Problem Books in Mathematics

Series Editor

Peter Winkler
Department of Mathematics
Dartmouth College
Hanover, NH
USA

More information about this series at https://link.springer.com/bookseries/714

Laszlo Csirmaz · Zalán Gyenis

Mathematical Logic

Exercises and Solutions

 Springer

Laszlo Csirmaz [ID]
Institute of Information
Theory and Automation
Prague, Czech Republic

Zalán Gyenis [ID]
Institute of Philosophy
Jagiellonian University
Kraków, Poland

ISSN 0941-3502   ISSN 2197-8506   (electronic)
Problem Books in Mathematics
ISBN 978-3-030-79012-7   ISBN 978-3-030-79010-3   (eBook)
https://doi.org/10.1007/978-3-030-79010-3

Mathematics Subject Classification: 03B05, 03B10, 03C10, 03C20, 03C62

This Springer imprint is published by the registered company Springer Nature Switzerland AG
The registered company address is: Gewerbestrasse 11, 6330 Cham, Switzerland

PREFACE

Delivering a Mathematical Logic course is always a challenge. Students enroll the course expecting to learn what "logical thinking" is and what the infallible rules of mathematical and rational thinking are. Instead, they get boring stuff like Zorn's lemma, torsion-free divisible Abelian groups, or dense linear ordering. They probably have heard of the infamous incompleteness theorem, and are eager to know how it destroys mathematics and rational thinking in general. Instead, they get the technicalities of Gödel's theorems and a page-long list of conditions under which they apply. The Mathematical Logic course covers a large and diverse segment of mathematics, and a significant part comes in the form of *problem-solving*. The weekly assignments play another important role: a crucial part of learning mathematics is gaining intuition about how the various concepts operate and interact, which, much like learning to drive a car, cannot be done without hands-on experience, and trial and error.

Problems in this volume have been collected over more than 30 years of teaching undergraduate students Mathematical Logic at Eötvös Loránd University, Budapest. The problems come in great variety: routine applications of a newly introduced technique, checking whether the conditions of a particular theorem are really necessary, extending or finding the limitations of various methods, to amusing puzzles and interesting applications of established results. They range from easy questions and riddles to proving hard theorems when all the necessary ingredients are—hopefully—available.

Several of the problems are part of the "mathematical folklore": well-known ones often used in teaching that everyone changes or twists slightly to fit their taste and the problem to illustrate. They are like good jokes or anecdotes that one keeps telling to new guests at a dinner party, although no one is sure exactly where they have come from. Others are extensions, details, or crucial points of the often hard and ingenious proofs of major theorems. Still others are based on solutions submitted by our students where an unexpected, clever method was used, or where the proposed solution had an interesting flaw or omission. And, some of the problems originated from intriguing questions from our students.

Chapters 1–4 contain problems from supporting fields: set-theoretical constructions, interesting applications in (perfect information) games, basic (and not so basic) results in formal languages, and recursion theory. Mathematical Logic proper is covered in Chapters 5–11 with problems in propositional and multi-valued logic, compactness and derivation; basic properties of first-order logic, derivation, compactness and completeness, elementary equivalence, and the ultraproduct technique. Chapter 10 covers arithmetics and incompleteness, and Chapter 11 touches on two advanced topics: the insufficiency of the Ehrenfeucht–Fraïssé game, and the zero-one law of random (universal) graphs.

The book concludes with solutions to all of the problems. We strongly encourage the reader to try to solve the problems before reading the solution included. Someone pointed out that there is a difference between doing push-ups and watching someone doing them (however fortunate it would be if it was otherwise!), which also seems to apply to mental exercise. The problems are organized such that ideas and techniques from previous ones can often be used to solve the next problem, so looking at the problems and their solutions leading up to the current one can be a good way of getting some inspiration.

One of the first homework problems in the Mathematical Logic course is a famous *sorosites* created by Lewis Carroll, so let us start our collection with it.

THE PIGS AND BALLOONS PUZZLE

The following facts are known:

(1) All, who neither dance on tight ropes nor eat penny-buns, are old.
(2) Pigs, that are liable to giddiness, are treated with respect.
(3) A wise balloonist takes an umbrella with him.
(4) No one ought to lunch in public who looks ridiculous and eats penny-buns.
(5) Young creatures, who go up in balloons, are liable to giddiness.
(6) Fat creatures, who look ridiculous, may lunch in public, provided that they do not dance on tight ropes.
(7) No wise creatures dance on tight ropes, if liable to giddiness.
(8) A pig looks ridiculous, carrying an umbrella.
(9) All who do not dance on tight ropes, and who are treated with respect are fat.

Show that no wise young pigs go up in balloons.

Acknowledgements The work of the first author for compiling and preparing the material has been funded by the GACR grant number 19-04579S. The second author wishes to acknowledge the project no. 2019/34/E/HS1/00044 financed by the National Science Centre, Poland.

Prague, Czech Republic   Laszlo Csirmaz
Kraków, Poland   Zalán Gyenis

CONTENTS

SPECIAL SET SYSTEMS

1

A *set system* is a collection of subsets of a set X. Frequently X is called the *ground set* or *base set* of the set system. Set systems are often denoted by calligraphic letters such as $\mathcal{F}$ or $\mathcal{A}$.

1.1 Definition (Families of sets). For sets X, Y we define the following notions:

- $\wp(X)$ is the set of all subsets of X, called the *powerset* of X.
- $[X]^\kappa$ is the family of all subsets of X of cardinality κ.
- $[X]^{<\kappa}$ is the family of all subsets of X of cardinality less than κ. In particular, $[X]^{<\omega}$ is the family of all finite subsets of X.
- $^Y X$ is the family of all functions $f : Y \to X$, i.e., $\mathrm{dom}(f) = Y$ and $\mathrm{ran}(f) \subseteq X$.

Notation. The set of natural numbers, integers, rationals, and real numbers are denoted by $\mathbb{N}, \mathbb{Z}, \mathbb{Q}$, and $\mathbb{R}$, respectively.

Zorn's lemma is an indispensable tool which will be used through this book.

1.2 Definition. A subset Q of the partially ordered set P is *totally ordered*, or *chain*, if every two elements in Q are comparable.

1.3 Theorem (Zorn's lemma). Every partially ordered set P, in which every chain has an upper bound in P, contains a maximal element.

1.1 BASIC CONSTRUCTIONS

1.4 Definition. The set system $\mathcal{F}$ is *almost disjoint* if any two different members of $\mathcal{F}$ intersect in a finite set: for $A, B \in \mathcal{F}, |A \cap B| < \omega$.

1. Show that there is an almost disjoint family of cardinality continuum on any (infinite) countable set.

© The Author(s), under exclusive license to Springer Nature Switzerland AG 2022
L. Csirmaz and Z. Gyenis, *Mathematical Logic,* Problem Books
in Mathematics, https://doi.org/10.1007/978-3-030-79010-3_1

1.5 Definition. The set system $\mathcal{F}$ of subsets of X is *maximal* with respect to property Φ, if $\mathcal{F}$ has property Φ, but for every $A \notin \mathcal{F}$, $\mathcal{F} \cup \{A\}$ does not have property Φ.

2. (a) Let $\mathcal{F}$ be a countable almost disjoint family on an infinite set X. Can $\mathcal{F}$ be maximal, i.e., no subset of X can be added to $\mathcal{F}$ so that it remains almost disjoint?

(b) Invent a condition which ensures that a countable almost disjoint family satisfying this condition is not maximal.

(c) Show that on a countable set one can always find an almost disjoint family of cardinality $\aleph_1$.

1.6 Definition. $\mathcal{F}$ is *κ-almost disjoint*, if for different $A, B \in \mathcal{F}$, $|A \cap B| < \kappa$.

3. Let κ be a regular infinite cardinal. Show that there is a κ-almost disjoint family $\mathcal{F}$ of cardinality κ^+.

4. Let κ be an infinite cardinal, and $\mathcal{F} \subseteq \wp(\kappa)$ be an almost disjoint family. Show that $|\mathcal{F}| \leq \kappa^\omega$. Show that there exists an almost disjoint family of this size.

1.7 Definition. The family $\mathcal{F}$ of subsets of X is *independent* if picking finitely many different A_i's from $\mathcal{F}$ and 0–1 numbers ε_i, the intersection of the sets $A_i^{\varepsilon_i}$ is not empty (here $A_i^1 = A_i$ and A_i^0 is $X \smallsetminus A_i$).

5. Prove that there is an independent family of subsets of ω of size 2^ω.

Hint. Consider $A_r = \{p : p$ is a polynomial with rational coefficient and $p(r) > 0\}$. Alternatively, for an element A of a set family on ω consider the set of finite subsets of ω that intersect A.

6. Prove that on an infinite set of cardinality κ there is an independent family of size 2^κ.

Hint. Assume that the members of the family are indexed by subsets of κ. Construct the elements of a member of the family from finite subsets of κ as the functions which map all subsets of such a finite subset to $\{0, 1\}$.

1.8 Definition. $\mathcal{F}$ has the *finite intersection property* (FIP) if the intersection of finitely many elements of $\mathcal{F}$ is never empty. In particular, the empty set cannot be an element of $\mathcal{F}$ in this case.

7. Let A be an infinite set, X be the collection of finite subsets of A, moreover for every $a \in A$ let $X_a = \{x \in X : a \in x\}$. Show that $\mathcal{U} = \{X_a : a \in A\}$ has the finite intersection property.

1.2 COUNTEREXAMPLES

8. Let $|X| = \kappa$ be an infinite cardinal. Does there exist an FIP family $\mathcal{F} \subseteq \wp(X)$ with $|\mathcal{F}| = \kappa$ so that every element of X is contained in finitely many elements of $\mathcal{F}$ only?

9. Let X be an infinite set. Construct an infinite family of infinite subsets of X so that the intersection of any two members of the family has infinitely many elements, while the intersection of any three elements is empty.

10. For each $n \geq 2$ construct a family $\mathcal{F}$ of subsets of the infinite set X so that the intersection of at most n elements from $\mathcal{F}$ is always infinite (non-empty), but the intersection of any $n + 1$ elements is always empty.

11. Give a property Φ and a family $\mathcal{F}$ of subsets of some set X having property Φ such that $\mathcal{F}$ cannot be extended to be maximal with respect to Φ.

1.3 SET SYSTEMS OF FUNCTIONS

12. (a) Construct a family F of continuum many functions from ω to ω such that for any two (different) $f, g \in F$, the set $\{i \in \omega : f(i) = g(i)\}$ is finite.

(b) Construct a family as in (a) with the additional property that all functions in F take all values in ω.

13. Let A_n be a finite set of cardinality at least 2^n. Construct a family $F \subseteq \prod_{n<\omega} A_n$ of continuum many elements such that for any two different $f, g \in F$ the set $|f \cap g|$ is finite.

14. Construct continuum many permutations of ω so that every two different permutations coincide on at most finitely many places.

15. Let κ be an infinite cardinal. Show that there exists a family F of functions $\kappa \to \kappa$ such that $|F| = \kappa^+$, and for any $f, g \in F$, $|\{\xi < \kappa : f(\xi) = g(\xi)\}| < \kappa$.

16. Let κ be an infinite cardinal. If f and g are different functions from κ to κ, then $D(f, g)$ is the set where they differ:

$$D(f, g) = \{\xi < \kappa : f(\xi) \neq g(\xi)\}.$$

Construct a family F of functions of cardinality 2^κ such that the family

$$\{D(f, g) : f, g \in F\}$$

has the finite intersection property.

17. Let κ be an infinite cardinal. Show that there are 2^κ many permutations of κ (i.e., one-to-one mapping of κ into itself) so that the family $\{D(f, g) : f, g \in F\}$ has the finite intersection property.

18. Let $\mathcal{D} \subseteq [\kappa]^\kappa$ be a κ-almost disjoint system. Show that there is family F of functions $\kappa \to \kappa$ of cardinality $|\mathcal{D}|$ such that for any two different functions $f, g \in F$, $|\{\xi < \kappa : f(\xi) = g(\xi)\}| < \kappa$.

19. Let κ be an infinite cardinal. Give a set X of cardinality κ and 2^κ many functions $X \to \omega$ so that no matter how finitely many of the functions are chosen, say $f_1, \ldots, f_n \in \mathcal{F}$, one can always find an $x \in X$ such that $f_1(x), \ldots, f_n(x)$ are all different.

20. Give a sequence of length ω_1 of functions $f_\alpha : \omega \to \omega$ such that f_β dominates f_α whenever $\alpha < \beta < \omega_1$, that is, for all but finitely many $n \in \omega$ we have $f_\alpha(n) < f_\beta(n)$.

1.4 FILTERS

Let $\langle P, \leq \rangle$ be a partially ordered set. In this section we assume that P has a maximal element but has no minimal element.

1.9 Definition. $\mathcal{F} \subseteq P$ is a *filter* if (a) whenever $p \in \mathcal{F}$ and $q \geq p$ then $q \in \mathcal{F}$; and (b) when p and q are in $\mathcal{F}$ then there is an $r \in \mathcal{F}$ for which $r \leq p$ and $r \leq q$.
The subset $D \subseteq P$ is *dense* if for every $p \in P$ one can find a $q \in D$ such that $q \leq p$.

21. Is it true that every filter on P can be extended to a maximal filter?

22. Is it true that if $q \in P$ is not maximal then there is a filter which avoids q?

23. Construct a partially ordered set P and dense subsets of P such that every filter on P avoids at least one of the dense subsets.

24 (Rasiowa–Sikorski Theorem). Let $\{D_i : i \in \omega\}$ be dense subsets of a partially ordered set P. Show that there is a filter $\mathcal{F}$ which intersects all D_i.

1.10 Definition. Let G be a group. A family $\mathcal{F}$ of subgroups of G is a *normal filter* if the following properties hold:
- $\mathcal{F}$ is not empty and the one-element subgroup is not in $\mathcal{F}$;
- $H_1, H_2 \in \mathcal{F}$ then $H_1 \cap H_2 \in \mathcal{F}$;
- if $H_1 \in \mathcal{F}$ and H_1 is a subgroup of H_2 then $H_2 \in \mathcal{F}$;
- if $H \in \mathcal{F}$ and $g \in G$ arbitrary group element, then $gHg^{-1} \in \mathcal{F}$.

25. Show that every normal filter can be extended to be a maximal one.

26. Is it true that if H is a non-trivial proper subgroup of G then there is a maximal normal filter containing H?

27. Let U be an infinite set and G be the symmetric group of U, that is, elements of G are the permutations of U and the group operation is the composition. For a finite $A \subset U$ the subgroup H_A is the pointwise stabilizer of A: $H_A = \{\pi \in G : \pi(a) = a \text{ for all } a \in A\}$. Show that $\mathcal{F} = \{H : H_A$ is a subgroup of H for some finite $A\}$ is a normal filter.

1.5 ULTRAFILTERS

> **1.11 Definition.** $\mathcal{F} \subseteq \wp(X)$ is a *filter* over X if the following properties hold.
> - $\emptyset \notin \mathcal{F}$ and $\mathcal{F} \neq \emptyset$.
> - If $A, B \in \mathcal{F}$ then $A \cap B \in \mathcal{F}$ (closed under intersection).
> - If $A \in \mathcal{F}$ and $A \subset B \subseteq X$ then $B \in \mathcal{F}$ (closed upward).
>
> Each filter has the FIP. The subfamily $\mathcal{B} \subseteq \mathcal{F}$ *generates* the filter $\mathcal{F}$ if each element of $\mathcal{F}$ is a superset of some element in $\mathcal{B}$: $\mathcal{F} = \{A \subseteq X : A \supseteq B \text{ for some } B \in \mathcal{B}\}$. We say that $\mathcal{F}$ is *trivial* or *principal* if it is generated by a single set, otherwise it is *non-trivial*.

> **1.12 Definition.** $\mathcal{F}$ is an *ultrafilter* over X if it is a filter and for each $A \subseteq X$ exactly one of A and $X - A$ is in $\mathcal{F}$.

28. Show that a set system which is maximal with respect to the finite intersection property is an ultrafilter.

29. Show that every set system with the finite intersection property can be extended to a maximal one.

30. Show that X is finite if and only if all ultrafilters on X are trivial. Prove that no non-trivial ultrafilter can contain a finite set.

31. Let $\mathcal{U}$ be an ultrafilter on X, and let $X = \bigcup_{i \in n}^* X_i$ be a finite partition of X. Show that there exists exactly one $i \in n$ such that $X_i \in \mathcal{U}$. What can be said about non-finite partitions?

32. Suppose $\emptyset \notin \mathcal{F} \subseteq \wp(X)$ satisfies the following assumption: If $X = \bigcup_{i \in n}^* X_i$ is a finite partition, then there is *exactly one $i \in n$* such that $X_i \in \mathcal{F}$. Does it follow that $\mathcal{F}$ is an ultrafilter?

33. For an infinite cardinal κ show that there are 2^{2^κ} many ultrafilters over κ.

> **Hint.** Use Problem 6.

34. Let $\mathcal{U}$ be an ultrafilter on ω and π be a permutation of ω (i.e., one-to-one function from ω onto ω). For $X \subseteq \omega$, πX is the image of X, and $\pi \mathcal{U}$ is the family of the images of the elements of $\mathcal{U}$. Show that for some non-trivial π we have $\pi \mathcal{U} = \mathcal{U}$.

35. Are there two non-trivial ultrafilters $\mathcal{U}$ and $\mathcal{V}$ on ω such that $\pi\mathcal{U} \neq \mathcal{V}$ for every permutation π of ω?

36. If $\mathcal{U}_1,\dots,\mathcal{U}_n$ are non-trivial ultrafilters on ω, then there is some infinite, co-infinite $A \in \mathcal{U}_1 \cap \dots \cap \mathcal{U}_n$.

37. Let $\mathcal{F} \subseteq \wp(X)$ be a family of subsets of X, and $Y \subseteq X$. The *trace of $\mathcal{F}$ on Y* is

$$\{Z \cap Y : Z \in \mathcal{F}\}.$$

Show that the trace of an ultrafilter on Y is either an ultrafilter, or the family of all subsets of Y.

Let $\mathcal{U}$ be an ultrafilter on X, and let $\mathcal{V} = \{Y \subseteq X :$ the trace of $\mathcal{U}$ on Y is an ultrafilter$\}$. Does $\mathcal{V}$ have some special property?

38. Let $f : \omega \to \omega$ be an injection, i.e., $f(i) \neq f(j)$ whenever $i \neq j$. Let $\mathcal{U}$ be an ultrafilter over ω, and $\mathcal{V} = \{f^{-1}(A) : A \in \mathcal{U}\}$. Find a necessary and sufficient condition for $\mathcal{V}$ to be an ultrafilter.

> **1.13 Definition.** A subset $\mathcal{U}$ of a Boolean algebra $\mathbf{B}$ is an *ultrafilter* if for all $a, b \in \mathcal{U}$ and $c \in \mathbf{B}$ we have $a \wedge b \in \mathcal{U}$; $a \vee c \in \mathcal{U}$, and exactly one of c or $-c$ is in $\mathcal{U}$.

39. Establish a connection between ultrafilters on $\mathbf{B}$ and homomorphisms $f : \mathbf{B} \to 2$.

> **Notation.** $\mathbf{B}^*$ denotes the set of ultrafilters on the Boolean algebra $\mathbf{B}$ and for $x \in \mathbf{B}$ we let $N_x = \{\mathcal{U} \in \mathbf{B}^* : x \in \mathcal{U}\}$.

40 (Stone's Representation Theorem). Prove that each Boolean algebra is isomorphic to a suitable subalgebra of the power set of a set.

> **Hint.** Find a suitable subalgebra of $\wp(\mathbf{B}^*)$.

41. Show that $\{N_x : x \in \mathbf{B}\}$ generates a zero-dimensional, compact and Hausdorff topology on $\mathbf{B}^*$.

42. Prove that $\mathcal{U} \subset \mathbf{B}$ is trivial if and only if it is an isolated point of $\mathbf{B}^*$. Does it follow that every infinite $\mathbf{B}^*$ contains a non-trivial ultrafilter?

> **1.14 Definition.** Let X be a topological space and $\mathcal{F}$ an ultrafilter over X. $p \in X$ is a *limit point of $\mathcal{F}$* if every neighborhood of p belongs to $\mathcal{F}$.

43. Give an example of a topological space X and a non-trivial ultrafilter $\mathcal{F}$ over X such that (a) $\mathcal{F}$ has a limit point; (b) $\mathcal{F}$ does not have a limit point.

44. Let X be a topological space and $\mathcal{F}$ be an ultrafilter on X. Show that

(a) X is Hausdorff if and only if each convergent ultrafilter has a unique limit point.

(b) X is compact if and only if every ultrafilter has some limit point.

45. Let $\mathcal{W}$ be a family of subsets of ω. The set $X \subseteq \omega$ is in the *closure of* $\mathcal{W}$ if for all finite $F \subseteq \omega$ there exists a $Y \in \mathcal{W}$ such that $X \cap F = Y \cap F$.

(a) What is the closure of an ultrafilter $\mathcal{U}$ over ω?

(b) Show that all ultrafilters $\mathcal{U}$ on ω are *dense in itself*, i.e., all elements of $\mathcal{U}$ are in the closure of the other elements of $\mathcal{U}$.

46. Give a topological space T and an open cover $\mathcal{O}$ of T which has no minimal subcover: from any subcover one can leave an open set so that it remains a cover.

> **1.15 Definition.** Let $\mathcal{U}$ be a non-principal ultrafilter on ω and $(x_n)_{n<\omega}$ be a bounded sequence of real numbers. The *ultralimit* of the sequence is $x \in \mathbb{R}$ if for all $\varepsilon > 0$ we have $\{n \in \omega : |x - x_n| < \varepsilon\} \in \mathcal{U}$.

47. Let A_n $(n < \omega)$ be a partition of ω into infinite sets. Construct an ultrafilter $\mathcal{U}$ on ω such that for all $n \in \omega$, $A_n \notin \mathcal{U}$ and no $X \in \mathcal{U}$ meets each A_n in a finite set.

48. Prove that the ultralimit x of every bounded sequence $(x_n)_{n<\omega}$ of real numbers exists and is unique. Show that x is a (traditional) limit of a subsequence of $(x_n)_{n<\omega}$. Is there always an $X \in \mathcal{U}$ such that the subsequence $(x_i)_{i \in X}$ converges to x?

49. Let $\mathcal{U}$ be a non-trivial ultrafilter on ω and consider the set

$$A = \left\{ \sum_{i \in X} \frac{1}{2^{i+1}} : X \in \mathcal{U} \right\}.$$

Can A be Lebesgue measurable? If yes, what is the measure of A?

> **1.16 Definition.** Let G be an (infinite) Abelian group. For subsets of G define $A + B = \{a + b : a \in A \text{ and } b \in B\}$, and $A - a = \{x - a : x \in A\}$. If $\mathcal{U}$ and $\mathcal{V}$ are ultrafilters on G, then let
>
> $$\mathcal{U} + \mathcal{V} = \{X \subseteq G : \{a \in G : X - a \in \mathcal{U}\} \in \mathcal{V}\}.$$

50. Which Abelian groups G admit a non-trivial translation-invariant ultrafilter $\mathcal{U}$, i.e., for all $a \in G$, $X \in \mathcal{U}$ iff $a + X \in \mathcal{U}$?

51. Is it true that $\mathcal{U} + \mathcal{V}$ is always an ultrafilter?

52. Is it always the case that $\mathcal{U} + \mathcal{V} = \mathcal{V} + \mathcal{U}$? What about $(\mathcal{U} + \mathcal{V}) + \mathcal{W} = \mathcal{U} + (\mathcal{V} + \mathcal{W})$?

53. Is it true that for every $\mathcal{U}$ and $\mathcal{V}$ one can find a $\mathcal{W}$ with (a) $\mathcal{U} + \mathcal{W} = \mathcal{V}$; (b) $\mathcal{W} + \mathcal{U} = \mathcal{V}$?

54 (Ramsey theorem). All k-element subsets of the natural numbers ω are colored by one of n different colors. We want to show that there is an infinite homogeneous set for some color (i.e., an infinite $A \subseteq \omega$ all k-element subsets of which share the same color). To this end, let U be a non-trivial ultrafilter on ω, and let the color of the $k-1$-element subset $\{i_1, \ldots, i_{k-1}\}$ be e, if for ultrafilter-many i_k, the color of the k-element set $\{i_1, \ldots, i_{k-1}, i_k\}$ is e.

Complete the proof.

55 (Sierpiński, $2^\kappa \nrightarrow (3)_\kappa^2$). Color edges of the complete graph with ω many colors (a) on ω_1 vertices (b) on continuum many vertices so that no triangle becomes homogeneous.

56 (Sierpiński, $2^\omega \nrightarrow (\aleph_1)_2^2$). The complete graph G is defined on the set of real numbers $\mathbb{R}$. The edges of G are colored by two colors as follows. Let $\prec$ be a well-ordering on $\mathbb{R}$, and color the edge between the points $x < y$ red if x precedes y in the well-ordering, and color it blue otherwise. Show that a homogeneous subset of G is at most countable.

> **1.17 Definition.** An ultrafilter $\mathcal{U}$ on κ is *regular* if there exists $E \subset \mathcal{U}$ such that $|E| = \kappa$ and for all $\xi \in \kappa$ the set $\{e \in E : \xi \in e\}$ is finite.
>
> An ultrafilter $\mathcal{U}$ on κ is *uniform* if for all $A \in \mathcal{U}$ we have $|A| = \kappa$.

57. For an infinite cardinal κ, construct a regular ultrafilter on κ.

> **Hint.** Replace κ with $[\kappa]^{<\omega}$ and for $X \in [\kappa]^{<\omega}$ consider the sets
>
> $$e_X = \left\{ Y \in [\kappa]^{<\omega} : X \subseteq Y \right\}.$$

58. (a) Show that every regular ultrafilter is uniform, as well.

(b) Prove that each non-trivial ultrafilter on ω is regular.

59. (a) Construct a non-trivial, non-uniform ultrafilter on every uncountable cardinal.

(b) Construct a non-regular ultrafilter on every uncountable cardinal.

60. Show that there are 2^{2^κ} regular ultrafilters over an infinite κ.

> **Hint.** Use problem 5.

GAMES AND VOTING

2

2.1 GAMES

> **2.1 Definition.** A *strategy* is a function which tells from all previous moves what the next move should be. A *positional strategy* is a strategy where the suggested move depends on the position of the game only and not on the way previous moves led to that position.

61. Let $\mathcal{F} \subseteq \wp(\omega)$ be a collection of subsets of ω and play the game $G(\mathcal{F})$ as follows. Two players, I and II, take turns choosing finitely many previously unchosen elements of ω. After infinitely many moves, player I picked $A \subseteq \omega$ and player II picked $B \subseteq \omega$. If A is in $\mathcal{F}$, then player I wins, otherwise, player II wins.

Which player can have a winning strategy in the following games?

(a) $\mathcal{F}$ contains all infinite sets $A \subseteq \omega$ such that every element of A is odd.

(b) $\mathcal{F}$ is the collection of all sets containing at least twenty primes.

(c) $\mathcal{F}$ contains all sets that contain all but finitely many of the even numbers.

62. Let U be a fixed non-trivial ultrafilter on ω. Two players, I and II, are playing the following game. They pick alternately elements of ω in ω many steps; no number can be picked twice. When the game ends, the set of numbers picked by I is A, and that of picked by II is B. I wins the game if $A \in U$, and II wins if $B \in U$, otherwise the game is a draw.

(a) Show that II cannot have a winning strategy.

(b) Show that I does not have a winning strategy.

63. Show that in the ultrafilter game of Problem 62 neither player has a strategy which guarantees a draw. In other words, if either player plays by a strategy, the other can win against it.

> **Hint.** If they play three games simultaneously, II can win at least one of them.

© The Author(s), under exclusive license to Springer Nature Switzerland AG 2022 9
L. Csirmaz and Z. Gyenis, *Mathematical Logic*, Problem Books
in Mathematics, https://doi.org/10.1007/978-3-030-79010-3_2

64. Let $A \subseteq [0, 1]$. Players I and II play the following game. They pick the digits of the real number $0.a_1 a_2 a_3 \dots$ alternately; odd digits are determined by I, even digits by II. I wins if the formed number is an element of A, and II wins if it is not.

(a) Suppose II has no winning strategy (for a suitable A). Then I can play so that after his steps II still has no winning strategy. Give a set A so that I playing this strategy loses.

(b) Show that if A is open or closed, then some player has a winning strategy.

(c) Prove that for countable A, Player II has a winning strategy.

(d)* Construct a set A so that no player has a winning strategy.

> **Hint.** Enumerate all the possible strategies and define the set A by transfinite recursion.

65. Let B be the complete infinite binary tree. Players I and II agree on a collection A of (full) branches of B, and place a token at the root of B. During the game the players move alternately. Player I can move the token higher along any branch to a new node, or leave it where it is. Player II must move the token exactly one level higher. When the game finishes, the token travels along a branch. Player I wins if this branch is in A.

(a) Show that if A is countable then II has a winning strategy.

(b)* Show that if II has a winning strategy then A is necessarily countable.

> **Warning.** The strategy might not be positional: the suggested move may depend on all previous moves, not only on the node where the token is.

(c) If I has a winning strategy then $|A|$ has cardinality continuum.

(d) Assume A has the property that whenever $x \in A$ then there is an initial segment s of x such that every full extension of s belongs to A. Show in this case that some player has a winning strategy.

66. I and II are playing a generalized 5-in-a-row game (Gomoku). Given are a set B and a collection $\mathcal{F}$ of finite subsets—the "winning sets"—of B. The players claim elements of B alternately in ω steps. The first to move is I, and he wins the game if he can claim all elements of some winning set in $\mathcal{F}$.

An n-move strategy for II is a function which tells, for each $k \leq n$, the field II should occupy given the sequence of the first k fields occupied by I in the first k moves. We know that for all n there is a strategy S_n for II of at least n-move such that if II plays according to S_n then I does not win before the n-th move.

Let U be a non-trivial ultrafilter on ω, and let II play as follows. Assume it is the k-th move, and I occupied the fields $b_1, \dots, b_k$ so far in this order. For each field $b \in B$ let I_b be the set of all indices $n \in \omega$ for which the strategy S_n advises b as the k-th move after playing $b_1, \dots, b_k$. If I_b is in U for some $b \in B$, then II should occupy b. If none of them is in U, then II can occupy any field of her choice.

Show that playing so, II never loses the game.

67. Let A_i be propositional variables for $i \in \omega$, and Γ be a set of propositional formulas using these variables. Players I and II set the truth value of the variables alternately. I wins the game if at the end all elements of Γ evaluate to true, otherwise the winner is II.

(a) Show that either I or II has a winning strategy.

(b) Construct a game where Γ is satisfiable, but II wins the game.

2.2 VOTING

> In a firm the board of directors are voting on a proposal which has several variants. Each member casts his or her vote on one of the variants. A *voting scheme* is a function $v: V^M \to V$ where V is the set of the variants and M is the set of board members. The vote for all board members is an element of V^M, and v tells the outcome for all possible distribution of votes. The scheme v is *fair* if it has the following properties:
> - if every vote goes to the same alternative, then the result is that alternative;
> - if every vote is changed, then the outcome, as determined by v, changes as well.
>
> Consider the following family $\mathcal{F}$ of subsets of M: if the outcome (according to v) of a particular voting is x, then the set of those members who voted for x is in $\mathcal{F}$.

68. Suppose $|V| \geq 3$, and let v be a fair voting scheme.

(a) Show that $\mathcal{F}$ is an ultrafilter on M.

(b) Show that if M is finite then every fair voting scheme is "autocratic," i.e., there is a single board member whose vote determines the outcome.

(c) Show that neither (a) nor (b) remains true if $|V| = 2$.

69. Describe all fair voting schemes when V is at least 3 and finite, but M is an infinite set.

> **2.2 Definition.** Let $G_i = (V_i, E_i)$ be graphs for $i \in I$. The *product* of these graphs is defined on the vertex set $V = \prod_{i \in I} V_i$, and two vertices in V are connected if for each $i \in I$ their i-th coordinates are connected in G_i.

70. Consider the product $K_n^I = \prod_{i \in I} K_n$, where K_n is the complete graph on $n \geq 3$ vertices. Show that the chromatic number of this product is n, and describe all correct n-colorings of K_n^I.

FORMAL LANGUAGES AND AUTOMATA

3

An *alphabet* is a finite set Σ of symbols. A *word* is a finite sequence of symbols, including the empty sequence which is denoted by λ. The set of words over Σ is denoted by Σ^*, and it can be considered as the universe of the free semigroup generated by Σ with λ as the unit element. The *length* of w, denoted as $|w|$, is the number of the symbols in it. The *concatenation* of two words is denoted by writing them next to each other. The *reverse* of the word $w = x_1 \ldots x_n$ is $w^R = x_n \ldots x_1$.

3.1 Definition (Language). A (formal) *language over* Σ is a subset of Σ^*.

3.2 Definition (Operations on languages). On languages over the alphabet Σ the usual operators are the following:

- *Union*: $L_1 \cup L_2$ is the set-theoretical union of L_1 and L_2. Frequently this operation is written as $L_1|L_2$.
- *Concatenation*: $L_1 L_2 = \{w_1 w_2 : w_1 \in L_1, w_2 \in L_2\}$.
- *Kleene star*: $L^* = \{\lambda\} \cup L \cup L^2 \cup L^3 \cup \ldots$.

3.1 REGULAR LANGUAGES AND AUTOMATA

3.3 Definition (Regular languages). The family of regular languages over the alphabet Σ is defined recursively as follows:

- $\emptyset$ and $\{\lambda\}$ are regular.
- For each $a \in \Sigma$, $\{a\}$ is regular.
- If A, B are regular, then so are $A|B$, AB, and A^*.

Generation of regular languages are described by *regular expressions*.

3.4 Definition (Regular expressions). Given a finite alphabet Σ we define the set of regular expressions as follows:

- $\emptyset$, λ, and a for each $a \in \Sigma$ are regular expressions.
- If σ and τ are regular expressions, so are $(\sigma|\tau)$, $(\sigma\tau)$, and $(\sigma)^*$.

© The Author(s), under exclusive license to Springer Nature Switzerland AG 2022   13
L. Csirmaz and Z. Gyenis, *Mathematical Logic,* Problem Books
in Mathematics, https://doi.org/10.1007/978-3-030-79010-3_3

Whenever it is clear we omit the parentheses. For example, $a|b^*$ is the language $\{a, \lambda, b, bb, bbb, \ldots\}$, while $(a|b)^*$ is the set of all words over the alphabet $\{a, b\}$.

3.5 Definition (Deterministic Finite Automaton, DFA). A *deterministic finite automaton* over the finite alphabet Σ is a tuple $\mathcal{A} = \langle Q, S, F, \delta \rangle$, where Q is the non-empty finite set of *states*, $S \in Q$ is the *starting state*, $F \subseteq Q$ is the set of *halting states*, and $\delta: Q \times \Sigma \to Q$ is the *transition function*.

Any word $w \in \Sigma^*$ acts on $\mathcal{A}$ as follows. Start from the initial state S, and apply the transition function to the actual state on the next symbol (from left to right) of w to get the next state. The word w is *accepted* by $\mathcal{A}$ if w moves the automaton into a final state (i.e., an element of F). The language *generated* or *accepted* by $\mathcal{A}$ is

$$L(\mathcal{A}) = \{w \in \Sigma^* : w \text{ is accepted by } \mathcal{A}\}.$$

71. (a) Show that every finite language L is generated by some DFA.

(b) Show that every finite language is regular.

72. Let L_1, L_2 be regular. Does $L_1^* \cap L_2^* = (L_1 \cap L_2)^*$ always hold?

73. Write the following languages over the alphabet $\Sigma = \{a, b\}$ using regular expressions.

(a) The set of strings that have at least one b.

(b) The set of strings that have at most one b.

(c) The set of strings that end in 3 consecutive b's.

In each case design a DFA which accepts the given language.

74. Let L be the set of words $w \in \{a, b\}^*$ that contain even number of a's and odd number of b's.

(a) Give a DFA that generates L.

(b)* Give a regular expression for L.

3.6 Theorem (Pumping lemma). If the language L is generated by a DFA, then there exist $p \geq 0$ (depending on L only) such that for all $uwv \in L$ with $|w| \geq p$, there exists a partition $w = xyz$ such that $|xy| \leq p$, $|y| > 0$ and for all $i \geq 0$, $uxy^i zv \in L$. Here u and v might be empty.

75. Prove the Pumping lemma 3.6.

76. Prove that no DFA accepts the language

$$L = \{w \in \{a, b\}^* : |w| = k^2 \text{ for some } k\}.$$

Hint. Use the Pumping lemma 3.6.

77. Determine all subsets $A \subseteq \omega$ for which the language $\{a^n : n \in A\}$ is generated by some DFA.

78. (a) Prove that $L = \{a^n b^n : n \in \omega\}$ is not accepted by any DFA.

(b) Determine all functions $f : \omega \to \omega$ such that the language $L_f = \{a^n b^{f(n)} : n \in \omega\}$ is accepted by a DFA.

> **3.7 Definition** (Non-deterministic Finite Automaton, NFA). A *non-deterministic finite automaton* over the finite alphabet Σ is a tuple $\mathcal{N} = \langle Q, S, F, \delta \rangle$ similar to DFA, with the exception that the transition function $\delta : Q \times \Sigma \to \wp(Q)$ gives a non-empty set of possible next states. The automaton accepts a word if it has such a run where it arrives at an accepting state.

79. Prove that a language L is accepted by some DFA if and only if L is accepted by some NFA.

80. We are given two DFA's $\mathcal{A}$ and $\mathcal{B}$ accepting the languages L_A and L_B over the alphabets Σ_A and Σ_B. Create an automaton which accepts

(a) the union $L_A \cup L_B$,

(b) the intersection $L_A \cap L_B$,

(c) the concatenation $L_A L_B$,

(d) the complement $\Sigma_A^* \smallsetminus L_A$, and

(e) the Kleene star L_A^*.

> **3.8 Theorem** (Kleene). Regular languages are precisely the languages accepted by DFA's.

81. Show that every regular language can be accepted by a DFA.

82.* Show that every language accepted by a DFA is regular.

> **Hint.** For each pair $q_1, q_2 \in Q$ let L_{q_1,q_2} be the language accepted by the automaton with initial state q_1 and accepting state q_2. Use induction on the number of states to show that all these languages are regular.

83. Give an example for a *non-regular* language that satisfies the conclusions of Pumping lemma 3.6.

> **Hint.** $L_1 = \{ca^i b^j : i, j \in \omega\}$ is regular while $L_2 = \{ca^n b^n : n \in \omega\}$ is not, see Problem 78(a). Adjust L_2 so that $L \cap L_1 = L_2$, while L can be pumped.

84. Let L be regular. Show that $\{w^R : w \in L\}$ is regular.

> **3.9 Definition.** Let $\mathcal{A} = \langle Q, S, F, \delta \rangle$ be a DFA. The equivalence relation $\sim$ on Q is a *congruence* if the followings hold:
> - If $a \sim b$ and a is a halting state, then so is b.

- If $a \sim b$ and $a \xrightarrow{\alpha} a'$ and $b \xrightarrow{\alpha} b'$ are transitions, then $a' \sim b'$.

Having a congruence one can define the *quotient automaton* $\mathcal{A}/\sim$ in the usual way.

85. Show that for a congruence $\sim$ the quotient $\mathcal{A}/\sim$ and $\mathcal{A}$ accept the same language.

3.10 Definition. An automaton $\mathcal{A}$ is *connected* if every state in $\mathcal{A}$ can be reached from the initial state by some word.

86. Suppose the DFA $\mathcal{A}$ and $\mathcal{B}$ accept the same language. For two states q_1 and q_2 in $\mathcal{A}$ define $q_1 \sim q_2$ if there are two words w_1 and w_2 so that w_1 sends the initial state of $\mathcal{A}$ to q_1, and w_2 sends it to q_2, while in $\mathcal{B}$ both words send the initial state to the same state. Prove that for connected $\mathcal{A}$ and $\mathcal{B}$ the equivalence relation $\sim$ is always a congruence.

87. Let $\mathcal{A}$ be a DFA with the smallest possible number of states which accepts the language L. Show that $\mathcal{A}$ is determined uniquely.

88. Give an algorithm which, from an automaton $\mathcal{A}$, generates another automaton $\mathcal{B}$ so that $\mathcal{B}$ accepts the same language and has the smallest possible number of states.

89. Let L be a language and define $s(n) = |L \cap \Sigma^n|$. Show that if L is regular, then there is a natural number d and constants $c_1, \ldots, c_d$ such that

$$s(n+d) = c_1 s(n+d-1) + \cdots + c_{d-1} s(n+1) + c_d s(n).$$

90. Find a regular language L such that the $|L \cap \Sigma^n|$ sequence is the Fibonacci numbers $1, 1, 2, 3, 5, 8, \ldots$

91. Let Σ be a finite alphabet and $a \notin \Sigma$. For a language L over the alphabet $\Sigma \cup \{a\}$ write

$$L_a = \{w \in \Sigma^* : w_1 a w a w_2 \in L \text{ for some } w_1, w_2 \in (\Sigma \cup \{a\})^*\}.$$

Prove that L_a is regular provided L is such.

3.2 WHEN THE CONTEXT DOES NOT MATTER

3.11 Definition (Context-free language). Let V be a finite set of nonterminal characters, Σ be a finite set of terminal characters, and $S \in V$ be the start symbol. The empty word λ is considered as a terminal. Context-free languages are generated from the start symbol S by finitely many production rules of the form $A \to \alpha$, where A is a single nonterminal symbol, and α is a non-empty word of terminal and/or nonterminal symbols.

Example. With $\Sigma = \{a, b\}$ the rules $S \to aSa$, $S \to bSb$, $S \to \lambda$ generate the word $aabbaa$ as follows:

$$S \to aSa \to aaSaa \to aabSbaa \to aabbaa$$

.

92. Prove that the following languages are context-free:

(a) $L = \{w : w \in \{(,)\}^*, w$ has matching parentheses$\}$.
(b) $L = \{ww^R : w \in \{a, b\}^*\}$.
(c) $L = \{w : w \in \{a, b\}^*$, the number of a's and the number of b's in w are different$\}$.

93. Which are (i) regular, (ii) context-free among the languages below:

(a) $\{a^n b^n : n \geq 0\}$,
(b) $\{a^n b^n c^j : n \geq 0, \ j \geq 0\}$,
(c) $\{(ab)^n : n \geq 2\}$,
(d) $\{(ab)^n (bc)^j : n \geq 2, \ j \geq 0\}$, and
(e) $\{b^n a^m b^{2n} : n \geq 0, \ m \geq 0\}$.

94. Prove that all regular languages are context-free.

95. Prove that context-free languages are closed under union, concatenation, and Kleene star.

96. Show that every context-free language has a generating rule set in which the right-hand side of each rule $A \to \alpha$

- does not contain the start symbol S;
- does not contain λ with the only possible exception $S \to \lambda$.

Example. The rule set $S \to aSa$, $S \to bSb$, $S \to \lambda$ generates the same language as the set $S \to \lambda$, $S \to T$, $T \to aTa$, $T \to bTb$, $T \to aa$, $T \to bb$.

3.12 Theorem (Pumping lemma). For every context-free language L there is an integer $p \geq 1$ such that every word $s \in L$ of length at least p can be written in the form $s = uvwxy$ such that $|vx| \geq 1$, $|vwx| \leq p$, and $uv^n wx^n y \in L$ for all $n \geq 0$.

97. Prove Theorem 3.12.

98. Show that $L = \{a^n b^n c^n : n \geq 0\}$ is not context-free.

99. Prove that the class of context-free languages is

(a) not closed under intersection,
(b) not closed under complementation, and
(c) is the class closed under taking intersection with regular languages?

3.13 Definition. Let $\mathcal{F}$ be a family of languages over the alphabet $|\Sigma| \geq 2$. We say that $U \subseteq \Sigma^*$ is *universal* for $\mathcal{F}$ if for each $L \in \mathcal{F}$ there is a prefix $w = w_L \in \Sigma^*$ such that $x \in L$ iff $wx \in U$.

100. (a) Is there a regular language that is universal for regular languages for the fixed alphabet Σ?

(b) What about context-free languages?

(c) Show that for every countable collection $\mathcal{F}$ of languages over Σ there is a universal language.

(d) Can a countable set $\mathcal{F}$ of languages always be extended to $\mathcal{F}^+$ such that it contains a language universal for $\mathcal{F}^+$?

RECURSION THEORY

4

4.1 Definition (Operators on functions). Let Ω be the set of all functions $f : \omega^n \to \omega$. An *operator on* Ω assigns an element of Ω to certain tuples from Ω. The following operators will be considered:

- The *composition operator* Comp is defined on the tuple of functions $(g, h_1, \ldots, h_\ell)$ if $\mathrm{dom}(g) = \omega^\ell$, and $\mathrm{dom}(h_i)$ is the same ω^m for each i. In this case

$$\mathrm{Comp}(g, h_1, \ldots, h_\ell) : \vec{x} \mapsto g(h_1(\vec{x}), \ldots, h_\ell(\vec{x})).$$

- The μ *operator* can be applied to the single function $g \in \Omega$ if it is defined on $\omega^{\ell+1}$ with $\ell \geq 1$ and for every $\vec{x} \in \omega^\ell$ there is a $u \in \omega$ such that $g(\vec{x}, u) = 0$. In this case $\mu(g) \in \Omega$ returns the smallest such a u:

$$\mu(g) : \vec{x} \mapsto \min\{u \in \omega : g(\vec{x}, u) = 0\}.$$

- The *primitive recursion operator* PrRec can be applied to the pair (g, h) if g is defined on ω^ℓ and h is defined on $\omega^{\ell+1}$ with $\ell \geq 1$. In this case $\mathrm{PrRec}(g, h)$ is the uniquely defined function f on ω^ℓ which satisfies, for all $\vec{x} \in \omega^{\ell-1}$ and $n \geq 0$,

$$f(\vec{x}, 0) = g(\vec{x}, 0),$$

and

$$f(\vec{x}, n+1) = h(\vec{x}, n, f(\vec{x}, n)).$$

4.1 PRIMITIVE RECURSIVE FUNCTIONS

4.2 Definition. The set of *primitive recursive* functions is the smallest subset of Ω that contains the initial functions below, and is closed for the Comp and PrRec operators:

- constant zero function $\mathbf{0}(n) = 0$,
- successor function $S(n) = n + 1$,
- projections $U_i^k(\vec{x}) = x_i$, where $\vec{x} = \langle x_1, \ldots, x_k \rangle$.

© The Author(s), under exclusive license to Springer Nature Switzerland AG 2022   19
L. Csirmaz and Z. Gyenis, *Mathematical Logic,* Problem Books
in Mathematics, https://doi.org/10.1007/978-3-030-79010-3_4

101. Show that the following functions are primitive recursive:

(a) constant functions,

(b) predecessor: $\delta(x) = \max\{0, x-1\}$,
sign: $\mathrm{sgn}(x) = \min\{x, 1\}$,
limited subtraction: $x \dot- y = \max\{0, x-y\}$,

(c) addition, multiplication, exponentiation,

(d) factorial $n!$.

102. Show that the following functions are primitive recursive:

$$K_<(x, y) = \begin{cases} 0 & \text{if } x < y, \\ 1 & \text{otherwise.} \end{cases} \qquad E(x, y) = \begin{cases} 1 & \text{if } x = y, \\ 0 & \text{otherwise.} \end{cases}$$

103. (a) If $f(\vec{x}, y)$ is primitive recursive, then so are $\sigma_f(\vec{x}, y) = \sum_{i \leq y} f(\vec{x}, i)$ (bounded sum) and $\pi_f(\vec{x}, y) = \prod_{i \leq y} f(\vec{x}, i)$ (bounded product).

(b) Bounded minimization with primitive recursive bound: if $h(\vec{x}, y)$ and $g(\vec{x})$ are primitive recursive, then so is

$$f(\vec{x}) = \min\{y < g(\vec{x}) : h(\vec{x}, y) = 0\},$$

which is the least y such that $h(\vec{x}, y) = 0$ if there is such a $y < g(\vec{x})$, and $g(\vec{x})$ otherwise.

4.3 Definition. The subset (relation) $A \subseteq \omega^n$ is *primitive recursive* if its characteristic function is primitive recursive.

104. (a) Primitive recursive relations are closed under Boolean operations (and, or, complement).

(b) The comparison relations ($=, \neq, \leq$, etc.) are primitive recursive.

(c) If $A \subseteq \omega^{n+1}$ is primitive recursive, then so are the relations when applying *bounded quantifiers*:

$$\{\langle \vec{x}, n \rangle : (\exists i < n)\langle \vec{x}, i \rangle \in A\},$$
and
$$\{\langle \vec{x}, n \rangle : (\forall i < n)\langle \vec{x}, i \rangle \in A\}.$$

105 (Definition by cases). Suppose $A_1, \ldots, A_k$ are primitive recursive relations such that for all $\vec{x} \in \omega^n$ exactly one of them holds. If $h_1, \ldots, h_k$ are primitive recursive functions, then so is

$$f(\vec{x}) = \begin{cases} h_1(\vec{x}) & \text{if } \vec{x} \in A_1 \\ h_2(\vec{x}) & \text{if } \vec{x} \in A_2 \\ \vdots & \vdots \\ h_k(\vec{x}) & \text{if } \vec{x} \in A_k \end{cases}$$

106. Show that the integer division $[x/y]$ is primitive recursive.

107. (a) Let $p(n)$ be the n-th prime ($p(0) = 2$, $p(1) = 3$, etc.). Show that $p(n)$ is primitive recursive.

(b) Show that the exponent of the n-th prime in the prime factorization of x is a primitive recursive function of x and n.

108. For a function $g : \omega \to \omega$ define

$$g^{(0)}(x) = x, \quad g^{(1)}(x) = g(x), \quad \text{and} \quad g^{(i+1)}(x) = g(g^{(i)}(x)).$$

(a) Show that if g is primitive recursive then so is $g^{(x)}(x)$.

(b) Let $g_0(x) = 2$ and $g_{k+1}(x) = (g_k(x))^x$. Show that $g_x(x)$ is primitive recursive.

109. For a given function g let us define the functions A_i for $i \geq 0$ as follows:

(i) $A_0(x) = g(x)$ and

(ii) $A_{i+1}(x) = A_i(A_i(\dots A_i(x) \dots))$, where we iterate A_i exactly $i + 1$ times.

Write $A(x) = A_x(x)$.

(a) Prove that $A(x)$ is primitive recursive provided $g(x)$ is such.

(b) Let $g(x) = x + 1$. Give an estimate for $A(10)$.

(c) Let $g(x) = (x+1)^{x+1}$. Give an estimate for $A(5)$.

> **4.4 Definition** (Ackermann function, R. Péter version).
>
> $$\begin{aligned} A(0, m) &= m + 1 \\ A(n+1, 0) &= A(n, 1) \\ A(n+1, m+1) &= A(n, A(n+1, m)). \end{aligned}$$

110. For each primitive recursive $f(n_1, \dots, n_k)$ there is an N such that

$$f(n_1, \dots, n_k) < A(N, n_1 + \dots + n_k).$$

111. Show that $A(m, m)$ is not primitive recursive.

> **4.5 Definition.** Fix $n \in \omega$ and let $\mathcal{F} \subseteq \{f : f : \omega^n \to \omega\}$. The function $U : \omega^{n+1} \to \omega$ is *universal for the family* $\mathcal{F}$ if for every $f \in \mathcal{F}$ there is an $i \in \omega$ such that $U(i, \bar{x}) = f(\bar{x})$. Such an i is the U-*index*, or just the *index* of f.

112. Prove that there is no primitive recursive $U : \omega^{n+1} \to \omega$ which is universal for the set of primitive recursive n-variable functions.

113. Show that there is a function from $\mathbb{R}^2$ to $\mathbb{R}$ which is universal for the $\mathbb{R} \to \mathbb{R}$ continuous functions. Show that this universal function cannot be continuous.

4.2 RECURSIVE FUNCTIONS

> **4.6 Definition.** The set of *recursive functions* is the smallest subset of Ω (Definition 4.1) that contains the initial functions below, and is closed for the Comp and the μ operators:
> - addition, multiplication, and the $K_<$ function (see Problem 102) and
> - the projection functions U_i^k for $1 \le i \le k$.

114. Show that the following functions are recursive:

(a) constant zero and constant one functions;

(b) sign and limited subtraction.

> **4.7 Definition.** The subset (relation) $A \subseteq \omega^n$ is *recursive* if its characteristic function is recursive.

115. Assume $A \subseteq \omega^{n+1}$ is recursive, and for each $\vec{x} \in \omega^n$ there is an $i \in \omega$ with $\langle \vec{x}, i \rangle \in A$. Show that the function assigning the minimal such i to $\vec{x}$ is recursive. This function is written as $\mu\{i : \langle \vec{x}, i \rangle \in A\}$.

116. (a) Recursive relations are closed under Boolean operations.

(b) The comparison relations ($=, \ne, \le$, etc.) are recursive.

(c) Recursive relations are closed for bounded quantifiers.

117. The following functions are recursive:

(a) integer division $[x/y]$ (zero if y is zero),

(b) integer part of the square root of x: $[\sqrt{x}]$,

(c) $\text{rem}(x, y) = \min\{z \in \omega : y \text{ is a divisor of } (x - z)\}$.

118. Show that after changing finitely many values of a recursive function f, it remains recursive.

119. Prove that every function $f : \omega \to \omega$ is a pointwise limit of recursive functions.

120. Show that there is a non-recursive function $f : \omega \to 2$.

121. Let $f : \omega \to \omega$ with $\text{ran}(f) = \omega$ be recursive and let $f^{-1}(y)$ be the least x such that $f(x) = y$. Prove that f^{-1} is recursive.

122. Let $f_0, f_1, \ldots$ be unary recursive functions. Give an example for such a sequence where $g(i) = f_i(i)$ is not recursive.

123 (Definition by cases, see Problem 105). Suppose $A_1, \ldots, A_k$ are recursive relations such that for all $\vec{x} \in \omega^n$ exactly one of them holds. If $h_1, \ldots, h_k$ are recursive functions, then so is

$$
f(\vec{x}) = \begin{cases}
h_1(\vec{x}) & \text{if } \vec{x} \in A_1 \\
h_2(\vec{x}) & \text{if } \vec{x} \in A_2 \\
\vdots & \vdots \\
h_k(\vec{x}) & \text{if } \vec{x} \in A_k
\end{cases}
$$

124. Let $K(u) = u \dotminus [\sqrt{u}]^2$, and $L(u) = ([\sqrt{u}] \dotminus K(u)) \dotminus 1$. Show that for each pair $(x, y) \in \omega^2$ there is an u with $K(u) = x$ and $L(u) = y$, and the function assigning the minimal such u to the pair (x, y) is recursive.

> **4.8 Definition** (Gödel's β function). $\beta(m, b, i) = \mathrm{rem}(m, b(i + 1) + 1)$.

125 (Gödel). Show that for any sequence of natural numbers $r_0, \ldots, r_{n-1}$ there exists m and b such that $\beta(m, b, i) = r_i$ for all $0 \le i < n$.

> **4.9 Definition** (Sequence coding). The unary function $\mathrm{Len}(u)$ and the binary function $\mathrm{Elem}(u, i)$, also written as $(u)_i$, *codes every sequence*, if for every $n \ge 0$ and sequence $\langle r_0, \ldots, r_{n-1} \rangle \in \omega^n$ there is a $u \in \omega$ such that $\mathrm{Len}(u) = n$ and $\mathrm{Elem}(u, i) = r_i$ for $i < n$. The *code of the sequence* is the minimal $u \in \omega$ with this property.

126. (a) Show that there are coding functions.

(b) Show that there are recursive coding functions with the additional property that $\mathrm{Elem}(u, i) < u$ when $i < \mathrm{Len}(u)$.

(c) If Len and Elem are recursive, then the set of sequence codes is recursive.

(d) If Len and Elem are recursive, then so is the "append" function $u \frown z$, which returns the code of the sequence where z is appended to the sequence coded by u.

127. Show that there are recursive coding functions Len^* and Elem^* with Elem^*
$(u, i) > u$ for some code $u \in \omega$ and $i < \mathrm{Len}^*(u)$.

128. Show that the coding functions $\mathrm{Len}(u)$, $\mathrm{Elem}(u, i)$, and $u \frown z$ based on Gödel's β function are primitive recursive.

> **Remark.** Problems 129–130 indicate an alternative method to code *sets of integers*. The method can be turned into sequence coding using the same trick which is used to define functions in set theory: an ordered pair (a, b) is the set $\{\{a\}, \{a, b\}\}$ and the sequence $\langle r_0, \ldots, r_{n-1} \rangle$ is the set containing the ordered pairs (i, r_i).

129. Show that the following functions are recursive:

(a) $f_a(x) = 1$ if x is a power of 3, and $f_a(x) = 0$ otherwise.

(b) $f_b(x)$ is the smallest number greater than x such that its base 3 expansion is of the form $10 \cdots 00$.

(c) $f_c(x, y) = z$, where the base 3 expansion of z is the concatenation of that of x and y.

(d) $f_d(x) = 1$ if the base 3 expansion of x starts with a block of 2's followed by a zero digit.

130. The number $x \in \omega$ "contains" $y \in \omega$ if the base 3 representations of x and y, as digit sequences, are u and v, respectively; u starts with a "separator" sequence s and contains the sequence svs, while v does not contain s. The separator sequence s is a non-empty sequence of 2's followed by a zero.

(a) Show that "x contains y" is a recursive relation.

(b) For every $y_1, \ldots, y_n$ there is an x which "contains" all y_i and nothing more.

(c) Given $\vec{y} \in \omega^n$ the minimal x which "contains" all y_i is a recursive function of $\vec{y}$.

131. Let $p(n)$ be the n-th prime. Define the code of the sequence $\langle r_0, \ldots, r_{n-1} \rangle$ be $u = p_n \cdot p_0^{r_0} \cdots p_{n-1}^{r_{n-1}}$. Show that the corresponding $\mathsf{Len}(u)$, $\mathsf{Elem}(u, i)$, and $u ^\frown z$ functions are primitive recursive.

> **Notation.** From this point on $\mathsf{Len}(u)$ and $\mathsf{Elem}(u, i) = (u)_i$ are fixed recursive coding functions as in Definition 4.9 with the additional property that $\mathsf{Elem}(u, i) < u$ for $i < \mathsf{Len}(u)$. The corresponding "append" function $u ^\frown z$ is also recursive by Problem 126.
> In subsequent problems no other property of these functions should be used: these statements are valid for arbitrary choice of the coding functions.

132. Show that the code of the empty sequence is 0.

133. Show that the following functions and relations are recursive.

(a) If $\mathsf{Len}(u) > 0$ then $f_a(u)$ is the code of the sequence of length $\mathsf{Len}(u) - 1$ where the last element is dropped from the sequence coded by u; otherwise $f_a(u)$ is the code of the empty sequence.

(b) $R_b(u, v) \subseteq \omega^2$ is true if the sequence coded by u is an initial segment of the sequence coded by v;

(c) if u codes the sequence $\langle r_0, \ldots, r_{n-1} \rangle$ then $f_c(u)$ is the code of a sequence of length $\mathsf{Len}(u)$ whose i-th element is the code of the initial segment $\langle r_0, \ldots, r_{i-1} \rangle$.

134. Show that 2^k is recursive.

> **Hint.** Show that the set of codes of the sequences $\langle 1, 2, \ldots, 2^{n-1} \rangle$ is recursive.

135. Show that the set of recursive functions is closed for the primitive recursive operator PrRec: if g and h are recursive and PrRec can be applied to them, then $\mathsf{PrRec}(g, h)$ is recursive. In particular, every primitive recursive function is recursive.

136. Suppose that the recursive f takes all of its values infinitely often. Let $g(i) = n$ if n is the $(i + 1)$-st place where $f(n) = i$ (e.g., $g(0)$ is the first n so that $f(n) = 0$). Prove that g is recursive.

137. Suppose f is recursive with $\limsup f = \infty$. Construct an increasing recursive g such that $f(i) = g(i)$ infinitely often.

138. Let f be a recursive function. If $u \in \omega$ encodes the sequence $\langle r_0, \ldots, r_{n-1} \rangle$, then let $g(u) = f(r_0) \cdot f(r_1) \cdots f(r_{n-1})$. Show that g is recursive.

> Primitive recursion defines the value of a function from the value it takes at the previous place. *Course-of-values recursion,* or simply *recursion,* allows all previous values to be used. It is done technically by passing the (code of the) sequence of the previous values to the defining function.

> **4.10 Definition.** The *recursive operator* Rec applied to the function $G(\vec{x}, u)$ returns the uniquely defined function $f(\vec{x}, n)$ which satisfies, for all $n \geq 0$,
>
> $$f(\vec{x}, n) = G(\vec{x}, u_n),$$
>
> where u_n is the code of the sequence $\langle f(\vec{x}, 0), \ldots, f(\vec{x}, n-1) \rangle$.

139. Show that recursive functions are closed for the Rec operator.

140. Take primitive recursive coding functions where the append function is primitive recursive as well (see Problem 131). Show that primitive recursive functions are closed for the Rec operator.

> **Remark.** See Problem 148 for the condition that the append function is primitive recursive.

141. Show that there is a function $H: \omega \to \omega$ which grows faster than every recursive function f in the sense that

$$\lim_{n \to \infty} \frac{f(n)}{H(n)} = 0.$$

142. Show that there is a recursive function $f(n, k)$ such that every sequence of length $\leq n$ with elements $\leq k$ has a code $\leq f(n, k)$.

143. Let $f(k, r)$ denote the least natural number n such that each coloring of the edges of the complete graph on n vertices with r colors contains a monochromatic complete subgraph of size k. Prove that f is recursive.

144. Show that the function $C(x, y)$ which returns the code of the concatenation of the sequences coded by its arguments is recursive.

> **Warning.** The natural idea to give the primitive recursion-like definition $C(x, 0) = x$, $C(x, y\frown z) = C(x, y)\frown z$, and then quote the recursive operator Rec, does not work.

145. Show that the function which reverses the order of the coded sequence is recursive.

146.* Show that the Ackermann function (Definition 4.4) is recursive.

> **Remark.** An immediate consequence of this problem is that $A(m, m)$ is a recursive function which is not primitive recursive (Problem 111). Another way to prove that there is a recursive but not primitive recursive function is indicated in Problem 179.

> **4.11 Definition.** The *graph of the function f* is the set
>
> $$\{\langle \vec{x}, f(\vec{x}) \rangle : \vec{x} \in \mathrm{dom}(f)\}.$$

147. (a) Show that if the graph of f is a recursive relation, then f is recursive.
(b) Show that there is a function f whose graph is primitive recursive, but f is not primitive recursive.

148. There are primitive recursive coding functions Len^* and Elem^* such that the corresponding append function is not primitive recursive, while for each $n \geq 1$ the function which at $(x_0, \ldots, x_{n-1})$ takes the code of the sequence $\langle x_0, \ldots, x_{n-1} \rangle$ is primitive recursive.

> **4.12 Definition.** The subset $A \subseteq \omega^n$ is *recursively enumerable*, or simply *enumerable*, if it is either empty, or there are unary recursive functions $f_1, \ldots, f_n$ such that $A = \{\langle f_1(i), \ldots, f_n(i) \rangle : i \in \omega\}$.

149. The non-empty $A \subseteq \omega^n$ is enumerable if and only if there is a unary recursive function f such that every $f(i)$ is a code of a sequence of length n, and $A = \{\langle (f(i))_0, \ldots, (f(i))_{n-1} \rangle : i \in \omega\}$.

> **Notation.** By abuse of notation we write $u \in A$ for $u \in \omega$ and $A \subseteq \omega^n$ to mean that u is a code of length n and $\langle (u)_0, \ldots, (u)_{n-1} \rangle \in A$. While this is ambiguous in the $n = 1$ case, the meaning should be clear from the context. Using this notation Problem 149 claims that a non-empty enumerable set is the range of a unary recursive function.

150. If A and B are enumerable, then so is $A \times B$. In particular, ω^n is enumerable.

151. The range of every recursive function (of several variables) is enumerable.

152. Let g be recursive and $A = \{\vec{x} : g(\vec{x}) = 0\}$. Show that A is enumerable.

153. (a) Every recursive set $A \subseteq \omega^n$ is enumerable.
(b) If A is recursive and infinite, then it is enumerable by an injective recursive function.
(c) Any infinite enumerable set can be enumerated by an injective recursive function.

154. The union and the intersection of two enumerable sets are enumerable.

155. Let $A \subseteq \omega$ be enumerable, and f be a recursive function. Show that $B = \{\vec{x} : f(\vec{x}) \in A\}$ is also enumerable.

156. If the graph of f is enumerable, then f is recursive.

157. $A \subseteq \omega^n$ is recursive if and only if both A and its complement are enumerable.

158. Let $A, B \subset \omega$ be disjoint, recursively enumerable but not recursive sets. Prove that $A \cup B$ is not recursive.

159. Every infinite recursively enumerable set has an infinite recursive subset.

160. Show that there is an infinite set which has no infinite recursive subset.

4.3 PARTIAL RECURSIVE FUNCTIONS

4.13 Definition (Operators on partial functions). Let Ω^* be the set of all *partial* functions mapping integers to integers, that is, all functions f with $\mathrm{dom}(f) \subseteq \omega^n$ and $\mathrm{ran}(f) \subseteq \omega$. Functions $f, g \in \Omega^*$ are equal if they have the same number of arguments, the same domain, and $f(\vec{x}) = g(\vec{x})$ for all $\vec{x} \in \mathrm{dom}(f)$. Partial functions with empty domain but different arity are considered to be different.

Operators Comp and μ are extended to functions in Ω^* as follows.
- Comp is defined on the tuple $(g, h_1, \ldots, h_\ell)$ if $\mathrm{dom}(g) \subseteq \omega^\ell$ and $\mathrm{dom}(h_i) \subseteq \omega^m$ for all i; $\vec{x} \in \omega^m$ is in the domain of the composite function if $\vec{x} \in \mathrm{dom}(h_i)$ for all i and $\langle h_1(\vec{x}), \ldots, h_\ell(\vec{x}) \rangle \in \mathrm{dom}(g)$, and the value is $g(h_1(\vec{x}), \ldots, h_\ell(\vec{x}))$.
- the μ *operator* can be applied to every $g \in \Omega^*$ with arity $\ell + 1 \geq 2$. $\mu(g)$ has arity ℓ and $\vec{x} \in \mathrm{dom}(\mu(g))$ if there is an $i \in \omega$ such that $\langle \vec{x}, j \rangle \in \mathrm{dom}(g)$ for all $j \leq i$ and $g(\vec{x}, i) = 0$. The value of $\mu(g)$ at $\vec{x}$ is the minimal such as i.

4.14 Definition. The set of *partial recursive* functions is the smallest subset of Ω^* that contains the initial functions below, and is closed for Comp and μ operators:
- addition, multiplication, and the $K_<$ function;
- the projection functions U_i^k for $1 \leq i \leq k$.

161. Show that every recursive function is partial recursive.

Notation. (a) For a function $f \in \Omega^*$ we write $f(\vec{x}) = \downarrow$ to indicate that $\vec{x} \in \mathrm{dom}(f)$, and $f(\vec{x}) = \uparrow$ to indicate that $\vec{x} \notin \mathrm{dom}(f)$.

(b) $f \in \Omega^*$ is *total* if $f \in \Omega$, i.e., $\mathrm{dom}(f) = \omega^n$ for some n.

162. Let $f : \omega^n \to \omega$ be a recursive function and $A \subseteq \omega^n$ be a recursive relation. Show that

$$h(\vec{x}) = \begin{cases} f(\vec{x}) & \text{if } \vec{x} \in A, \\ \uparrow & \text{otherwise} \end{cases}$$

is partial recursive.

163. Show that every function with finite domain is partial recursive.

164. Let f be recursive and g be partial recursive. Define h so that h is defined where g is, and at those places it takes the same value as f does. Show that h is partial recursive.

165. Prove that for all partial recursive f there is a partial recursive g which takes the same values as f but takes all of these values infinitely often.

166. Recall (Definition 4.11) that the graph of $f \in \Omega^*$ is the relation $\{\langle \vec{x}, f(\vec{x}) \rangle : \vec{x} \in \text{dom}(f)\}$. Suppose that the graph of $f \in \Omega^*$ is recursively enumerable. Show that f is partial recursive.

167.[*] Show that the graph of every partial recursive function is recursively enumerable.

168. Let $A \subseteq \omega$ be recursive, and f_1 and f_2 be unary partial recursive functions. Write

$$g_1(x) = \chi_A(x) \cdot f_1(x) + \chi_{\neg A}(x) \cdot f_2(x),$$

and

$$g_2(x) = \begin{cases} f_1(x) & \text{if } x \in A, \\ f_2(x) & \text{otherwise.} \end{cases}$$

Give examples for f_1 and f_2 when g_1 and g_2 are different functions. Verify that both g_1 and g_2 are partial recursive.

169. Prove that the following are equivalent for any set $A \subseteq \omega$:

(a) A is recursively enumerable,
(b) A is the domain of a partial recursive function,
(c) A is the range of a partial recursive function, and
(d) the function

$$f(x) = \begin{cases} 0 & \text{if } x \in A, \\ \uparrow & \text{otherwise} \end{cases}$$

is partial recursive.

170. Prove that all total partial recursive functions are, in fact, recursive.

> **Remark.** The result of the operators Comp and μ can be total even if some of the arguments are not total functions. This problem claims that a total

partial recursive function can be generated by the Comp and μ operators in a way that all intermediate functions are total.

4.4   CODING

Coding, or *arithmetization*, is a powerful technique which assigns natural numbers to mathematical objects in a way that preserves most of the structural properties. The code of the object o, denoted here by $\alpha(o)$, is typically the code (in the meaning of Definition 4.9) of a sequence of natural numbers reflecting how o is built up from other objects.

The technique relies on coding sequences of natural numbers. The additional property of the recursive coding functions Len and Elem that elements of a sequence are smaller than its code implies that objects which take part in the definition of another object will have smaller codes. Consequently "structural induction" on the objects translates to an application of course-of-values recursion (Definition 4.10). While not essential, it significantly simplifies the treatment.

No other intrinsic property of the coding functions will be (or should be) used. In particular, choosing primitive recursive coding functions, all functions and relations claimed here to be recursive are actually primitive recursive, see the remark after Theorem 4.16.

Notation. To ease the burden of complicated notations, $\langle x_0, \ldots, x_{n-1} \rangle$ will denote *both* the sequence itself (as an element of ω^n) *and* the code of this sequence (as an element of ω). The exact meaning should be clear from the context.

4.4.1   REGULAR EXPRESSIONS

A regular expression over the finite alphabet Σ defines a regular language (Definition 3.4). Elements of the alphabet Σ are identified with the natural numbers from 0 to $|\Sigma| - 1$, and words over Σ are (codes of) the sequences formed from them, giving a natural coding of Σ^*. The code $\alpha(\sigma)$ of the regular expression σ is defined by induction along its definition as follows:

$$
\begin{array}{lll}
\alpha(\lambda) = \langle 0 \rangle & & \text{empty word,} \\
\alpha(a) = \langle 1, \alpha(a) \rangle & & \text{singletons for } a \in \Sigma, \\
\alpha(\sigma|\tau) = \langle 2, \alpha(\sigma), \alpha(\tau) \rangle & & \text{union,} \\
\alpha(\sigma\tau) = \langle 3, \alpha(\sigma), \alpha(\tau) \rangle & & \text{concatenation,} \\
\alpha(\sigma^*) = \langle 4, \alpha(\sigma) \rangle & & \text{Kleene star.}
\end{array}
$$

171. (a) Show that different regular expressions have different codes.

(b) Show that the set of regular expression codes is recursive.

172. We say that $u \in \omega$ encodes the finite language $L \subset \Sigma^*$ if elements of u encode just the elements of L. Show that there is a recursive function $f(n, u, v)$ such that if u encodes L_1, v encodes L_2, then $f(n, u, v)$ encodes all words of length at most n in (a) $L_1 | L_2$ (union), (b) $L_1 L_2$ (concatenation).

173. Show that there is a recursive function $f(n, u)$ such that if u encodes the finite language L, then $f(n, u)$ encodes all words of length at most n in L^*.

174. Let $\Sigma^{<n}$ be the set of all words in Σ^* of length smaller than n. Show that there is a recursive function $f(u, n)$ such that if σ is a regular expression, then $f(\alpha(\sigma), n)$ encodes $\sigma \cap \Sigma^{<n}$.

175. Show that there is a recursive function $f(x, y)$ such that for every regular expression σ, the word w is in the generated language if and only if $f(\alpha(\sigma), \alpha(w)) = 1$.

4.4.2 PRIMITIVE RECURSIVE FUNCTIONS

The *code* of a primitive recursive function (Definition 4.2) is a description of how it is built up from the initial functions using the Comp and PrRec operators. The code reflects the definition and not the function: if f is primitive recursive, then so is $f + 0$, which, as a function, is the same, but has a different code. For the sake of convenience, the first element of the code is the arity, the second element is an indication of the applied operator.

$$\alpha(0) = \langle 1, 0, 0 \rangle,$$
$$\alpha(S) = \langle 1, 0, 1 \rangle, \qquad \left. \right\} \quad \text{initial functions,}$$
$$\alpha(U_i^k) = \langle k, 0, 2, i \rangle,$$
$$\alpha(f) = \langle m, 1, \alpha(g), \alpha(h_1), \dots, \alpha(h_\ell) \rangle \quad \text{for } f = \text{Comp}(g, h_1, \dots, h_\ell),$$
$$\alpha(f) = \langle \ell, 2, \alpha(g), \alpha(h) \rangle \qquad \text{for } f = \text{PrRec}(g, h).$$

176. Show that the set of function codes is recursive.

4.15 Definition. The sequence $u = \langle d_0, \dots, d_{n-1} \rangle$ is a *justified computation*, if (a) each d_i codes a triplet $\langle c_i, x_i, y_i \rangle$, where $c_i = \alpha(f_i)$ is a function code, x_i is the code of the sequence specifying the arguments of f_i, and y_i is value of f_i at the given arguments; (b) if f_i is an initial function then y_i is computed correctly; (c) if f_i is an application of Comp or PrRec, then all necessary computations justifying that the correct value of the function is y_i are coded earlier in the sequence.

177. Show that the set of justified computations is recursive.

178. Show that there are recursive functions $W_n(i, \bar{x})$ which are universal (see Definition 4.5) for the n-variable primitive recursive functions.

179. Show that $W_1(x, x) + 1$ is a recursive function which is not primitive recursive.

180. Define the unary primitive recursive functions $f_i(x)$ for $i = 0, 1, \ldots$ as follows. $f_0(n) = 2^n$; $f_{i+1}(0) = 1$, and $f_{i+1}(n+1) = f_i(f_{i+1}(n))$. Show that the function $i \mapsto \alpha(f_i)$ which assigns the code of f_i to i is recursive.

4.4.3 PARTIAL RECURSIVE FUNCTIONS

Partial recursive functions are coded analogously to that of primitive recursive functions. As codes for the two function types will never be mixed, there is no need to make their codes to be different.

$$\left.\begin{aligned}
\alpha(+) &= \langle 2, 0, 0 \rangle, \\
\alpha(\cdot) &= \langle 2, 0, 1 \rangle, \\
\alpha(K_<) &= \langle 2, 0, 2 \rangle, \\
\alpha(U_i^k) &= \langle k, 0, 3, i \rangle,
\end{aligned}\right\} \quad \text{initial functions,}$$

$$\alpha(f) = \langle m, 1, \alpha(g), \alpha(h_1), \ldots, \alpha(h_\ell) \rangle \quad \text{for } f = \text{Comp}(g, h_1, \ldots, h_\ell),$$

$$\alpha(f) = \langle \ell, 2, \alpha(g) \rangle \qquad\qquad\qquad \text{for } f = \mu(g).$$

181. Show that the set of the codes of partial recursive functions is recursive.

182. There is a recursive function f such that for every partial recursive function g the partial recursive function with code $f(\alpha(g))$ has the same domain as g, and takes zero everywhere.

183. There is a unary recursive function f such that $f(i)$ is the code of the unary recursive function which always returns i.

184. There is a recursive function $f(c, i)$ such that if g is a two-variable partial recursive function, then $f(\alpha(g), i)$ is the code of a one-variable partial recursive function g_i such that $g_i(x) = g(x, i)$ for all $x \in \omega$.

185. Extend the definition of justified computation (Definition 4.15) with the μ operator, and show that justified computations form a recursive relation.

186. Show that the graph of a partial recursive function is recursively enumerable (see Problem 167).

4.16 Theorem (Kleene's normal form theorem). There are recursive functions $G(t)$ and $H_n(e, \vec{x}, u)$ such that for each n-variable partial recursive function g there is an e such that $g(\vec{x}) = G(\mu\{u : H_n(e, \vec{x}, u) = 0\})$.

187. Prove Theorem 4.16.

Remark. The coding functions Len, Elem and $u^\frown z$ are primitive recursive, see Problem 128. Along the proof only course-of-values recursion (Definition 4.10) is used, which gives primitive recursive function from a primitive recursive one. Consequently the stronger statement that G and H_n can be chosen to be primitive recursive is also true.

188. Using Theorem 4.16 show that the graph of a partial recursive function is enumerable (see Problem 167).

189. Construct a partial recursive function $H(x, y)$ with the property that for each partial recursive $g(x, y)$ there is a recursive $\tilde{g}(x)$ such that $g(x, y) = H(\tilde{g}(x), y)$.

Hint. Let e be the code of g. The function $x \mapsto \langle e, x \rangle$ is recursive.

190.[*] Suppose $H(x, y)$ is as in Problem 189. Show that for each unary partial recursive function h there is an $m \in \omega$ such that $H(h(m), y) = H(m, y)$ (see also Problem 220).

4.5 UNIVERSAL FUNCTION

191. Prove that there is no total recursive function $U(i, \bar{x})$ that is universal for the set of total recursive n-ary functions (see Definition 4.5).

4.17 Theorem (Kleene)**.** For each $n \in \omega$ there is a partial recursive function $U_n : \omega^{1+n} \to \omega$ such that

- U_n is universal for the n-variable partial recursive functions;
- the functions U_n form a coherent family: there are recursive functions $S_n^m : \omega^{1+m} \to \omega$ for $m, n > 0$ such that

$$U_{m+n}(i, x_1, \ldots, x_m, y_1, \ldots, y_n) = U_n(S_n^m(i, x_1, \ldots, x_m), y_1, \ldots, y_n).$$

The second item is known as the *s-m-n theorem*.

192. Prove Theorem 4.17

Notation. If $g(\bar{x}) = U_n(i, \bar{x})$, then i is the *index* of g, or *U-index* if the universal function is not clear from the context. Following the common practice, $\varphi_i(x)$ abbreviates $U_1(i, x)$, $\varphi_i(x, y)$ abbreviates $U_2(i, x, y)$, etc.

For the rest of the section U_n and S_n^m are arbitrary but fixed functions satisfying Kleene's theorem. Problems below ask to show properties which are shared by every such family of coherent universal functions.

193. Show that for some recursive function $r(i)$, if i is the index of $g(x, y)$, then $r(i)$ is the index of $g(y, x)$.

194. Show that there is a recursive function r such that whenever the unary partial recursive g has index i, then $r(i)$ is the index of a partial recursive g' for which we have $\mathrm{dom}(g) = \mathrm{dom}(g')$ and the only value g' takes is 0 (see Problem 182).

195. Show that there is a recursive $r(i, j)$ such that $\varphi_{r(i,j)}(x) = \varphi_i(x) + \varphi_j(x)$.

196. Prove that there is a recursive function $c(i, j)$ with the property

$$\varphi_{c(i,j)}(x) = \varphi_i(\varphi_j(x)).$$

197. Let $A \subseteq \omega$ be recursive and f_1, f_2 be partially recursive functions. Show that there is a recursive $r(x)$ such that

$$\varphi_{r(i)}(x) = \begin{cases} f_1(x) & \text{if } i \in A, \\ f_2(x) & \text{otherwise.} \end{cases}$$

198. For each partial recursive $g(x, y)$ one can find a recursive $\tilde{g}(x)$ such that $g(x, y) = \varphi_{\tilde{g}(x)}(y)$ (see Problem 189).

199. Prove that for all partial recursive $f(x)$ there is a total recursive $g(x)$ such that $\varphi_{f(i)}(x) = \varphi_{g(i)}(x)$.

200. Show that there is a recursive function $F(j)$ such that if the partial recursive $f(x)$ has index j, then $F(j)$ is an index of a (total) recursive $g(x)$ such that $\varphi_{f(i)}(x) = \varphi_{g(i)}(x)$.

201. Show that there is a recursive function f so that $\varphi_{f(i)}(x)$ is the constant i function.

202. Show that there is a recursive function $H(i, x, y, u)$ such that $\varphi_i(x) = y$ if and only if $H(i, x, y, u) = 0$ for some $u \in \omega$.

203. (a) Prove that there is a recursive function h such that for all $i \in \omega$, $\mathrm{ran}(\varphi_i) = \mathrm{dom}(\varphi_{h(i)})$.
(b) Prove that there is a recursive function h such that $\mathrm{dom}(\varphi_i) = \mathrm{ran}(\varphi_{h(i)})$, and $\varphi_{h(i)}$ is total whenever $\mathrm{dom}(\varphi_i)$ is not empty.

204. (a) Show that there is a partial recursive function which is *not* the restriction of any recursive function to its domain.
(b) Show that there is such a partial recursive function with a $\{0, 1\}$ range.
(c) Let f be partial recursive with infinite domain and suppose there is a recursive function g such that $g(n)$ is the n-th element of $\mathrm{dom}(f)$. Show that f is a restriction of a recursive function.

205. Show that there is an infinite, recursively enumerable set $A \subseteq \omega$ for which the function f defined by the primitive recursion

$$f(0) = \min\{i : i \in A\}$$
$$f(n+1) = \min\{i : i > f(n) \ \wedge \ i \in A\}$$

is not recursive.

206. Give an example for a recursively enumerable set which is not recursive.

207. Show that there is a partial recursive function which takes (a) somewhere (b) infinitely often a larger value than any recursive function.

208.[*] Suppose the partial recursive f has infinite domain. Prove that there is a total recursive g which takes a larger value than f infinitely many times.

> **Notation.** The function f *dominates* g if, with finitely many exceptions, $f(x) > g(x)$ for all $x \in \operatorname{dom}(f) \cap \operatorname{dom}(g)$.

209. Show that there is a total function $f : \omega \to \omega$ which dominates all recursive functions. Can this f be recursive?

210. Show that no partial recursive function with infinite domain dominates all recursive functions.

211. Show that there is a partial recursive function which is not dominated by any recursive function.

212. Let A be recursively enumerable and suppose for all $i \in A$ the function $\varphi_i(x)$ is total recursive. Show that there is a total recursive g which dominates all φ_i for $i \in A$.

213. Given a counterexample to the following claim: "Suppose $R \subseteq \omega^2$ is a recursive relation such that for every x there is at least one, but finitely many y with $(x, y) \in R$. The function $g(x) = \max\{y : (x, y) \in R\}$ is recursive."

> **Remark.** If one takes minimum instead of maximum, then g is clearly recursive.

4.6   DECIDABILITY

> **4.18 Definition.** A set $A \subseteq \omega^n$ is *decidable* if it is a recursive relation, and *undecidable* otherwise. The words *recursive* and *decidable* are used as synonyms.

> The notion of decidability extends to other mathematical disciplines using *coding*: a question about certain mathematical objects is decidable if the set of the codes for the positive instances is recursive. In other words, just looking at the code, one can tell whether the answer to the question is yes or no. While this definition depends on the coding, in practice, it is quite robust and gives equivalent notions of decidability for a large variety of "natural" coding.

214. The problem of whether a word matches a regular expression is decidable.

215. Show that the set $K = \{i \in \omega : \varphi_i(i) = \downarrow\}$ is undecidable.

> **Remark.** The problem for which K is the solution set is known as the *diagonal halting problem*, and can be phrased as "for which i does the computation of the partial function with code i at input i halt?"

216. Show that K is recursively enumerable.

217. Show that if A is recursively enumerable, then there is a recursive function f such that $i \in A$ iff $f(i) \in K$.

218. Show that $\{i \in \omega : \varphi_i$ is total$\}$ is not recursively enumerable (hence not decidable).

219. Show the condition whether the μ operator is applicable for a two-variable function is undecidable. More precisely, the set $\{i \in \omega : \varphi_i(x, u)$ is total, and for every x there is a u such that $\varphi_i(x, u) = 0\}$ is not recursive.

> **Remark.** This result explains why there is no coding for recursive functions: there is no (recursive) way to tell whether the μ operator is applicable.

> **4.19 Theorem** (Fixed point theorem). For each recursive function $g(x)$ there exists $m \in \omega$ such that $\varphi_m \equiv \varphi_{g(m)}$.

220. (a) Prove Theorem 4.19.

> **Hint.** If i is the index of h, then $h(i) = \varphi_i(i)$. Set $m = h(i)$ for some h.

(b) What happens if g is partial recursive?

221. Show that there is an $m \in \omega$ such that $\varphi_m(x) \equiv m$ (interpreted as "φ_m returns its own code").

222. Give an $i \in \omega$ such that $\varphi_i(x) = i + x$.

223. There is a recursive function $m(i)$ which returns a fixed point for φ_i.

224. (a) Let f and g be two-variable recursive functions. Show that there are integers i and j such that $\varphi_i = \varphi_{f(i,j)}$ and $\varphi_j = \varphi_{g(i,j)}$.
(b) Show that for unary recursive functions f and g, in general, it is impossible to find an i such that $\varphi_{f(i)} = \varphi_{g(i)}$.
(c) What happens if, in addition, every integer is taken by f and g?

> **4.20 Definition.** The *fixed point set* of the recursive function f is $\mathrm{Fix}(f) = \{n \in \omega : \varphi_n = \varphi_{f(n)}\}$. It is never empty by Theorem 4.19.

225. Prove that $\mathrm{Fix}(f)$ is infinite.

226. There is no recursive A such that both A and its complement are fixed point sets.

227. Let $A \subset \omega$ be recursive and let f be the recursive function such that

$$\varphi_{f(i)}(x) = \begin{cases} \varphi_i(x) & \text{if } i \in A \\ \uparrow & \text{otherwise.} \end{cases}$$

Is it always true that $\text{Fix}(f) = A$?

228. Show that the Ackermann function (Definition 4.4) is recursive (cf. Problem 146).

> **Hint.** The idea is to find the Ackermann function in the form $\varphi_n(x, y)$ for a suitable index n. Write the defining equations as
>
> $$g(n, x, y) = \begin{cases} y + 1 & \text{if } x = 0, \\ \varphi_n(x, 1) & \text{if } x > 0 \text{ and } y = 0, \\ \varphi_n(x \doteq 1, \varphi_n(x, y \doteq 1)) & \text{if } x > 0 \text{ and } y > 0, \end{cases}$$
>
> and apply Theorem 4.19.

229. Consider the ternary function $B(x, y, z)$ defined by the recursion

$$B(x, y, 0) = x + y,$$

$$B(x, 0, z + 1) = \begin{cases} 0 & \text{if } z = 0, \\ 1 & \text{if } z = 1, \\ x & \text{if } z > 1, \end{cases}$$

$$B(x, y + 1, z + 1) = B(x, B(x, y, z + 1), z).$$

Show that $B(x, y, 1) = x \cdot y$ and $B(x, y, 2) = x^y$. Prove that B is recursive.

> **Remark.** $B(x, y, z)$ is the original version of Ackermann's function.

> **4.21 Theorem** (Rice's theorem). The subset $A \subseteq \omega$ is an *index property*, if $\varphi_i \equiv \varphi_j$ implies $(i \in A \Leftrightarrow j \in A)$. If an index property is decidable, then it is either empty, or it is the whole ω.

230. Prove Rice's theorem 4.21.

231. Which one of the following sets is decidable? Which one is recursively enumerable?

(a) $A = \{i \in \omega : \varphi_i(0) = \downarrow\}$,

(b) $B = \{i \in \omega : \text{dom}(\varphi_i) = \emptyset\}$,

(c) $C = \{i \in \omega : \varphi_i(x) = x^2\}$,

(d) $D = \{i \in \omega : 0 \notin \text{dom}(\varphi_i)\}$,

(e) $E = \{i \in \omega : 0 \in \text{ran}(\varphi_i)\}$,

(f) $F = \{i \in \omega : \varphi_i(x) = x\}$.

232. Show that every partial recursive function has infinitely many indices.

233. Show that there is a universal partial recursive function $W(i,x)$ so that some unary partial function has a single W-index.

234. Find two recursive functions f and g such that $\lim_{n\to\infty} \frac{f(n)}{g(n)} = \alpha$ exists, but the function $d(n)$ that gives the n-th digit of the binary expansion of the fractional part of α is not recursive.

235. Are there recursive functions f, g and a real number α such that $|\frac{f(n)}{g(n)} - \alpha| < \frac{1}{n}$ while the function $d(n)$ that gives the n-th decimal digit of α is not recursive?

4.7 RECURSIVE ORDERS

> **4.22 Definition.** An ordinal α is called *recursive* if there is a recursive set $A \subset \omega$ and a recursive linear ordering $\prec$ on A such that $\langle A, \prec \rangle$ is isomorphic to $\langle \alpha, < \rangle$.

236. Prove that the *infinite* ordinal α is recursive if and only if there is a recursive ordering $\lhd$ *on* ω such that $\langle \omega, \lhd \rangle$ is isomorphic to $\langle \alpha, < \rangle$.

237. (a) Show that $\omega + 1$ and $\omega + \omega$ are recursive ordinals.

(b) Show that ω^2 is a recursive ordinal.

(c) Prove that ω^ω is a recursive ordinal.

238. Prove the following statements.

(a) If α is a recursive ordinal and $\beta < \alpha$, then β is a recursive ordinal.

(b) If α and β are recursive ordinals, then so is $\alpha \cdot \beta$.

(c) If α is a recursive ordinal, then $\alpha + 1$ is a recursive ordinal as well.

(d) There is a minimal, *countable*, non-recursive, limit ordinal.

> **Remark.** This ordinal is called the *Church–Kleene* ordinal, denoted by ω_1^{CK}.

239. Prove or disprove the following statements:

(a) The supremum of a recursive sequence of recursive ordinals is a recursive ordinal. More precisely, if $I \subset \omega$ is a recursive set and α_i is a recursive ordinal for each $i \in I$, then $\sup_{i \in I} \alpha_i$ is a recursive ordinal.

(b) Suppose $f, g : \omega \to \omega$ are recursive functions such that $\varphi_{f(i)}$ is the characteristic function of $A_i \subset \omega$, and $\varphi_{g(i)}$ is the characteristic function of a well-ordering $<_i$ on A_i. Assume $\langle A_i, <_i \rangle$ has order-type α_i. Then $\sup_{i \in \omega} \alpha_i$ is a recursive ordinal.

240. Prove that ω_1^{CK} is at least as large as $\omega^{\omega^{\omega^{\cdot^{\cdot^{\cdot}}}}}$.

PROPOSITIONAL CALCULUS

5

5.1   FORMULAS

5.1 Definition (Syntax). Propositional formulas are elements of the free algebra generated by the set of propositional variables V and constants $\top, \bot$ in the Boolean-type $\langle \vee, \wedge, \neg \rangle$.

The following abbreviations will be used:

$$\varphi \to \psi \quad \text{instead of} \quad \neg\varphi \vee \psi,$$
$$\varphi \leftrightarrow \psi \quad \text{instead of} \quad (\varphi \to \psi) \wedge (\psi \to \varphi).$$

5.2 Definition (Evaluation). Given the set V of propositional variables, an *evaluation* is a function $f : V \to \{\top, \bot\}$, which tells the *truth value* of each variable. The evaluation, in fact, maps the generators of the free algebra of formulas into the two-element Boolean algebra $\mathbf{B} = \{\top, \bot\}$, thus f automatically extends to a homomorphism from all formulas to $\mathbf{B}$.

5.3 Definition (Satisfiability). A formula φ is *satisfiable* if for some evaluation f, $f(\varphi) = \top$. A set F of propositional formulas is satisfiable if for some evaluation f, all elements of F have value $\top$. F is *contradictory* if it is not satisfiable.

5.4 Definition (Tautology). A formula φ is a *tautology* if for each evaluation f we have $f(\varphi) = \top$. φ is *refutable* if it is not a tautology.

241. We have 11 tokens and 10 boxes. The value of the propositional variable $A_{i,j}$ is $\top$ if the i-th token is in the j-th box and $\bot$ otherwise. Write formulas expressing the following claims:

(a) the first token is in at most one box;

(b) the second token is in at least three boxes;

(c) the first five tokens are in the first three boxes;

(d) pigeonhole principle: it is impossible that all tokens are in one or more boxes, but each box contains at most one token.

L. Csirmaz and Z. Gyenis, *Mathematical Logic,* Problem Books in Mathematics, https://doi.org/10.1007/978-3-030-79010-3_5

242. Let G be a finite graph with V as the set of vertices. For $v \in V$ and $e = 1, 2, 3, 4$ the propositional variable $A_{v,e}$ means that the vertex v is colored by color e. Write formulas which express the facts that each vertex has exactly one color and vertices connected by an edge have different colors.

243. Let G be a finite directed graph. For vertices v and w the propositional variable $A_{v,w}$ indicates whether there is a directed edge from v to w or not. Construct a formula which expresses that there exists a vertex from which every other vertex can be reached in at most two steps.

244. Let $|G| = 100$, $|B| = 100$. The value of the propositional variable A_{gb} is $\top$ if the girl $g \in G$ knows the boy $b \in B$ and $\bot$ otherwise. Construct a single formula which expresses

(a) every girl knows exactly one boy and the other way around;

(b) all subsets of girls know at least as many boys as many girls are in that subset.

> **5.5 Definition.** The function f is *Boolean* if $\mathrm{dom}(f) = \{0, 1\}^n$ for some $n \geq 1$, and $\mathrm{ran}(f) \subseteq \{0, 1\}$.

245. Show that Boolean functions can be expressed using the traditional "and," "or," and "not" operators only.

246. Show that for some n-ary Boolean function the shortest such expression in Problem 245 has length $\geq 2^{n/10}$.

247. Ternary functions are defined on $\{0, 1, 2\}^n$ and take values in $\{0, 1, 2\}$. Show that every ternary function can be expressed from the following functions using composition only: $x \wedge y = \min(x, y)$, $x \vee y = \max(x, y)$, $\curvearrowright x = x + 1 \pmod{3}$, and $\neg x = 2 - x$.

248. Show that all Boolean functions can be obtained from the following functions by taking compositions only:

(a) $f_1(x, y) = xy \pmod 2$, $f_2(x) = 1 - x$.

(b) $f_1(x, y) = xy + x \pmod 2$, $f_2(x) = 1 - x$.

(c) Show that the following two functions do not generate all Boolean functions: $f_1(x, y) = xy \pmod 2$, $f_2(x, y) = x + y \pmod 2$.

249. (a) Does there exist a single Boolean function from which all other Boolean functions can be obtained by composition? How many such binary functions exist?

> **Warning.** No constants can be used in the composition.

(b) What is the case for the collection of real functions $\mathbb{R}^n \to \mathbb{R}$?

> **5.6 Definition.** A *k-valued logical function* is a function defined on $\{0, 1, \ldots, k-1\}^n$ which has values in $\{0, 1, \ldots, k-1\}$.

250. Let $k \geq 2$. The function set F consists of $+$, $\cdot$ (modulo k), characteristic functions of single elements of $\{0, \ldots, k-1\}$, and constant functions:

$$F = \{+, \cdot, \chi_a, a : 0 \leq a < k\}.$$

Show that F generates all k-valued logical functions.

251. Show that there exists a single k-valued logical function which generates all other such functions (including the constant ones) by taking compositions only.

> **Hint.** Merge the functions in Problem 250 into a single function with four variables.

252.[*] Prove that all k-valued functions can be obtained by compositions from the binary function $F(x, y) = \max\{x, y\} + 1 \pmod{k}$.

253. **A** is a *free algebra* over the set of generators $G \subseteq \mathbf{A}$, if every embedding of G to an algebra **B** of the same type can be extended to a homomorphism from **A** to **B**. Let $\mathbf{A}_1$ and $\mathbf{A}_2$ be two free algebras over the same generator set G. Show that they are isomorphic.

> **5.7 Definition.** A *congruence* on an algebra **A** is a partition of **A** for which there exists a homomorphism $\varphi : \mathbf{A} \to \mathbf{B}$ collapsing elements of the same class. A congruence is *smaller* than another one if the former is a refinement of the latter.

254. Show that there are smallest and largest congruences. Show that any two congruences have a smallest upper bound and greatest lower bound.

255. **L** is the five element non-modular lattice as pictured. Draw the congruence lattice of **L**.

$$\mathbf{L} =$$

256. Let ϑ be an equivalence relation on the set X different from $\Delta = \{\langle x, x \rangle : x \in x\}$ and $\nabla = X \times X$. Is it always possible to define an algebra on X whose congruences are exactly ϑ, Δ, and ∇?

> **5.8 Theorem** (Compactness—weak form). Let Σ be a set of propositional formulas. If every finite subset of Σ is satisfiable, then Σ is also satisfiable.

257. Prove Theorem 5.8.

> **Hint.** Consider the problem as a design for a voting system (Chapter 2.2). Every finite subset of Σ has a vote on what the value of the propositional variable $v \in V$ (or a formula φ) should be. Try to find a voting system which is compatible with the evaluation of formulas.

258 (Compactness of k-valued logic). Let V be a set of k-valued logical variables taking values in $\{0, \ldots, k-1\}$. The elements of Σ are $\varphi_\alpha(v_{\alpha_1}, \ldots, v_{\alpha_n})$ where φ_α is a k-valued logical function. Σ is satisfiable if there is an evaluation of the variables V such that each function in Σ has value 0.

Show that if every finite subset of Σ is satisfiable, then Σ is also satisfiable.

259. Let Σ be a collection of (multivariate) polynomials over a finite field. Show that if every finite subset of Σ has a solution (there are values for the variables which make every polynomial in the set vanish), then the whole set has a solution.

260. Give a counterexample to the previous problem when the polynomials are over (a) real numbers; (b) integers; (c) complex numbers.

> **5.9 Definition.** φ is a *semantical consequence* of Σ, written as $\Sigma \vDash \varphi$, if every evaluation that satisfies Σ also satisfies φ.

> **5.10 Theorem** (Compactness—strong form). If $\Sigma \vDash \varphi$, then $\Sigma' \vDash \varphi$ for some finite subset $\Sigma' \subseteq \Sigma$.

261. Prove Theorem 5.10.

> **Hint.** Use the fact that $\Sigma \vDash \varphi$ if and only if $\Sigma \cup \{\neg\varphi\}$ is not satisfiable.

262. Let V be a set of propositional variables, F be the set of formulas using variables from V only, and let X consist of all evaluations of V. A *topology* is defined on X by stipulating $U \subseteq X$ to be open if, for some appropriate set $\Gamma \subseteq F$ of formulas, U consists of those evaluations which make some element of Γ true.

(a) Show that this is a topological space indeed. Show that this space has a basis consisting of sets which are simultaneously closed and open (clopen sets).

(b) Show that the compactness theorem is equivalent to the fact that this space is compact.

(c) What separation properties among T_0, T_1, T_2, and T_3 this space has?

(d) Does this space have the M_1 or M_2 property? Is it separable? Is it metrizable?

(e) Prove that X is homeomorphic to $^V2$, the product of discrete topology.

263 (Interpolation theorem). Suppose $\varphi \to \psi$ is a tautology. Show that there exists a formula ϑ, which might contain only those variables which occur both in φ and ψ for which both $\varphi \to \vartheta$ and $\vartheta \to \psi$ are tautologies.

264 (Lindenbaum algebra). Let V be a set of propositional variables, and F be the set of all formulas using variables from V only. Define $\varphi_1 \equiv \varphi_2$ if $f(\varphi_1) = f(\varphi_2)$ for all homomorphisms (i.e., evaluations) f. Let $[\varphi]$ denote the equivalence class of φ. Define $[\varphi_1] \leq [\varphi_2]$ if $f(\varphi_1) \leq f(\varphi_2)$ always, where, as you might expect, $\bot < \top$. Show that $\leq$ is a partial ordering, and the equivalence classes with this ordering form a distributive lattice.

> **Remark.** Actually, this lattice is a Boolean algebra. It has a minimal and maximal element, and the complement of the class $[\varphi]$ is $[\neg\varphi]$.

265 (Erdős–deBruijn theorem). Suppose that all finite subgraphs of G can be colored by $n < \omega$ colors. Show that then the whole graph can also be colored by n colors.

266 (Erdős–deBruijn theorem—alternative proof). Suppose that all finite subgraphs of the graph G can be colored by k colors. Using Zorn's lemma show that G can be extended to a maximal such graph H by adding new edges (i.e a graph to which no more edges can be added without causing some finite subgraph to require more than k colors). Show that *non*-adjacency in H is an equivalence relation. Prove that H (and hence G) can be colored by k colors.

267. Consider the following graphs. Pick n different points arbitrarily in the d-dimensional Euclidean space and connect two points if their distance is exactly one. Suppose all of these graphs can be properly colored by k colors. Show that in this case all points of the space can also be colored by k colors so that points at distance 1 are colored differently.

268. Show that the points of the plane can be colored by 9 colors so that any two points at distance 1 are colored differently.
   Show that you need at least four colors to do that.

269. Let $k < \omega$ be fixed. Suppose that *edges* of each finite induced subgraph of G can be colored by two colors such that it has no homogeneous subgraph of size k. Show that G can also be colored in such a way.

270. Let G and B be two (possibly infinite) sets, the set of girls and boys, respectively. Suppose that every subset A of G together knows at least as many boys from B as the cardinality of A, and the same is true for B. Give an example where, in spite of this, G and B cannot be matched.

271. Assume, in addition to the conditions of Problem 270, that every girl knows finitely many boys from B only, and vice versa. Show that in this case there exists a matching between G and B.

> **Hint.** Use the compactness theorem. You'll need a special combinatorial lemma.

5.11 Definition. A *tournament* is a directed graph where between any two vertices there is exactly one edge. A vertex v in a tournament is a *king* if every other vertex can be reached from v by a (directed) path of length at most two.

272. (a) Show that in every finite tournament there is a king.

(b) Give an infinite tournament without a king.

273. (a) Show that for every large enough n there is a tournament on n vertices where everyone is a king.

(b) Does there exist an infinite tournament where everyone is a king?

274. Suppose that in an infinite tournament every vertex has finite "in-degree" (i.e., only finitely many edges are directed toward this vertex). Show that there is a king in it.

Hint. Use the compactness theorem.

275. Prove or disprove that a finite tournament has a source (vertex of in-degree zero) if and only if it has a sink (vertex of out-degree zero). What is the case for infinite tournaments?

5.2 DERIVATION

5.2.1 RESOLUTION METHOD

Given a set Σ of propositional formulas, the resolution method determines whether Σ is refutable or not. This yields strong completeness since $\Sigma \vDash \varphi$ if and only if $\Sigma \cup \{\neg\varphi\}$ is refutable.

5.12 Definition. A *conjunctive normal form* is a formula of the form

$$(\pm v_{0,0} \vee \cdots \vee \pm v_{0,n-1}) \wedge \cdots \wedge (\pm v_{v,0} \vee \cdots \vee \pm v_{k,n-1})$$

where $v_{i,j}$ is a propositional variable and $\pm v$ is either v or $\neg v$ (see Problem 245).

276. Prove that every formula can be transformed into conjunctive normal form using the following steps until they cannot be applied:

Step 1. If φ has a subformula of the form $\neg(\psi \vee \vartheta)$, then replace it by $(\neg\psi) \wedge (\neg\vartheta)$. If φ has a subformula of the form $\neg(\psi \wedge \vartheta)$, then replace it by $(\neg\psi) \vee (\neg\vartheta)$.

Step 2. Replace any subformula of φ of the form $\neg(\neg\psi)$ by ψ.

Step 3. If φ has a subformula of the form $\psi \vee (\vartheta_0 \wedge \vartheta_1)$ or $(\vartheta_0 \wedge \vartheta_1) \vee \psi$, then replace it by $(\psi \vee \vartheta_0) \wedge (\psi \vee \vartheta_1)$.

Show that if starting from φ we get φ^*, then $\vDash \varphi \leftrightarrow \varphi^*$.

> **5.13 Definition.** A *literal* is either a propositional variable or its negation; a *clause* is a finite set of literals; and $\square$ denotes the empty clause. The clause set $\mathcal{C}$ is *satisfiable*, if there is an evaluation such that every clause in $\mathcal{C}$ contains at least one true literal. $\mathcal{C}$ is *refutable* if it is not satisfiable.

> If $\square \in \mathcal{C}$ then $\mathcal{C}$ is refutable (as no true literal can be in $\square$). If $\mathcal{C} = \emptyset$, then $\mathcal{C}$ is satisfiable.

> **5.14 Definition** (Resolvent). Let c_0, c_1 be clauses, ℓ be a literal so that $\ell \in c_0$ and $\neg \ell \in c_1$. The *resolvent* of c_0 and c_1 with respect to ℓ is
>
> $$\mathbb{R}(c_0, c_1, \ell) = (c_0 \smallsetminus \{\ell\}) \cup (c_1 \smallsetminus \{\neg \ell\})$$

277. Let c be a one step resolvent of $\mathcal{C}$. Then $\mathcal{C}$ is satisfiable if and only if $\mathcal{C} \cup \{c\}$ is satisfiable.

> **5.15 Definition** (Resolution method). Let $\mathcal{C}$ be a set of clauses and c be a clause. c is *derivable* from $\mathcal{C}$ by the resolution method, denoted as $\mathcal{C} \vdash^R c$, if there is a sequence of clauses $c_0, c_1, \ldots, c_{n-1}$ ending with $c_{n-1} = c$ so that each c_i is a one step resolvent from the set $\mathcal{C} \cup \{c_0, \ldots, c_{i-1}\}$.

> **5.16 Theorem** (Completeness of Resolution). $\mathcal{C}$ is refutable if and only if $\mathcal{C} \vdash^R \square$.

278. What is the set of clauses arising from this formula:

$$\neg(A \vee \neg((B \wedge \neg C) \vee (\neg B \wedge C))) \vee \neg(\neg A \vee \neg((B \wedge C) \vee (\neg B \wedge \neg C)))$$

279. Deduce the following tautology using the resolution method:

$$\neg(\neg A \vee B) \vee (\neg(\neg C \vee A) \vee (\neg C \vee B))$$

280. Using the resolution method prove that the following formula is a tautology:

$$((A \rightarrow (B \rightarrow C)) \wedge (A \rightarrow (B \rightarrow D))) \rightarrow ((A \rightarrow B) \rightarrow (A \rightarrow (C \wedge D)))$$

281 (Deduction lemma). Let ℓ be a literal, and $\{\ell\}$ be a one-element clause. Suppose that for the clause c differing from $\{\ell\}$ we have $\mathcal{C}, \{\ell\} \vdash^R c$. Then either $\mathcal{C} \vdash^R c$ or $\mathcal{C} \vdash^R c \cup \{\neg \ell\}$.

282. Let $\mathcal{C}$ be a maximal set of clauses from which the empty clause cannot be derived using resolution. Show that for each propositional variable v exactly one of $\{v\} \in \mathcal{C}$ or $\{\neg v\} \in \mathcal{C}$ hold.

283. Prove Theorem 5.16.

284. For a set of clauses C and a clause c not containing both a literal and its negation the following two statements are equivalent:

(i) $C \vdash^R c'$ for some clause $c' \subseteq c$ and

(ii) If an evaluation satisfies every element of C, then it also satisfies c.

285. Solve Lewis Carroll's sorosites given in the preface.

286. Assume clauses can be *multisets* rather than sets, meaning that in a clause the same literal may occur repeatedly. The resolvent $\mathbb{R}(c_0, c_1, \ell)$ discards one instance of ℓ and one instance of $\neg\ell$ from $c_0 \cup c_1$. Give an example of a refutable C for which $C \not\vdash^R \square$. Where does the proof of the deduction lemma in Problem 281 fail for multisets?

5.2.2   GENTZEN'S NATURAL DEDUCTION

5.17 Definition (Sequent). A pair $\Gamma \vdash \Delta$ is a *sequent* if Γ and Δ are finite sets of propositional formulas. Curly brackets and the union sign are omitted, thus $\Gamma, \varphi \vdash \varphi, \Delta, \psi$ is the sequent $\Gamma \cup \{\varphi\} \vdash \Delta \cup \{\varphi, \psi\}$.

5.18 Definition (Natural deduction). A *Gentzen-style derivation* is a tree of sequents that is formed using axioms at the leaves and inference rules at nodes.
Axioms: $\Gamma, \varphi \vdash \varphi, \Delta$ (the two sides have a common formula).
Inference rules:

$$(\neg \vdash) \quad \frac{\Gamma \vdash \varphi, \Delta}{\Gamma, \neg\varphi \vdash \Delta} \qquad\qquad (\vdash \neg) \quad \frac{\varphi, \Gamma \vdash \Delta}{\Gamma \vdash \neg\varphi, \Delta}$$

$$(\vee \vdash) \quad \frac{\Gamma, \varphi \vdash \Delta \quad \Gamma, \psi \vdash \Delta}{\Gamma, \varphi \vee \psi \vdash \Delta} \qquad\qquad (\vdash \vee) \quad \frac{\Gamma \vdash \varphi, \psi, \Delta}{\Gamma \vdash \varphi \vee \psi, \Delta}$$

$$(\wedge \vdash) \quad \frac{\Gamma, \varphi, \psi \vdash \Delta}{\Gamma, \varphi \wedge \psi \vdash \Delta} \qquad\qquad (\vdash \wedge) \quad \frac{\Gamma \vdash \varphi, \Delta \quad \Gamma \vdash \psi, \Delta}{\Gamma \vdash \varphi \wedge \psi, \Delta}$$

287. Construct a Gentzen-style proof for the following tautologies:

(a) $A \to (A \vee B)$.

(b) $A \to (B \to (A \wedge B))$.

(c) $(A \to B) \to ((A \to (B \to C)) \to (A \to C))$.

288. Show that the sequent $\emptyset \vdash \varphi$ is derivable if and only if φ is a tautology.

5.2.3   HILBERT-TYPE DERIVATION

A Hilbert-type derivation is specified by a given set of formulas, called *axioms*, and a collection of *inference rules*. An inference rule is a (partial) function that assigns a formula (the conclusion) to a finite set of other formulas (called premises).

5.19 Definition (Hilbert-type derivation). A Hilbert-type *derivation* of φ from Σ is a finite sequence of formulas $\varphi_1, \ldots, \varphi_n$ ending in $\varphi_n = \varphi$ such that each formula in the sequence is either

- an element of Σ, or
- an axiom, or
- the conclusion of an inference rule where all premises of the rule are in the sequence before the formula.

We write $\Sigma \vdash \varphi$ to indicate that there is a Hilbert-type proof of φ from Σ.

Remark. For a fixed set of axioms and inference rules the Hilbert-type derivation has the following properties:

 (i)   *monotonicity:* $\Sigma_0 \subset \Sigma$, $\Sigma_0 \vdash \varphi$ implies $\Sigma \vdash \varphi$.
 (ii) *compactness:* if $\Sigma \vdash \varphi$, then for some finite $\Sigma' \subset \Sigma$ we have $\Sigma' \vdash \varphi$.
 (iii) *transitivity:* if $\Sigma_0 \vdash \varphi$ and $\Sigma_1 \vdash \Sigma_0$, then $\Sigma_1 \vdash \varphi$.

The main issue is to find a set of axioms and inference rules which leads to a *sound* (if $\Sigma \vdash \varphi$ then $\Sigma \vDash \varphi$) and *complete* (if $\Sigma \vDash \varphi$ then $\Sigma \vdash \varphi$) proof system. These sets depend on the available propositional variables and are typically specified by *schemes*: to get an axiom or an inference rule from the scheme the Greek letters (metavariables) in the scheme are to be replaced by propositional formulas (generated by the given set of propositional variables) so that the same formula is replaced for each occurrence of the same letter.

Only one type of inference rule will be used, the *modus ponens*, abbreviated as MP.

5.20 Definition (Modus Ponens).

$$\text{MP:} \quad \frac{\varphi \quad \varphi \to \psi}{\psi}$$

Here the two premises are φ and $\varphi \to \psi$, they are above the line, and the conclusion, ψ is below.

289. Suppose all axioms are tautologies. Show that $\vdash$ is sound.

290. Suppose all tautologies are axioms. Show that $\vdash$ is complete.

Hint. Use the Compactness Theorem 5.10.

The following problems indicate how to find a sound and complete proof system with *finitely many* axiom schemes. The schemes will be unveiled one by one. These axioms form neither a nice, nor an independent set, however they suffice for our purposes.

291 (Syntactic deduction lemma). $\Sigma, \psi \vdash \varphi$ if and only if $\Sigma \vdash \psi \to \varphi$.

> **Hint.** ($\Rightarrow$) Use induction on the length of the derivation of which φ is the last member. Introduce the following axiom schemes:
>
> Ax_1 $\varphi \to (\psi \to \varphi)$
> Ax_2 $\varphi \to \varphi$
> Ax_3 $(\psi \to (\vartheta \to \varphi)) \to ((\psi \to \vartheta) \to (\psi \to \varphi))$.
>
> The next problem formalizes the "proof by contradiction" technique: assume that the statement to be proved is false and derive a contradiction.

292. $\Sigma, \neg\varphi \vdash \bot$ if and only if $\Sigma \vdash \varphi$.

> **Hint.** Use the axiom schemes
>
> Ax_4 $(\neg\varphi \to \bot) \to \varphi$
> Ax_5 $\varphi \to (\neg\varphi \to \bot)$.

> **Notation.** The set Σ of propositional formulas is *syntactically consistent*, if $\Sigma \nvdash \bot$.

> Let Σ be consistent. By the compactness of $\vdash$ (and by Zorn's lemma) it can be extended to a maximal consistent subset set of propositional formulas. In the next problems assume Σ is maximal consistent, that is, $\varphi \notin \Sigma$ just in case $\Sigma, \varphi \vdash \bot$. Maximality also implies that $\varphi \in \Sigma$ whenever $\Sigma \vdash \varphi$.

293. If Σ is maximal consistent, then for every φ exactly one of φ and $\neg\varphi$ is in Σ.

294. Let Σ be maximal consistent. Define the valuation $f_\Sigma : V \to \{\top, \bot\}$ by

$$f_\Sigma(v) = \begin{cases} \top & \text{if } v \in \Sigma, \\ \bot & \text{if } \neg v \in \Sigma. \end{cases}$$

This definition is justified by Problem 293. Show that for every formula

$$f_\Sigma(\varphi) = \begin{cases} \top & \text{if } \varphi \in \Sigma, \\ \bot & \text{if } \neg\varphi \in \Sigma. \end{cases}$$

> **Hint.** Use induction on the complexity of φ. For negation use Problem 293. For $\vee$ and other connectives new axioms are needed, such as
>
> Ax_6 $\varphi \to (\varphi \vee \psi)$
> Ax_7 $\psi \to (\varphi \vee \psi)$
> Ax_8 $\neg\varphi \to (\neg\psi \to \neg(\varphi \vee \psi))$.

295 (Weak completeness). For any set Σ of propositional formulas Σ is satisfiable if and only if $\Sigma \nvdash \bot$.

5.21 Theorem (Strong completeness of propositional logic)**.**

$$\Sigma \vDash \varphi \quad \text{iff} \quad \Sigma \vdash \varphi$$

296. Prove Theorem 5.21.

Remark. The exact set of axioms depends on the set of propositional variables, but more importantly, on the logical operators which can be used in the formulas. According to Definition 5.1 $\varphi \to \psi$ is only a shorthand for $\neg \varphi \vee \psi$, thus all axioms and inference rules should be "expanded" to a formula without this connective. For example, the $\to$-free form of axiom schemes Ax_4 and Ax_5 are

Ax_4' $\neg(\neg\neg\varphi \vee \bot) \vee \neg\varphi$
Ax_5' $\neg\varphi \vee (\neg\neg\varphi \vee \bot)$.

The method outlined in the above problems works for other functionally complete sets of logical operators, see Problem 298.

Remark. The Hilbert-style derivation does not allow manipulating subformulas, in particular, one cannot replace a subformula with another "provably equivalent" subformula and expect to get a valid derivation (but see Problem 299). For example, deleting the double negation in Ax_4' gives a scheme which is too weak to prove the statement in Problem 292.

297. Provide the derivations claimed below using schemes $Ax_1 - Ax_8$:

(a) $\varphi \to \psi,\ \psi \to \vartheta \vdash \varphi \to \vartheta$,

(b) $\varphi,\ \neg\varphi \vdash \psi$,

(c) $\varphi \to (\psi \to \vartheta) \vdash \psi \to (\varphi \to \vartheta)$,

(d) $\vdash \neg\neg\varphi \to \varphi$,

Hint. In (b) use that $\varphi \to \psi$ is a shorthand for $(\neg\varphi) \vee (\psi)$.

298. Let us restrict propositional logic in the following sense: besides propositional variables the only connectives are $\to$ and $\neg$. What axiom schemes should we make use of to be able to deduce each tautology with MP as the only rule of inference?

Remark. In *Principia Mathematica* Bertrand Russel and Alfred Whitehead proved that for propositional formulas using $\to$ and $\vee$ only the following four axiom schemes together with MP are sufficient to derive all tautologies:

RW1 $\varphi \vee \varphi \to \varphi$
RW2 $\varphi \to \varphi \vee \psi$
RW3 $\varphi \vee \psi \to \psi \vee \varphi$
RW4 $(\varphi \to \psi) \to (\vartheta \vee \varphi \to \vartheta \vee \psi)$.

Another frequently used complete set of axiom schemes for the case when the logical connectives are $\neg$ and $\to$ only (covered in Problem 298) is the following:

- $\alpha \to (\beta \to \alpha)$
- $(\alpha \to (\beta \to \gamma)) \to ((\alpha \to \beta) \to (\alpha \to \gamma))$
- $(\neg\alpha \to \neg\beta) \to ((\neg\alpha \to \beta) \to \alpha)$.

299. Suppose φ is a subformula of Φ and replace each occurrence of φ by ψ. Denote the obtained formula by Ψ. Prove that $\vdash \varphi \leftrightarrow \psi$ implies $\vdash \Phi \leftrightarrow \Psi$.

300. Let us suppose that a propositional variable may take three different values: $\{0, *, 1\}$. The effect of operations $\neg$ and $\vee$ are given by two tables, such as $\neg 0 = 1$, $\neg * = *$, $\neg 1 = 0$. A formula is *all-1* if its value is always 1 for all possible values of the propositional variables.

(a) What condition the $\vee$ table should satisfy so that from all-1 formulas only all-1 formulas can be derived using MP? What happens if $\neg * = 1$?

(b) What are the possible values of the four axioms above if $1 \vee x = x \vee 1 = 1$, $0 \vee x = x \vee 0 = x$, $* \vee * = *$, and the definition of $\neg$ is the first or the second one?

(c) Find tables for $\neg$ and $\vee$ so that among the four axioms only $\varphi \to \varphi \vee \psi$ is not an all-1 formula.

301. Show that none of the four axioms of RW1, RW2, RW3, RW4 can be derived from the others, i.e., the system is independent.

Hint. Use three truth values $\{0, 1, *\}$ and define tables for $\vee$ and $\neg$ so that from all-1 formulas only all-1 formulas can be derived, and exactly three of the axioms are all-1.

5.3 CODING

With countably many propositional variables there is a natural coding (see Section 4.4) of propositional formulas. Suppose the logical operators are $\neg$, $\vee$, and $\bot$ only. The *code* of the propositional formula φ, denoted as $\alpha(\varphi)$, can be defined along the complexity of the formulas as follows:

$$\begin{aligned}
\alpha(v_i) &= \langle 0, i \rangle \quad \text{for each propositional variable } v_i \in V, \\
\alpha(\bot) &= \langle 1 \rangle, \\
\alpha(\neg\varphi) &= \langle 2, \alpha(\varphi) \rangle, \\
\alpha(\varphi \vee \psi) &= \langle 3, \alpha(\varphi), \alpha(\psi) \rangle.
\end{aligned}$$

As $(u)_i < u$ the code of any subformula of φ is smaller than the code of φ. For a set Σ of propositional formulas we identify Σ and $\{\alpha(\varphi) : \varphi \in \Sigma\}$, so we can say that Σ is recursive, or recursively enumerable.

302. Show that the following sets are recursive:

(a) (the codes of) all propositional formulas;

(b) formulas generated by an axiom scheme;

(c) the set of triplets $\langle \alpha(\varphi), \alpha(\varphi \to \psi), \alpha(\psi) \rangle$.

303. Let g be a recursive function, and define the evaluation f by $f(v_i) = \top$ iff $g(i) \neq 0$. Show that there is a recursive function G such that $G(\alpha(\varphi)) \neq 0$ iff $f(\varphi) = \top$.

304. Show that the set of tautologies is recursive. In other words, "tautology" is a decidable property.

305. Let Σ be a recursive formula set. The binary relation "x is a derivation of y from Σ" holds for two natural numbers x and y if elements of the sequence (coded by) x just code a valid Hilbert-type derivation from Σ of the formula (coded by) y. Show that this relation is recursive (decidable).

306. (a) Let Σ be recursive. Show that the set of its consequences $\{\varphi : \Sigma \vdash \varphi\}$ is recursively enumerable.

(b) Show that the same conclusion holds when Σ is recursively enumerable only.

307. Give an example for a recursive Σ such that the set of its consequences is not recursive (i.e., it is undecidable).

6

FIRST-ORDER LOGIC

6.1 BASICS

6.1 Definition (Similarity type). The *similarity type*, or signature, or alphabet is a collection of constant, relation, and function symbols with their fixed arities. Similarity types are usually denoted by τ.

The cardinality of τ, denoted as $|\tau|$, is the cardinality of its symbols. $\tau_1 \subseteq \tau_2$ means that τ_1 can be got from τ_2 by deleting some of the symbols. τ_1 and τ_2 are *compatible*, if there is a type extending both, the minimal such type is denoted by $\tau_1 \cup \tau_2$. The smallest type is the empty type which contains no symbols at all.

The countable set of *individual variables*, or just *variables*, is denoted by X. Variables are typically denoted by x, y, z, possibly with indices.

6.2 Definition (Terms). The set $E(\tau)$ of τ-type *terms* (expressions) is the free algebra generated from the variable and constant symbols using the functions symbol in τ.

308. Show that $\omega \leq |E(\tau)| \leq |\tau| \cdot \omega$.

6.3 Definition (First-order formula). An *atomic formula* of type τ is a sequence of symbols either of the form $t_1 = t_2$ with $t_1, t_2 \in E(\tau)$ (equality); or $r(t_1, \ldots, t_n)$ for some n-place relation symbol $r \in \tau$ and terms $t_1, \ldots, t_n \in E(\tau)$ (relation).
The set of τ-type *first-order formulas* $F(\tau)$ is the smallest set satisfying the following conditions:

- every atomic formula is in $F(\tau)$,
- if $\varphi, \psi \in F(\tau)$ then so are $(\varphi) \vee (\psi)$, $\neg(\varphi)$, and $\exists x(\varphi)$, where $x \in X$ is a variable symbol.

Notation. As the equality symbol $=$ may occur in formulas, equality of formulas φ and ψ (as symbol sequences) will be denoted by $\varphi \equiv \psi$.

© The Author(s), under exclusive license to Springer Nature Switzerland AG 2022
L. Csirmaz and Z. Gyenis, *Mathematical Logic*, Problem Books
in Mathematics, https://doi.org/10.1007/978-3-030-79010-3_6

In the formulas logical operators and constructs not mentioned in Definition 6.3 will also be used. Similar to the Propositional Logic discussed in Chapter 5, they can be considered as either abbreviations, or, with some additional work, the treatment can be extended to handle them. In particular,

$$\forall x \varphi \quad \text{is an abbreviation for} \quad \neg \exists x \neg \varphi.$$

309. Show that $|F(\tau)| = |\tau| \cdot \omega$.

The set of *free variables* of a formula is defined as follows. For an atomic formula φ, $V(\varphi)$ is the set of all variables occurring in φ. Otherwise $V(\neg \varphi) = V(\varphi)$, $V(\varphi \vee \psi) = V(\varphi) \cup V(\psi)$, and $V(\exists x \varphi) = V(\varphi) \smallsetminus \{x\}$. An occurrence of the variable x in φ is *bound* or *free* depending on whether it is discarded or not in the above procedure. The formula φ is *closed* if $V(\varphi) = \emptyset$, that is, has no free variables, and its (universal) *closure* is $\bar{\varphi} \equiv \forall x_1 \cdots \forall x_n \varphi$ where $V(\varphi) = \{x_1, \ldots, x_n\}$. The closure of a formula is not necessarily unique.

Notation. We write $\varphi(x, y, \ldots)$, or just $\varphi(\vec{x})$ to indicate that the free variables of φ are among the ones in the brackets.

6.4 Definition (Structure). For a given similarity type τ, a τ-*type structure* $\mathfrak{A}$ consists of a non-empty set A (universe, or underlying set) together with the *interpretation* of the symbols in τ. For each constant symbol $c \in \tau$, its interpretation, denoted by $c^{\mathfrak{A}}$, is an element of A; for an n-place relation symbol $r \in \tau$, $r^{\mathfrak{A}}$ is a subset of A^n; finally for an n-place function symbol $f \in \tau$, $f^{\mathfrak{A}} : A^n \to A$.

An *evaluation* of the variable symbols over the structure $\mathfrak{A}$ is a function $e : X \rightharpoonup A$. For an evaluation e, variable symbol $x \in X$ and element $a \in A$, $e(x/a)$ is the evaluation that takes the same values as e except for at $x \in X$, where it takes the value $a \in A$.

For a term $t \in E(\tau)$ and evaluation $e : X \to A$ the value of t computed in $\mathfrak{A}$ using $e(x)$ as value of the variable symbol x is denoted by $t^{\mathfrak{A}}[e] \in A$.

6.5 Definition (Semantics). For a τ-type structure $\mathfrak{A}$, evaluation e over $\mathfrak{A}$ and formula $\varphi \in F(\tau)$ we define $\mathfrak{A} \models \varphi[e]$ to be read as "φ is true in $\mathfrak{A}$ under the evaluation e" by induction on the complexity of φ as follows:

- $\mathfrak{A} \models (t_0 = t_1)[e]$ if $t_0^{\mathfrak{A}}[e] = t_1^{A}[e]$,
- $\mathfrak{A} \models r(t_1, \ldots, t_n)[e]$ if $\langle t_1^{\mathfrak{A}}[e], \ldots, t_n^{\mathfrak{A}}[e] \rangle \in r^{\mathfrak{A}}$,
- $\mathfrak{A} \models (\varphi \vee \psi)[e]$ if either $\mathfrak{A} \models \varphi[e]$ or $\mathfrak{A} \models \psi[e]$,
- $\mathfrak{A} \models (\neg \varphi)[e]$ if $\mathfrak{A} \not\models \varphi[e]$,
- $\mathfrak{A} \models (\exists x \varphi)[e]$ if for some $a \in A$ we have $\mathfrak{A} \models \varphi[e(x/a)]$.

If $\mathfrak{A} \vDash \varphi[e]$ for all evaluations e, then we write $\mathfrak{A} \vDash \varphi$, and say that $\mathfrak{A}$ *is a model for* φ, or φ is satisfied, or valid in $\mathfrak{A}$. For $\Sigma \subseteq F(\tau)$, $\mathfrak{A} \vDash \Sigma$ means $\mathfrak{A} \vDash \varphi$ for every $\varphi \in \Sigma$.

310. Let e_0 and e_1 be evaluations over $\mathfrak{A}$ so that whenever $x \in V(\varphi)$, then $e_0(x) = e_1(x)$. Show that $\mathfrak{A} \vDash \varphi[e_0]$ iff $\mathfrak{A} \vDash \varphi[e_1]$.

Notation. For a formula $\varphi(x_1, \ldots, x_n)$ and elements $a_1, \ldots, a_n \in A$ we write $\mathfrak{A} \vDash \varphi[a_1, \ldots, a_n]$, or just $\mathfrak{A} \vDash \varphi[\bar{a}]$ to mean that $\mathfrak{A} \vDash \varphi[e]$ for all (or for any by Problem 310) evaluation e with $e(x_i) = a_i$.

311. Let $\bar{\varphi}$ be the universal closure of φ. Prove that $\mathfrak{A} \vDash \varphi$ iff $\mathfrak{A} \vDash \bar{\varphi}$.

6.6 Definition (Substitution). For a formula $\varphi \in F(\tau)$, variable x and term $t \in E(\tau)$, $\varphi[x/t]$ is the formula obtained from φ by replacing each free occurrence of x with t. *Admissible substitutions* are defined by recursion on the complexity of φ as follows:
- if φ is atomic, then $\varphi[x/t]$ is admissible,
- $(\varphi \vee \psi)[x/t]$ is admissible if both $\varphi[x/t]$ and $\psi[x/t]$ are admissible,
- $(\neg\varphi)[x/t]$ is admissible if $\varphi[x/t]$ is such,
- $(\exists x\varphi)[x/t]$ is admissible (note that the substitution gives $\exists x\varphi$),
- $(\exists y\varphi)[x/t]$ is admissible if x and y are different, $\varphi[x/t]$ is admissible, and either x is not free in φ or y does not occur in t.

312. Give example for $\varphi \in F(\tau)$ and $t \in E(\tau)$ such that $\vDash \varphi$ while $\nvDash \varphi[x/t]$.

6.7 Theorem (Substitution lemma). If $\varphi[x/t]$ is an admissible, then

$$\mathfrak{A} \vDash (\varphi[x/t])[e] \quad \text{if and only if} \quad \mathfrak{A} \vDash \varphi[e(x/t^{\mathfrak{A}}[e])].$$

313. Let φ be a formula, t be a term and e be an evaluation over $\mathfrak{A}$. Give examples for all possible truth values of $\mathfrak{A} \vDash \varphi(x/t)[e]$ and $\mathfrak{A} \vDash \varphi[e(x/t^{\mathfrak{A}}[e])]$ when the substitution is (a) admissible (b) not admissible.

6.8 Definition. The formula $\varphi \in F(\tau)$ is *semantical consequence* of $\Gamma \subset F(\tau)$, in symbols $\Gamma \vDash \varphi$, if whenever $\mathfrak{A} \vDash \Gamma$, then $\mathfrak{A} \vDash \varphi$.
$\Gamma \vDash \Delta$ means that $\Gamma \vDash \varphi$ for each $\varphi \in \Delta$; and $\vDash \Delta$ means $\emptyset \vDash \Delta$, that is, all formulas in Δ are true in every τ-type structure.

314. Give Γ and φ such that $\Gamma \nvDash \varphi$ and $\Gamma \nvDash \neg\varphi$. Give $\mathfrak{A}$ such that $\mathfrak{A} \nvDash \varphi$ and $\mathfrak{A} \nvDash \neg\varphi$.

6.9 Theorem (Deduction lemma). Suppose $\psi \in F(\tau)$ is closed. Then

$$\Gamma \cup \{\psi\} \vDash \varphi \quad \text{if and only if} \quad \Gamma \vDash \psi \to \varphi$$

315. Give examples Γ, φ and ψ for all the possible values of $\Gamma \vDash \varphi \to \psi$ and $\Gamma \cup \{\varphi\} \vDash \psi$.

316. Semantic consequence can be defined in a slightly different way. Write $\Gamma \vDash^* \varphi$ if for every structure $\mathfrak{A}$ and evaluation e over $\mathfrak{A}$, whenever $\mathfrak{A} \vDash \vartheta[e]$ for all $\vartheta \in \Gamma$, then $\mathfrak{A} \vDash \varphi[e]$. Are $\Gamma \vDash \varphi$ and $\Gamma \vDash^* \varphi$ equivalent?

> **6.10 Definition.** Γ is *consistent* if there is a structure $\mathfrak{A}$ with $\mathfrak{A} \vDash \Gamma$, that is, Γ has a model. Otherwise, Γ is *inconsistent* or *contradictory*. As for all structures $\mathfrak{A}$ we have $\mathfrak{A} \nvDash \bot$, Γ is inconsistent iff $\Gamma \vDash \bot$.

317. Let Γ be maximal consistent, and φ and ψ be closed formulas. Show that

(a) $\Gamma \vDash \varphi$ if and only if $\varphi \in \Gamma$.

(b) $\varphi \lor \psi \in \Gamma$ if and only if $\varphi \in \Gamma$ or $\psi \in \Gamma$.

(c) $\varphi \notin \Gamma$ if and only if $\neg \varphi \in \Gamma$.

> **6.11 Definition.** The *theory* of a τ-structure $\mathfrak{A}$ is the formula set $\mathrm{Th}(\mathfrak{A}) = \{\varphi \in F(\tau) : \mathfrak{A} \vDash \varphi\}$.

318. Show that Γ is maximal consistent if and only if it is the theory of some structure.

319. Find a maximal consistent Γ and formula φ such that neither $\varphi \in \Gamma$, nor $\neg \varphi \in \Gamma$.

320. (a) An oracle can decide for every type and every set Γ of formulas whether Γ has a model or not, provided that no formula in Γ contains a function symbol (even if there is one in the type). How can the oracle be used to decide the consistency of an arbitrary formula set?

(b) The oracle further restricted its usage, and now no formula in Γ can contain the equality sign. Can the oracle still be used to decide the consistency of arbitrary formula sets?

321. Let $\mathfrak{A}$ and $\mathfrak{B}$ be two structures with the same similarity type. Find a closed formula φ such that $\mathfrak{A} \vDash \varphi$ and $\mathfrak{B} \vDash \varphi$, but $\nvDash \varphi$.

> **6.12 Definition** (Substructure). $\mathfrak{B}$ is a *substructure* of $\mathfrak{A}$, written as $\mathfrak{B} \subseteq \mathfrak{A}$, if $B \subseteq A$ and $c^{\mathfrak{B}} = c^{\mathfrak{A}}$ for each constant symbol $c \in \tau$, $r^{\mathfrak{B}} = r^{\mathfrak{A}} {\restriction} B^n$ for each n-place relation symbol $r \in \tau$, and $f^{\mathfrak{B}} = f^{\mathfrak{A}} {\restriction} B^n$ for each n-place function symbol $f \in \tau$.
> For $X \subseteq A$ the substructure *generated by* X is the smallest $\mathfrak{B} \subseteq \mathfrak{A}$ with $X \subseteq B$.

322. Construct a structure $\mathfrak{A}$ with universe $\mathbb{R}$ (the real numbers) so that

(a) $\mathfrak{A}$ has exactly one proper substructure with universe $\mathbb{Z}$ (the integers),

(b) $\mathfrak{A}$ has two proper substructures with universe the even and odd integers, respectively.

323. Prove that for every non-empty $X \subseteq A$ the substructure $\mathfrak{B}$ generated by X in $\mathfrak{A}$ exists and $|\mathfrak{B}| \leq \max(|X|, |\tau|, \omega)$.

324. (a) Show that the condition in Problem 323 that X is not empty is necessary: there are structures with no smallest substructure.

(b) Construct a structure $\mathfrak{A}$ with substructures $\mathfrak{B}_1$ and $\mathfrak{B}_2$ such that $\mathfrak{B}_1 \cup \mathfrak{B}_2$ is not a substructure of $\mathfrak{A}$.

325. Suppose $\mathfrak{B} \subseteq \mathfrak{A}$, $\varphi(\vec{x})$ is quantifier-free, and $\vec{a} \in B$. Prove that

$$\mathfrak{B} \vDash \varphi[\vec{a}] \quad \Longleftrightarrow \quad \mathfrak{A} \vDash \varphi[\vec{a}].$$

326. Let $\mathcal{F}$ be a collection of non-empty subsets of A such that (a) the intersection of any subcollection of $\mathcal{F}$, if not empty, belongs to $\mathcal{F}$, (b) the union of any increasing sequence of elements of $\mathcal{F}$ is in $\mathcal{F}$, and (c) $A \in \mathcal{F}$.

Is it true that there is a structure $\mathfrak{A}$ with universe A such that the collection of the underlying sets of its substructures is exactly $\mathcal{F}$?

> **Remark.** The underlying sets of the substructures of $\mathfrak{A}$ clearly satisfy conditions (a), (b), (c) above.

> **6.13 Definition.** The structures $\mathfrak{A}$ and $\mathfrak{B}$ are *isomorphic* if they have the same type, and there is a bijection $f : A \to B$ which preserves the interpretation of every symbol in τ. An isomorphism from $\mathfrak{A}$ to $\mathfrak{A}$ is called *automorphism*.

327. The structure $\mathfrak{A}$ is rigid if its only automorphism is the identity. Give a finite signature and construct (a) countably infinite (b) continuum size rigid structure.

6.2   EXPRESSING PROPERTIES

> **6.14 Definition.** A *graph* is a structure $\mathfrak{A}$ with a single binary relation E, that is, its similarity type is $\tau = \langle E \rangle$. Elements of the universe A are the nodes, and two nodes $u, v \in A$ are connected by an edge just in case the pair $\langle u, v \rangle$ is in $E^{\mathfrak{A}}$, the interpretation of the relation symbol E.

328. What are the axioms of graph theory? That is, find a formula set Γ such that every graph satisfies Γ, and if an $\langle E \rangle$-type structure satisfies Γ, then it is a graph.

329. Let $\langle \mathfrak{A}, E \rangle$ be a graph. Write formulas which express the fact that

(a) the graph is 3-regular (every vertex has degree 3),

(b) it is a star,

(c) contains no triangles.

(d) What property corresponds to the formula

$$\neg \exists x_1 \exists x_2 \exists x_3 \exists x_4 \left(E(x_1, x_2) \wedge E(x_2, x_3) \wedge E(x_3, x_4) \wedge E(x_4, x_1) \right).$$

330. Construct formulas or sets of formulas Γ such that for a graph $\mathfrak{A} \vDash \Gamma$ iff

(a) $\mathfrak{A}$ contains no cycles,

(b) the diameter of $\mathfrak{A}$ is 15,

(c) $\mathfrak{A}$ is a bipartite graph.

331. Construct a consistent graph formula φ such that if $\mathfrak{A} \vDash \varphi$ for a graph $\mathfrak{A}$ then $\mathfrak{A}$ is infinite.

> **6.15 Definition.** A (multiplicative) *group* is a structure of type $\langle e, \cdot, ^{-1} \rangle$ denoting the unit, product, and inverse, respectively, where the multiplication is associative, e is both left and right unit, and $^{-1}$ is both left and right inverse.

332. What are the formulas expressing the group properties?

333. Construct formula sets which express that the group is

(a) infinite,

(b) torsion-free (has no finite order elements),

(c) torsion-free divisible Abelian group.

(d) elementary p-group (the rank of every element is p).

334. Let $\mathfrak{A}$ be the symmetric group of the integers $\mathbb{Z}$, and $\pi \in A$ be the right shift, namely, the following permutation of $\mathbb{Z}$: $\pi(j) = j + 1$. Add this π as a constant symbol to the group signature. Construct a formula $\varphi(x)$ which is satisfied exactly by the shifts in $\mathfrak{A}$.

> **6.16 Definition.** An *ordered set* is a structure with a binary relation $\leq$ satisfying
>
> - $x \leq x$ (reflexivity)   ,
> - $x \leq y$ and $y \leq z$ then $x \leq z$ (transitivity)   ,
> - $x \leq y \wedge y \leq x \rightarrow x = y$ (antisymmetry)   ,
> - $x \leq y \vee y \leq x$ (total)   .
>
> For a *partial order* the last property is not required.

> Sometimes the name *total order* or *linear order* is used to emphasize that $\leq$ is not a partial order. In an ordered set $y \geq x$ means $x \leq y$, and $x < y$ means $x \leq y$ and $x \neq y$. Using this notation properties other than transitivity can be merged into stipulating that exactly one of $x < y$, $x = y$, and $y < x$ holds (trichotomy).

335. Write the axioms of total order using the strict inequality sign $<$.

336. Construct formulas using the partial order $\leq$ which express that

(a) any two elements have a least upper bound,

(b) any two elements have a greatest lower bound,

(c) there exist a minimal and a maximal element.

337. Let $\leq$ be a total ordering. Construct a formula which says that on one side of every element one can find other elements arbitrarily close, while there is a closest one on the other side. Does there exist an ordering with this property? Can this side be the same for all points in the ordering?

338. For each $n \in \omega$ construct a closed formula $\varphi_n \in F(<)$ such that φ_n is true in a linear order $\mathfrak{A}$ if and only if the ground set A has at least n elements, and φ_n uses only two variable symbols x and y.

> **6.17 Definition.** The similarity type of the *fields* is $\langle 0, 1, +. -, \cdot, / \rangle$, where, as usual, the $-$ sign can be used both as a unary and as a binary function symbol. Zero has no inverse, so, strictly speaking, the division is not a binary function. However, we can pretend that it is one by stipulating that when dividing by zero, the result is zero.

339. Construct formulas or formula sets in the language of fields which express the following properties of a field:

(a) Every polynomial with integer coefficients has a root.

(b) The characteristics of the field is 2.

(c) The field is of characteristic zero.

(d) -1 is not a sum of finitely many squares.

(e) The field is algebraically closed.

(f) Each polynomial of odd degree has a root.

340. Let τ be a similarity type, and $\approx$ be a new binary relation symbol.

(a) Create a formula set Δ which expresses that $\approx$ is an equivalence relation which is compatible with all relation and function symbols of τ.

(b) To get $\varphi^{\approx}$ from $\varphi \in F(\tau)$ replace the equality symbol everywhere by the new binary relation symbol $\approx$. Show that $\Gamma \vDash \varphi$ if and only if $\Gamma^{\approx} \cup \Delta \vDash \varphi^{\approx}$.

> **Notation.** When the intended interpretation of the symbols in the similarity type is clear from the context, a structure will be specified by listing the base set followed by the symbols of the type. Example: $\mathfrak{A} = \langle \omega, 0, 1, +, \leq \rangle$.

341. Let $\mathfrak{A} = \langle \omega, < \rangle$, that is the universe is ω and the similarity type τ contains only the binary relation symbol $<$ with the natural interpretation. Construct a formula $\varphi(x, y)$ such that $\mathfrak{A} \vDash \varphi[a, b]$ iff $a \geq b + 1024$, and apart from x and y the formula φ contains a single additional variable symbol.

342. Let $\mathfrak{A} = \langle \omega, \in \rangle$ where $i \in^{\mathfrak{A}} j$ holds if the $(i+1)$-st digit in the binary representation of j (counting from right) is 1. (For example, $0 \in^{\mathfrak{A}} j$ iff j is odd.) Which of the following formulas are true in $\mathfrak{A}$?

(a) $\exists x \forall y (y \notin x)$

(b) $\forall x \forall y \exists z \forall u (u \in z \leftrightarrow (u \in x \vee u \in y))$

(c) $\forall x \exists y \forall u (u \in y \leftrightarrow \forall v (v \in u \rightarrow v \in x))$

(d) $\exists x (\exists v (v \in x) \wedge \forall y (y \notin x \vee \exists v (v \in x \wedge v \neq y \wedge \forall z (z \in y \rightarrow z \in v))))$.

343. Let $\mathfrak{A} = \langle \omega, \in \rangle$ as in Problem 342. In set theory the axiom scheme of comprehension stipulates that for any formula $\varphi(x,y)$ the axiom

$$\forall x \exists y (\forall u \in x)(\exists v \varphi(u,v) \rightarrow (\exists v \in y)\varphi(u,v))$$

holds. Show that this scheme holds in $\mathfrak{A}$.

344. Construct formulas which define the following sets in $\mathfrak{A} = \langle \omega, 0, 1, +, \cdot, \leq \rangle$:

(a) primes (warning: 0 and 1 are *not* primes),

(b) powers of two,

(c) prime powers,

(d) numbers which can be written as the sum of at most nine perfect cubes.

345. Let $\mathfrak{A} = \langle \omega, 0, 1, +, \cdot, \leq \rangle$. Construct a formula which is true in $\mathfrak{A}$ if and only if the twin prime conjecture holds.

346. Let $\mathfrak{A} = \langle \omega, +, | \rangle$ where $|$ is the divisibility relation. Find formulas φ_1, φ_2, φ_3 such that the following holds:

(a) $\mathfrak{A} \vDash \varphi_1[a]$ iff $a = 0$,

(b) $\mathfrak{A} \vDash \varphi_2[a]$ iff $a = 1$,

(c) $\mathfrak{A} \vDash \varphi_3[a,b]$ iff $a = b^2$.

347. Let $\mathfrak{A} = \langle \mathbb{Z}, S, \cdot \rangle$, where $\mathbb{Z}$ is the set of integers, $S^{\mathfrak{A}}$ is the successor function and $\cdot^{\mathfrak{A}}$ is the usual multiplication of integers. Find a formula $\varphi(x,y,z)$ such that $\mathfrak{A} \vDash \varphi[a,b,c]$ iff $a + b = c$.

348. Let $\mathfrak{A} = \langle \mathbb{Z}, 1, +, | \rangle$, where $|$ is the divisibility relation. Construct formulas such that

(a) $\mathfrak{A} \vDash \varphi_1[j]$ iff $|j| > 1$,

(b) $\mathfrak{A} \vDash \varphi_2[j]$ iff $|j|$ is a prime,

(c) $\mathfrak{A} \vDash \varphi_3[j,u]$ iff $|u|$ is a power of the prime $|j|$,

(d) $\mathfrak{A} \vDash \varphi_4[j,u]$ iff $|x|$ is a prime and $|u| = j^2$.

349. Let $\mathfrak{A} = \langle \mathbb{Z}, +, | \rangle$ where $|$ is divisibility. Show that there is no formula $\varphi(x)$ such that $A \vDash \varphi[a]$ iff $a = 1$.

350. Let $\mathfrak{A} = \langle \mathbb{Z}, 1, +, | \rangle$ where $|$ is divisibility. Construct formulas such that

(a) $\mathfrak{A} \vDash \varphi_1[i,j]$ iff $j = \pm i^2$,

(b) $\mathfrak{A} \vDash \varphi_2[i,j]$ iff $j = i^2$,

(c) $\mathfrak{A} \vDash \varphi_3[i]$ iff $i \geq 0$.

| **Hint.** Use that $\mathrm{lcm}(i, i+1) = \pm i(i+1)$ and $\mathrm{lcm}(i-1, i) = \pm(i-1)i$.

351. $\mathbb{R}$ is the set of reals. Let $\mathbb{F} = \{f : f$ is a function from $\mathbb{R}$ to $\mathbb{R}\}$. Let $\mathfrak{A} = \langle \mathbb{R} \cup \mathbb{F}, R, F, +, \cdot, -, \leq, 0, 1, H \rangle$, where the interpretation of the unary relation symbol R is $\mathbb{R}$, that of F is $\mathbb{F}$, the restriction to $\mathbb{R}$ of the functions $+, \cdot, -$ and $\leq$ are the usual ones. The interpretation of 0 and 1 are the real numbers zero and one. Finally, the value of $H(f, x)$ is $f(x)$ if $f \in \mathbb{F}$ and $x \in \mathbb{R}$, and 0 otherwise.

Construct a formula $\varphi(x)$ that holds in $\mathfrak{A}$ exactly for continuous functions.

6.3 MODELS AND CARDINALITIES

352. The similarity type τ consists of the constant symbols $\{c_\xi : \xi < \lambda\}$ for an infinite cardinal λ. Let $\Gamma = \{c_\xi \neq c_\eta : \xi < \eta < \lambda\}$. For what λ does Γ have a unique (up to isomorphism) model of cardinality λ?

353. (a) Let τ be the empty similarity type. What is the number of non-isomorphic countable τ-structures?

Let τ contain (b) exactly one (c) exactly $n \in \omega$ unary relation symbols. What is the number of non-isomorphic countable τ-structures?

354. Let τ be a finite similarity type and κ be an infinite cardinal. Prove that there are at most 2^κ pairwise non-isomorphic τ-structures of cardinality κ. Find some τ such that there exist more than continuum many non-isomorphic countable τ-structures.

355. Give a finite type τ such that for each infinite κ there are exactly 2^κ pairwise non-isomorphic τ-type structures of cardinality κ.

356. (a) The type τ contains a single unary function symbol f. How many pairwise non-isomorphic countable τ-structures exist?
(b)* How many such structures are of cardinality $\kappa > \omega$?

357. Give a similarity type τ and a set $\Gamma \subseteq F(\tau)$ of formulas such that Γ has a unique model of cardinality κ (up to isomorphism) for each infinite κ, but it has at least three 3-element models.

358. Give a set Γ of formulas such that for a finite $n \in \omega$, Γ has an n-element model iff n is a prime.

359. Give Γ that has a unique countable model but has at least two non-isomorphic models of cardinality ω_1.

360. In the language of graphs give a set Γ of formulas expressing that each vertex has degree 3 and there are no cycles. What is the number of non-isomorphic models of Γ of cardinality (a) ω (b) ω_1?

361. In the similarity type of a binary relation symbol E let Γ be the theory expressing that E is an equivalence relation having infinitely many infinite equivalence classes. Write this set Γ. What is the number of non-isomorphic models of Γ of cardinality (a) $\aleph_0$ (b) $\aleph_1$ (c) $\aleph_2$ (d) $\aleph_\omega$?

362.[*] The similarity type τ consists of $k \in \omega$ unary relation symbols. Let φ be a formula of length n. Prove that if φ has a model, then it has a model of size at most $n \cdot 2^k$.

363. The similarity type τ consists of 2 unary relation symbols R and P. Prove that if Γ has a unique model of cardinality ω_1, then it has a unique countable model. Does the converse hold?

6.4   ORDERED SETS

> **6.18 Definition** (Dense ordering). An ordering is *dense* if between any two different elements there is a third one. Frequently we require that an ordering should not have *endpoints*. This means that there are neither smallest, nor largest elements.

> **6.19 Definition** (Discrete ordering). The ordering is *discrete* if whenever an element is not the last one, then there is a smallest among those which are larger than it; and if it is not the smallest, then there is a largest element among those which are smaller than it.

> **6.20 Definition** (Well ordering). In a well-ordering every non-empty subset (including the whole set) has a minimal element.

364. Prove that there are exactly 2^κ non-isomorphic linear orders on an infinite set of cardinality κ.

365. How many countable dense orderings are?

366. Show that one can color the rational numbers with countably many colors so that between any two rationals all colors occur. Show that any two such colorings are isomorphic.

367. Show that every countable ordered set can be embedded into the rationals in an order-preserving way.

368. Construct, with proof, continuum many pairwise non-isomorphic discrete ordering without endpoints on a countable set.

369. Construct at least two, but you might try 2^κ many pairwise non-isomorphic dense orderings without endpoints of cardinality κ, for any $\kappa > \omega$.

370. Construct a dense linear order $(A, <)$ without endpoints so that there is an $a \in A$ such that any order-preserving permutation of A keeps a fixed.

> **6.21 Definition.** Let $\mathfrak{A} = \langle A, \leq \rangle$ be a linear ordering without endpoints. If $a \in A$ is the intersection of the open intervals (a_α, b_α) with $a_\alpha <$

$a < b_\alpha$, then the *weight* of $a \in A$ is the minimal cardinality of such a collection of intervals.

371. Construct an ordering where there is an internal point which has no weight.

372. Show that the weight of a point is either 1 or an infinite cardinal.

373. Determine the weight of $\sqrt{2}$ in the usual ordering of the reals.

374. Find an ordering where all elements have weight ω_1.

375. For every infinite cardinal κ, give an ordering on κ such that all elements have countable weight.

6.22 Definition. Let $(P, \leq)$ be a partially ordered set. A (generalized) *sequence* is a mapping from an ordinal to P; the α-th member of the sequence s is denoted by s_α. The sequence is *increasing* if $s_\alpha \leq s_\beta$ whenever $\alpha < \beta$.

376. Construct a partially ordered set where every increasing sequence has an upper bound, but there is an (increasing) sequence which does not have a *least* upper bound.

377. Construct a partially ordered set in which every countable increasing sequence $s_0, s_1, \ldots$ has an upper bound, but there is no maximal element in the set (i.e., the conclusion of Zorn's lemma does not hold).

378. Suppose that every increasing sequence (of arbitrary length) in P has an upper bound. Does it follow that every totally ordered subset of P has an upper bound as well?

379. Every partial ordering can be extended to a total ordering.

6.5 CODING

6.23 Definition (Tautology). A formula $\varphi \in F(\tau)$ is a *tautology* if there is a tautological propositional formula ψ so that φ is the result when the propositional variables in ψ are replaced systematically by appropriate first-order formulas.

380. If $\varphi \in F(\tau)$ is a tautology, then $\vDash \varphi$.

Similar to coding propositional formulas in Section 5.3, there is a natural coding of first-order formulas when the similarity type τ is finite. Such coding can be defined along the complexity of terms and formulas. Let us start with the symbols of the similarity type τ:

$\alpha(c) = \langle 0, i \rangle$ $c \in \tau$ is the i-th constant symbol in τ,

$\alpha(r) = \langle 1, i, n_i \rangle$ r is the i-th relation symbol in τ with arity n_i,

$\alpha(f) = \langle 2, i, n_i \rangle$ f is the i-th function symbol in τ with arity n_i.

As τ is finite, only finitely many codes can start with 0, 1, or 2. The code of τ-type terms can be defined as

$\alpha(x_i) = \langle 3, i \rangle$ where x_i is the i-the variable symbol, and

$$\alpha(f(t_1, \ldots, t_n)) = \langle 4, \alpha(f), \alpha(t_1), \ldots, \alpha(t_n) \rangle.$$

The code of atomic and composite formulas can be defined similarly to the propositional case.

381. Let τ be a fixed finite similarity type. Show that the following sets, functions, and relations are recursive:

(a) the set of (the code of) τ-type terms, that of τ-type formulas,

(b) the relation which decides if x is free in φ,

(c) the code of closed formulas,

(d) the function which returns the universal closure of φ,

(e) the 3-place function which returns the result of the substitution $\varphi[x/t]$ as $\langle \alpha(\varphi), \alpha(x), \alpha(t) \rangle \mapsto \alpha(\varphi[x/t])$,

(f) the relation which decides whether the substitution $\varphi[x/t]$ is admissible.

382. Suppose τ contains countably many function symbols which are coded as triplets $\langle 2, i, n_i \rangle$. Show that the set of τ-type terms is not necessarily decidable.

383. Suppose τ contains the constant symbols 0 and 1 and the binary function symbol +. Define the terms $\pi_0 = 0$, $\pi_1 = 0 + 1$, and $\pi_{n+1} = \pi_n + 1$. Show that the function $n \mapsto \alpha(\pi_n)$ is recursive.

384. Show that the property "φ is a tautology" is decidable.

385. Suppose $\mathfrak{A}$ is a finite structure. Show that its theory is decidable.

FUNDAMENTAL THEOREMS

7

7.1 FIRST-ORDER DERIVATIONS

7.1.1 HILBERT-TYPE DERIVATION

Definition 5.19 of Hilbert-type derivation applies to first-order logic. Axioms and derivation rules are *schemes* where the metavariables are to be replaced by τ-type first-order formulas, occasionally satisfying additional requirements. Both the axioms and the derivation rules depend on the similarity type τ, which is arbitrary but fixed in this section.

The main goal is to find axioms and inference rules which lead to a sound and complete proof system. Two types of inference rules will be used: the *modus ponens*

$$\text{MP} \quad \frac{\varphi, \quad \varphi \to \psi}{\psi}$$

and *generalization*

$$\text{G} \quad \frac{\varphi}{\forall x \varphi}.$$

Axiom schemes will be unveiled along the course of establishing various properties of the proof system; we start with schemes $Ax_1 - Ax_8$ from propositional logic. Further axiom schemes will not be the minimal ones which allow to prove the required property. The reason is explained in Problem 399.

386. Let $\varphi \in \mathcal{F}(\tau)$ be a tautology as in Definition 6.23. Show that $\vdash \tau$.

> **Remark.** By this result tautologies can be used in a derivation as if they were axioms: the necessary derivation can always be supplied.

387. Suppose $\vDash \vartheta$ for all axioms ϑ. Show that $\Gamma \vdash \varphi$ implies $\Gamma \vDash \varphi$.

> **7.1 Theorem** (Syntactical deduction lemma). Suppose $\psi \in F(\tau)$ is a closed formula. Show that $\Sigma, \psi \vdash \varphi$ if and only if $\Sigma \vdash \psi \to \varphi$.

> **Remark.** This lemma is about derivability, while the similar statement in Theorem 6.9 is about semantical consequence.

© The Author(s), under exclusive license to Springer Nature Switzerland AG 2022
L. Csirmaz and Z. Gyenis, *Mathematical Logic,* Problem Books
in Mathematics, https://doi.org/10.1007/978-3-030-79010-3_7

388. Prove Theorem 7.1.

> **Hint.** Try the same method which was used in propositional logic (Problem 291). The only case not treated there is the inference rule G. For that case use the axiom scheme
>
> Ax_9 $(\forall x(\psi \to \varphi)) \to (\psi \to \forall x\varphi)$ assuming x is not free in ψ.

389. Let $\bar{\varphi}$ be the universal closure of φ. Show that $\Gamma \vdash \varphi$ iff $\Gamma \vdash \bar{\varphi}$.

> **Hint.** Use one of the schemes below:
>
> Ax_{10} $\varphi[x/t] \to (\exists x\varphi)$ assuming $\varphi[x/t]$ is an admissible substitution,
>
> Ax_{11} $(\forall x\varphi) \to \varphi[x/t]$ assuming $\varphi[x/t]$ is an admissible substitution.

390. If $\forall x\varphi$ is an abbreviation for $\neg\exists x\neg\varphi$, then $Ax_{10} \vdash Ax_{11}$.

> **Notation.** The set $\Sigma \subset F(\tau)$ is *syntactically consistent*, or s-consistent for short, if $\Sigma \nvdash \bot$.

391. Suppose $\Sigma \subset F(\tau)$ is maximal s-consistent. Then

(a) for every $\varphi \in F(\tau)$, $\Sigma \vdash \varphi$ iff $\varphi \in \Sigma$,

(b) for a closed $\varphi \in F(\tau)$ exactly one of φ or $\neg\varphi$ is in Σ,

(c) for closed formulas $\varphi, \psi \in F(\tau)$, $\varphi \vee \psi \in \Sigma$ iff either $\varphi \in \Sigma$ or $\psi \in \Sigma$.

392. Let Γ_α for $\alpha < \kappa$ be an increasing chain of s-consistent theories. Show that $\bigcup\{\Gamma_\alpha : \alpha < \kappa\}$ is syntactically consistent.

393. Prove that every syntactically consistent theory can be extended to a maximal one.

> **7.2 Definition.** $\Sigma \subset F(\tau)$ is a *Henkin theory* if $\Sigma \vdash \exists x\varphi(x)$ for a closed formula $\exists x\varphi(x)$, then there is a constant symbol $c \in \tau$ such that $\Sigma \vdash \varphi[x/c]$.

394. Let $\Sigma \subset F(\tau)$ be a Henkin theory.

(a) Show that there is a constant symbol in τ.

(b) For a function symbol f and constant symbols c_i there is a constant symbol c such that $\Sigma \vdash f(c_1, \ldots, c_n) = c$.

> **Hint.** Use Ex_1 and Ex_2 from the following set of *equality axioms*:
>
> Ex_1 $x = x$, $x = y \to y = x$, $(x = y \wedge y = z) \to x = z$.
>
> Ex_2 $\bigwedge_i x_i = y_i \to f(\bar{x}) = f(\bar{y})$ for each n-place function symbol $f \in \tau$.
>
> Ex_3 $\bigwedge_i x_i = y_i \to r(\bar{x}) \leftrightarrow r(\bar{y})$ for each n-place relation symbol $r \in \tau$.

395. Is there a Henkin theory which has a model, but has no model defined on the set of the constant symbols in τ where each constant symbol is interpreted as itself?

396. Suppose that for every closed formula $\exists x \varphi(x)$ if $\mathfrak{B} \models \exists x \varphi(x)$, then there is a constant symbol c such that $\mathfrak{B} \models \varphi[x/c]$. Let

$$A = \{c^{\mathfrak{B}} : c \in \tau \text{ is a constant symbol}\}$$

(a) Prove that A is the universe of a substructure $\mathfrak{A}$.

(b) Prove that for every formula $\psi \in F(\tau)$, $\mathfrak{B} \models \psi$ iff $\mathfrak{A} \models \psi$.

397. Suppose the similarity type τ does not contain function symbols. Let Γ be a maximal syntactically consistent Henkin theory. The universe of $\mathfrak{A}$ is the set of constant symbols of τ, and for a relation symbol $r \in \tau$ we let

$$\langle c_1, \ldots, c_n \rangle \in r^{\mathfrak{A}} \text{ iff } \Gamma \vdash r(c_1, \ldots, c_n).$$

Prove that if φ does not contain the equality symbol, then $\mathfrak{A} \models \varphi$ iff $\Gamma \vdash \varphi$.

398. Let $\Sigma \subset F(\tau)$ be a maximal syntactically consistent Henkin theory. For constant symbols $c_0, c_1 \in \tau$ define $c_0 \sim c_1$ if $\Sigma \vdash c_0 = c_1$. Show that $\sim$ is an equivalence relation.

399. The formula set $\Sigma \subseteq F(\tau)$ is τ-consistent if using τ-type axioms and inference rules only, $\bot$ cannot be derived from Σ. Let $\tau \subseteq \tau'$ (symbols in τ are also symbols in τ').

(a) Show that if Σ is τ'-consistent, then it is also τ-consistent.

(b) Assuming that $\tau' \smallsetminus \tau$ has constant symbols only, show that the converse is also true.

400. Let $c \in \tau$ be a constant symbol such that neither φ nor any formula in $\Sigma \subset F(\tau)$ contains c. Suppose $\Sigma \vdash \varphi[x/c]$. Show that in this case $\Sigma \vdash \forall x \varphi$.

401. Let $\Sigma \subset F(\tau)$ be a maximal syntactically consistent Henkin theory. Prove that Σ has a model.

> **Hint.** Elements of the ground set are equivalence classes of the constant symbols from τ.

402. Suppose $\Sigma \subset F(\tau)$ is syntactically consistent. Show that Σ can be embedded into a maximal s-consistent Henkin theory $\Sigma' \subset F(\tau')$ in a larger type τ' such that $\tau' \smallsetminus \tau$ has constant symbols only.

> **Hint.** Follow the method of Problem 393. You will need the scheme
>
> Ax_{12} $(\forall x \neg \varphi) \rightarrow \neg \exists x \varphi$.

> **7.3 Theorem** (Gödel's first completeness theorem). If $\Sigma \nvdash \bot$, then Σ has a model.

403. Prove Theorem 7.3.

7.4 Theorem (Gödel's second completeness theorem). The Hilbert-type derivation is strongly complete and sound, that is,

$$\Sigma \vDash \varphi \quad \text{iff} \quad \Sigma \vdash \varphi.$$

404. Prove Theorem 7.4.

405. During the proof of the completeness theorem we made use of the fact that there are an infinite number of variables. Where exactly did we use it?

406. Suppose neither Γ nor φ contain the equality symbol and $\Gamma \vdash \varphi$. Prove that in this case φ can be derived from Γ without using the equality axioms Ex_1–Ex_3.

407. Let $\tau \subset \tau'$ be similarity types and $\Gamma \subset F(\tau')$, $\varphi \in F(\tau)$. Assume $\Gamma \vdash \varphi$. Is it true that there is a derivation of φ from Γ that consists of τ-formulas and axioms only? What happens if $\Gamma \subset F(\tau)$?

7.1.2 RESOLUTION METHOD

The resolution method is a syntactical tool which—based on the written form of the formulas only—gives a condition whether a given set of formulas has a model or not. As $\Sigma \vDash \varphi$ if and only if $\Sigma \cup \{\neg\bar{\varphi}\}$ does not have a model, it is strongly complete.

7.5 Definition (Prenex normal form). A formula φ is in *prenex form*, if it starts with a block of $\forall$ and $\exists$ quantifiers to be applied to a quantifier-free formula.

408. Every formula can be converted to an equivalent formula in prenex form.

7.6 Definition (Skolem function). Let $\mathfrak{A}$ be a τ-type structure, and $\psi(x_1, \ldots, x_n) \in F(\tau)$ be a formula of the form $\psi(\vec{x}) \equiv \exists y \varphi(y, \vec{x})$. (If ψ has no free variables, then choose $\vec{x}$ to be a single variable differing from y.) The *Skolem function* belonging to the formula ψ is an n-variable function f_ψ defined on A^n such that for all $\vec{a} \in A^n$,

$$\mathfrak{A} \vDash (\exists y \varphi)[\vec{a}] \quad \text{implies} \quad \mathfrak{A} \vDash \varphi[f_\psi(\vec{a}), \vec{a}].$$

409. Let $\Sigma \subset F(\tau)$. For each formula $\psi(\vec{x}) \in \Sigma$ of the form $\exists y \varphi(y, \vec{x})$ add the function symbol f_ψ to τ, and replace ψ by $\varphi[y / f_\psi(\vec{x})]$. Let the extended type be τ^*, and the new formula set be Σ^*. Show that Σ has a model if and only if Σ^* has a model.

410. The oracle from Problem 320 can be asked whether or not a set of quantifier-free formulas has a model. Show that the oracle can be used to decide the consistency of arbitrary formula sets.

7.7 Definition (Herbrand structure). Let τ be a similarity type with at least one constant symbol, and let $K(\tau)$ be the free term algebra generated by the constant symbols of τ (that is, $K(\tau)$ is the set of variable-free terms). The τ-structure $\mathfrak{M}$ is a *Herbrand structure* if

- the universe of $\mathfrak{M}$ is $M = K(\tau)$,
- for a constant symbol $c \in \tau$ we have $c^{\mathfrak{M}} = c$,
- for an n-place function symbol $f \in \tau$ and terms $t_1, \ldots, t_n \in M$ we have
$$f^{\mathfrak{M}}(t_1, \ldots, f_n) = f(t_1, \ldots, t_n).$$

A Herbrand structure which is a model of a theory Γ is called the *Herbrand model* of Γ.

411. (a) In the definition of Herbrand structure τ was required to contain at least one constant symbol. Why?

(b) What is the cardinality of a Herbrand structure?

(c) Show that for each τ-term t and evaluation e over $\mathfrak{M}$, the value of $t^{\mathfrak{M}}[e]$ is the term $t[x_1/e(x_1), \ldots, x_n/e(x_n)]$.

(d) Give a formula set Γ which has no Herbrand model.

412. The similarity type τ contains at least one constant symbol. The formula set $\Gamma \subset F(\tau)$ is such that no formula in Γ contains quantifiers or equality symbols. Prove that if Γ has a model, then it has a Herbrand model. Show that both conditions on Γ are necessary.

413. Show that if $\Gamma \subseteq F(\tau)$ has a model, then it has a model of cardinality at most $\max(\omega, |\tau|)$. (For a more general statement see Problem 493.)

Hint. Replace the equality symbol by a binary relation symbol as in Problem 340, then consider the Herbrand model.

7.8 Definition. A (first-order) *literal* ℓ is either $r(t_1, \ldots, t_n)$ or its negation, where $r \in \tau$ is an n-variable relation symbol, and $t_1, \ldots, t_n$ are τ-terms. A *clause* is a finite set of literals. The τ-structure $\mathfrak{A}$ is a model for the clause set C if for every evaluation e and every clause $c \in C$ at least one literal of c evaluates to true. C is *satisfiable* if it has a model, and *refutable* otherwise.

414. Show that if the clause set C has a model, then it has a Herbrand model.

415. Given a formula set $\Sigma \subseteq F(\tau)$, describe how to convert it to a clause set C such that Σ has a model iff C has a model. Check that if Σ is finite, then C can also be finite.

7.9 Definition. A *substitution* σ maps variable symbols to τ-type terms. Applying σ to a literal ℓ means that all variable symbols in ℓ are replaced

> simultaneously by the terms supplied by σ. When applied to a clause, σ should be applied to all literals in the clause.

416. For each relation symbol $r \in \tau$ and variable-free terms $t_1, \ldots, t_n$ consider $r(t_1, \ldots, t_n)$ as a propositional variable. For a set $\mathcal{C}$ of first-order clauses let $\mathcal{C}^o = \{\sigma(c) : c \in \mathcal{C}$ and σ assigns variable-free terms to the variables$\}$. Show that the propositional clause set $\mathcal{C}^o$ is satisfiable if and only if $\mathcal{C}$ is satisfiable.

> By Problem 416 and Theorem 5.16, $\mathcal{C}$ has no model iff $\mathcal{C}^o \vdash^R \square$. Rather than using the fully substituted forms of the clauses from $\mathcal{C}$ in the resolution steps, the substitutions can be left partly unspecified (so called lazy substitution), and perform the resolution step on these partially substituted clauses. This method, however, handles clauses as multisets necessarily (after full substitution different literals may become identical), in which case the resolution method is not guaranteed to work, see Problem 286. Thus, next to the generalized resolution step, a reduction called *factoring* is necessary: it checks whether two different literals in a clause can be (partially) substituted to become identical, and if so, the substitution is applied to the whole clause, and the literals are merged.

7.2 COMPACTNESS AND OTHER PROPERTIES

7.2.1 COMPACTNESS

> **7.10 Theorem** (Compactness). $\Sigma \models \varphi$ if and only if there is a finite $\Gamma \subset \Sigma$ such that $\Gamma \models \varphi$.

417. Prove Theorem 7.10.

418. Show that if every finite subset of Σ has a model, then Σ has a model, too. (See also Theorem 9.4.)

419. Suppose $\Sigma \subset F(\tau)$ has arbitrary large finite models. Show that Σ has an infinite model.

420. If Σ has an infinite model, then it has arbitrary large models.

421. In all models $\mathfrak{A}$ of Γ the relation $\leq^{\mathfrak{A}}$ is a linear ordering. Show that if Γ has an infinite model, then it has a model in which there is a decreasing sequence of length ω_1.

422. The partial order E is *well founded* if each non-empty subset contains an E-minimal element. Suppose τ contains the relation symbol E, and in each model $\mathfrak{A}$ of $\Sigma \subset F(\tau)$, the partial order $E^{\mathfrak{A}}$ is well founded. Show that there is a $k \in \omega$ (depending only on Σ) such that in every model $\mathfrak{A}$ of Σ, every increasing sequence $a_1 \, E^{\mathfrak{A}} \, a_2 \, E^{\mathfrak{A}} \cdots$ has length at most k.

423. Let τ have two binary symbols: E for the edges of a graph (see Definition 6.14), and $C(x, y)$ with the intended meaning that x and y are in the same connected component. Let $\Sigma \subset F(\tau)$ so that models of Σ are graphs, and in every model $\mathfrak{A}$ and nodes $u, v \in A$, $\langle u, v \rangle \in C^{\mathfrak{A}}$ iff u and v are in the same connected component. Show that there is a $k \in \omega$ depending only on Σ such that every model of Σ has diameter less than k.

Notation. The composition of the binary relations R and S is

$$R \circ S = \{(x, z) : \text{there is a } y \text{ such that } (x, y) \in R \text{ and } (y, z) \in S\}.$$

R^0 is the identity relation, and $R^{n+1} = R \circ R^n$ for $n \geq 0$.

424. The similarity type of Γ contains the binary relation symbol R. Suppose for each $n \geq 1$ there is a model $\mathfrak{A}_n \vDash \Gamma$ in which $R^0 \cup R^1 \cup \cdots \cup R^n$ is not transitive. Show that there is an $\mathfrak{A} \vDash \Gamma$ and elements $a_i \in A$ such that $\mathfrak{A} \vDash a_i \, R \, a_{i+1}$ but for $i + 1 < j$ we have $\mathfrak{A} \nvDash a_i \, R \, a_j$.

425. Let τ be a similarity type that contains the binary relation symbol R and let $\Gamma \subset F(\tau)$ be a theory. Prove that the following statements are equivalent:

 (i) There is an n such that $R^0 \cup R^1 \cup \cdots \cup R^n$ is transitive in each model of Γ.
 (ii) There is a $\varphi(x, y)$ such that $\{(a, b) : \mathfrak{A} \vDash \varphi[a, b]\}$ is the transitive closure of R in each model $\mathfrak{A}$ of Γ.

426. Let $\Gamma \subset F(\tau)$ be a theory and suppose that $\{\varphi_n(x) : n < \omega\}$ are formulas such that $\Gamma \vDash \forall x(\varphi_n(x) \rightarrow \varphi_{n+1}(x))$ for all $n \in \omega$. Assume that every element of every model of Γ satisfies some φ_n. Prove that $\Gamma \vDash \forall x \varphi_n(x)$ holds for some $n \in \omega$.

427.[*] The set $\Delta \subseteq F(\tau)$ of closed formulas has the property that if $\delta_1, \delta_2 \in \Delta$, then so is $\delta_1 \vee \delta_2 \in \Delta$. Let $\Gamma \subseteq F(\tau)$ and suppose that for every pair of structures

$$\text{if } \mathfrak{A} \vDash \Gamma \text{ and } (\mathfrak{A} \vDash \delta \text{ implies } \mathfrak{B} \vDash \delta \text{ for every } \delta \in \Delta), \text{ then } \mathfrak{B} \vDash \Gamma. \qquad (\star)$$

Show that there is a $\Delta' \subseteq \Delta$ such that $\Gamma \vDash \Delta'$ and $\Delta' \vDash \Gamma$.

Hint. Let $\Delta' = \{\delta \in \Delta : \Gamma \vDash \delta\}$.

7.2.2 DEFINABILITY

428. Give theories $\Gamma_i \subset F(\tau)$ for $i \in \omega$ such that the union of any three of them is consistent, while the union of any four is not.

The next two theorems connect theories on two compatible types τ_1 and τ_2. They have profound applications and, along with the compactness theorem, they are considered to be the most fundamental properties of the first-order logic.

7.11 Theorem (Robinson's consistency theorem). Let $\Gamma_1 \subset F(\tau_1)$ and $\Gamma_2 \subset F(\tau_2)$. If there is no closed $\varphi \in F(\tau_1 \cap \tau_2)$ such that both $\Gamma_1 \vDash \varphi$ and $\Gamma_2 \vDash \neg\varphi$, then $\Gamma_1 \cup \Gamma_2$ is consistent.

7.12 Theorem (Craig's interpolation theorem). Let $\varphi \in F(\tau_1)$ and $\psi \in F(\tau_2)$ and suppose $\varphi \vDash \psi$. Then there is $\vartheta \in F(\tau_1 \cap \tau_2)$ such that

$$\varphi \vDash \vartheta \quad \text{and} \quad \vartheta \vDash \psi.$$

429. Prove Robinson's consistency theorem using Craig's interpolation and vice versa.

430. Let $\tau = \tau_1 \cap \tau_2$. Prove Craig's theorem for the special case when $\tau_2 \smallsetminus \tau$ contains constant symbols only.

431. Let τ_i be compatible types and $\Gamma_i \subset F(\tau_i)$ be consistent theories. Suppose each pair (Γ_i, Γ_j) satisfies the assumptions of Robinson's consistency theorem. Does it follow that $\bigcup_i \Gamma_i$ is consistent?

432. For $1 \le i \le n$ let τ_i be compatible types, $\Gamma_i \subset F(\tau_i)$ be consistent theories. Suppose that for every pair of disjoint subsets I and J of $\{1, \ldots, n\}$ there is no closed formula φ of type $(\bigcup_{i \in I} \tau_i) \cap (\bigcup_{j \in J} \tau_j)$ such that $\bigcup_{i \in I} \Gamma_i \vDash \varphi$ and $\bigcup_{j \in J} \Gamma_j \vDash \neg\varphi$. Show that $\bigcup_i \Gamma_i$ is consistent.

7.13 Definition. Let τ be a similarity type and P be a relation symbol not in τ. We say that $\Sigma(P) \subset F(\tau \cup \{P\})$ defines P *implicitly* if each τ-structure can have at most one extension to a $\tau \cup \{P\}$ structure to become a model of $\Sigma(P)$. If for some $\vartheta(\vec{x}) \in F(\tau)$ we have $\Sigma(P) \vDash \vartheta(\vec{x}) \leftrightarrow P(\vec{x})$, then we say that $\Sigma(P)$ defines P *explicitly*.

In other words, $\Sigma(P)$ defines P implicitly if $\mathfrak{A}, \mathfrak{B} \vDash \Sigma(P)$ and $\mathfrak{A} \restriction \tau = \mathfrak{B} \restriction \tau$ implies $P^{\mathfrak{A}} = P^{\mathfrak{B}}$. Yet in other words, if $\mathfrak{A} \vDash \Sigma(P) \cup \Sigma(P')$, then $\mathfrak{A} \vDash \forall \vec{x}(P(\vec{x}) \leftrightarrow P'(\vec{x}))$. Naturally, this is the situation when P is defined explicitly.

7.14 Theorem (Beth). The theory $\Sigma(P) \subset F(\tau \cup \{P\})$ defines P implicitly if and only if it defines P explicitly.

433. Prove Theorem 7.14 using Craig's interpolation theorem.

Hint. Add new constant symbols $\vec{c}$ to the type, and use that $\Sigma(P) \cup \Sigma(P') \vDash P(\vec{c}) \leftrightarrow P'(\vec{c})$.

434. Suppose $\Sigma(P) \subset F(\tau \cup \{P\})$ defines P explicitly and let $\mathfrak{A} \vDash \Sigma(P)$. Show that every automorphism f of $\mathfrak{A} \restriction \tau$ preserves $P^{\mathfrak{A}}$ as well.

Remark (L. Svenonius). Under the additional assumption that $\Sigma(P)$ is maximal consistent the converse statement is also true: if every automorphism of $\mathfrak{A} \restriction \tau$ preserves $P^{\mathfrak{A}}$, then P is defined explicitly.

7.15 Definition. Let $\mathfrak{A}$ be a structure. The subset $X \in A^n$ is *definable in* $\mathfrak{A}$ if there is a formula $\varphi(\vec{x})$ such that $X = \{\vec{a} \in A^n : \mathfrak{A} \vDash \varphi[\vec{a}]\}$. The subset X is *definable using parameters*, if there are parameters $\vec{p} \in A$ such that $X = \{\vec{a} \in A^n : \mathfrak{A} \vDash \varphi[\vec{x}, \vec{p}]\}$ for some formula $\varphi(\vec{x}, \vec{y})$. A function f is definable if its graph (Definition 4.11) is definable.

435. Suppose X is definable in $\mathfrak{A}$ from the parameters $\vec{p}$. Show that if an automorphism fixes $\vec{p}$ pointwise, then it fixes X setwise.

436. Prove that the field of real numbers is not definable in the field of complex numbers.

437. Show that the addition cannot be defined from multiplication in the structure $\mathfrak{A} = \langle \omega, \cdot \rangle$, that is, the set of triplets $\langle a, b, a+b \rangle \in \omega^3$ is not definable in $\mathfrak{A}$.

438. Prove that no formula $\varphi(x) \in F(\leq, +)$ defines the one-element set $\{1\}$ in the structure $\langle \mathbb{Q}, \leq, + \rangle$.

439. Let $\mathfrak{A} = \langle \omega, 0, S \rangle$ where the interpretation of S is the successor function. Prove that the set of even numbers cannot be defined in $\mathfrak{A}$.

440. Let $\mathfrak{A} = \langle \omega, 0, \leq, S \rangle$ where S is the successor function. Show that the addition cannot be defined in $\mathfrak{A}$.

> **Remark.** It is also true that multiplication cannot be defined in the structure $\langle \omega, 0, \leq, S, + \rangle$, see Problem 795. Surprisingly, the exponentiation (in fact, every recursive function) can be defined in $\langle \omega, 0, \leq, S, +, \cdot \rangle$, see Problem 833.

441. Give an example for a countable discrete ordering with initial point 0 and successor function S in which one can define, in multiple ways, a 2-place function f (the "addition") that satisfies

$$f(0, x) = f(x, 0) = x,$$
$$f(Sx, y) = f(x, Sy) = Sf(x, y),$$
$$x \leq y \leftrightarrow \exists z f(x, z) = y.$$

> **Hint.** Use Beth's definability theorem.

7.2.3 SEMANTICAL INTERPRETATION

442. Show that the constants $0^{\mathfrak{A}}$, $1^{\mathfrak{A}}$ and the ordering $\leq^{\mathfrak{A}}$ can be defined in the structure $\mathfrak{A} = \langle \omega, + \rangle$.

> **7.16 Definition.** The τ'-type structure $\mathfrak{B}$ is a *semantical substructure* of the τ-structure $\mathfrak{A}$ if the ground set B is a definable subset of $\mathfrak{A}$, and the interpretation of every τ'-symbol in $\mathfrak{B}$ is $\mathfrak{A}$-definable. The phrase

$\mathfrak{B}$ *can be semantically defined in* $\mathfrak{A}$, or simply $\mathfrak{B}$ can be defined in $\mathfrak{A}$, means that $\mathfrak{B}$ is isomorphic to a semantical substructure of $\mathfrak{A}$.

Semantical interpretation is the basic tool to establish undecidability of a theory, see Section 10.2.

443. Show that if $\mathfrak{C}$ can be defined in $\mathfrak{B}$, and $\mathfrak{B}$ can be defined in $\mathfrak{A}$, then $\mathfrak{C}$ can be defined in $\mathfrak{A}$.

444. $\mathfrak{N} = \langle \omega, +, \cdot \rangle$ can be defined semantically in the ring $\langle \mathbb{Z}, +, \cdot \rangle$.

445. Let $\mathfrak{A} = \langle \omega, e \rangle$ where $e^{\mathfrak{A}}(x, y) = x^y$. Show that $\mathfrak{N} = \langle \omega, +, \cdot \rangle$ can be defined semantically in $\mathfrak{A}$.

446. Consider the structure $\mathfrak{A} = \langle \omega, +, \mathsf{sq} \rangle$ where the unary relation $\mathsf{sq}^{\mathfrak{A}}$ is the set of perfect squares. Show that $\mathfrak{N} = \langle \omega, +, \cdot \rangle$ can be semantically defined in $\mathfrak{A}$.

447. A set is *hereditarily finite* if it is either empty, or have finitely many elements each of which is hereditarily finite. Let H be the set of hereditarily finite sets, and let $\mathfrak{A}$ be the structure $\langle H, \in \rangle$.

(a) Show that $\mathfrak{N} = \langle \omega, +, \cdot \rangle$ can be defined semantically in $\mathfrak{A}$.

(b) Let $f(i) = 2^i$. Show that $\mathfrak{A}$ can be defined semantically in $\langle \omega, +, \cdot, f \rangle$.

448. Let ρ be a binary relation symbol and f be a binary function symbol. Show that every $\langle f \rangle$-structure can be semantically defined in some $\langle \rho \rangle$-structure.

449. Show that any structure with a single binary relation can be semantically defined in some graph.

450. Show that there is graph in which the structure $\mathfrak{N} = \langle \omega, +, \cdot \rangle$ of the natural numbers can be defined semantically.

451. Show that every graph can be semantically defined in some lattice with minimal and maximal elements.

452. Let $\mathfrak{A} = \langle \mathbb{Z}, 1, +, | \rangle$ where $|$ is the divisibility relation. Show that $\mathfrak{N} = \langle \omega, +, \cdot \rangle$ can be semantically defined in $\mathfrak{A}$.

453 (A. Tarksi). Let G be the symmetric group of $\mathbb{Z}$ with an additional constant symbol π denoting the right shift $j \to j + 1$. Show that $\mathfrak{N} = \langle \omega, +, \cdot \rangle$ is semantically definable in G.

Hint. Among the shifts in G addition is the composition, and the divisibility relation is definable.

7.17 Theorem (J. Robinson). $\mathfrak{N} = \langle \omega, +, \cdot \rangle$ is semantically definable in the field of rational numbers $\langle \mathbb{Q}, 0, 1, +, \cdot \rangle$.

7.2.4 OMITTING TYPES

Notation. For a set of formulas we write $T(\vec{x})$ to indicate that the free variables of formulas $\varphi \in T$ are among $\vec{c} = \langle x_1, \ldots, x_n \rangle$.

7.18 Definition (Types). Let $\Sigma \subset F(\tau)$ be a consistent theory. A set $T(x_1 \ldots, x_n) \subset F(\tau)$ is an *n-type*, or simply a *type* of Σ, if for each finite subset $\{\psi_1, \ldots, \psi_k\} \subseteq T$, $\Sigma \cup \{\exists \vec{x}(\psi_1(\vec{x}) \wedge \cdots \wedge \psi_k(\vec{x}))\}$ is consistent.

The type $T(\vec{x})$ is *isolated* if there is a $\varphi(\vec{x}) \in F(\tau)$ such that $\Sigma \cup \{\exists \vec{x} \varphi(\vec{x})\}$ is consistent, and for all $\psi(\vec{x}) \in T$, $\Sigma \vDash \varphi(\vec{x}) \to \psi(\vec{x})$.

The type $T(\vec{x})$ is *realized in* $\mathfrak{A} \vDash \Sigma$ if for some $\vec{a} \in A^n$, $\mathfrak{A} \vDash \psi[\vec{a}]$ for all $\psi \in T$. If $T(\vec{x})$ is not realized in $\mathfrak{A}$, then $\mathfrak{A}$ *omits* $T(\vec{x})$.

454. Suppose τ contains countably many constant symbols c_i. Show that $T(x) = \{x \neq c_i\}$ is a Σ-type when Σ has an infinite model. What models of Σ omit the type $T(x)$?

455. Let τ contain the relation symbol $\leq$ for ordering, and $\psi_n(x, y)$ be the formula which expresses that there are at least n different elements between x and y. Which ordered structures omit the 2-type $T(x, y) = \{\psi_n(x, y) : n \in \omega\}$?

456. Give a theory Σ and a 1-type $T(x)$ such that for each $k \geq 0$ there is a model $\mathfrak{A}_k \vDash \Sigma$ in which $T(x)$ is realized by exactly k elements in $\mathfrak{A}_k$.

7.19 Theorem (Omitting types theorem). Let τ be a countable signature, $\Sigma \subset F(\tau)$ be consistent, and $T_i(\vec{x})$ ($i \in \omega$) be countably many non-isolated types for Σ. Then there is a countable model $\mathfrak{A} \vDash \Sigma$ which omits every $T_i(\vec{x})$.

457. Show that there is an uncountable type τ, a consistent theory $\Sigma \subset F(\tau)$, and a non-isolated type $T(x)$ such that no model of Σ can omit $T(x)$.

458. Let $\tau = \langle 0, \leq, c_i : i \in \omega \rangle$ and Σ be the set of all τ-formulas true in the model $\mathfrak{A} = \langle \omega, 0, \leq, c_i \rangle$ where the interpretation of c_i is $i \in \omega$. The type $T(x)$ is $\{c_i \leq x : i \in \omega\}$. It is clearly finitely satisfiable, and it is not isolated: if $\Sigma \cup \{\exists x \varphi(x)\}$ is consistent, then $\mathfrak{A} \vDash \exists x \varphi(x)$, and then Σ cannot prove that x is bigger than every c_i. By Theorem 7.19 there is a model $\mathfrak{B} \vDash \Sigma$ which omits $T(x)$. But this is impossible, as every model of Σ contains the interpretations of the constants c_i at the beginning, and there is nothing in between them.
Where is the error?

In the rest of this section the similarity type τ contains the constant symbol 0 and the unary function symbol S. The terms π_i for $i \in \omega$ are defined by $\pi_0 = 0$, $\pi_1 = S(0)$, and in general $\pi_{n+1} = S(\pi_n)$. Theories are required to guarantee that the terms π_i have different interpretations. It can be achieved, among others, by requiring that the following formulas are in every formula set considered:

$$\forall x(0 \neq S(x)), \quad \text{and} \quad \forall x \forall y(S(x) = S(y) \rightarrow x = y).$$

7.20 Definition. The theory $\Gamma \subset F(\tau)$ is *ω-consistent* if there is no formula $\varphi(x) \in F(\tau)$ such that $\Gamma \vDash \exists x \neg \varphi(x)$, while $\Gamma \vDash \varphi(\pi_i)$ for every $i \in \omega$. Γ is *ω-complete* if for every formula $\varphi(x) \in F(\tau)$,

$$\Gamma \vDash \varphi(\pi_0), \ \Gamma \vDash \varphi(\pi_1), \ \Gamma \vDash \varphi(\pi_2), \ \ldots$$

implies $\Gamma \vDash \forall x \varphi(x)$.

$\mathfrak{A}$ is an *ω-model* if the ground set is ω, the interpretation of 0 is zero, and $S^{\mathfrak{A}}$ is the successor function. $\pi_i^{\mathfrak{A}} = i$ for all $i \in \omega$ in any ω-model.

459. Let Γ be a consistent theory. Prove that

(a) If Γ is ω-complete, then Γ has an ω-model.

(b) If Γ has an ω-model, then Γ is ω-consistent.

7.21 Definition. The *ω-rule* infers the conclusion from infinitely many premises. φ is a formula φ with x as a free variable:

$$\frac{\varphi(\pi_0), \ \varphi(\pi_1), \ \varphi(\pi_2), \ \varphi(\pi_3), \ \ldots}{\forall x \varphi(x).}$$

460. Add the ω-rule as a rule of inference. Prove that a theory Γ is syntactically consistent (with the additional ω-rule) if and only if it has an ω-model.

461. We have seen that every syntactically consistent theory can be extended to a maximal one (Problem 393). Add the ω-rule as a rule of inference. Prove or disprove that every syntactically ω-consistent theory can be extended to a maximal one.

462. Add the ω-rule as a rule of inference. Show that the compactness property of the derivation does not hold anymore.

ELEMENTARY EQUIVALENCE

8

8.1 BASICS

> **8.1 Definition** (Elementary equivalence). The τ-type structures $\mathfrak{A}$ and $\mathfrak{B}$ are *elementarily equivalent*, written as $\mathfrak{A} \equiv \mathfrak{B}$ if $\mathfrak{A}$ and $\mathfrak{B}$ have the same theory $\mathrm{Th}(\mathfrak{A}) = \mathrm{Th}(\mathfrak{B})$, that is, for each $\varphi \in F(\tau)$, $\mathfrak{A} \vDash \varphi$ iff $\mathfrak{B} \vDash \varphi$.

463. Prove that isomorphic structures are elementarily equivalent.

464. Find two elementarily equivalent structures which are not isomorphic.

465.[*] If two structures are elementarily equivalent and one of them is finite, then they are isomorphic.

466. Prove or disprove: if $\mathfrak{A} \!\upharpoonright\! \sigma$ and $\mathfrak{B} \!\upharpoonright\! \sigma$ are isomorphic for all finite subtypes $\sigma \subseteq \tau$, then $\mathfrak{A}$ and $\mathfrak{B}$ are also isomorphic.

467. Show that among more than continuum many linear orderings there are at least two which are elementarily equivalent. Construct continuum many linear orderings such that no two of them are elementarily equivalent.

468. Give two not elementarily equivalent structures such that each is isomorphic to a substructure of the other.

469. Find an increasing chain of elementarily equivalent structures such that the union of this chain is not elementarily equivalent to any element of the chain.

470. Prove or disprove: two graphs are elementarily equivalent if and only if they have the same set of spanned finite subgraphs.

471. Do there exist elementarily equivalent graphs with different but finite chromatic numbers? See also Problem 499.

472. Let $\mathfrak{B}$ be a substructure of $\mathfrak{A}$ and e be an evaluation over $\mathfrak{B}$. Prove that

$$\mathfrak{B} \vDash \varphi[e] \quad \Leftrightarrow \quad \mathfrak{A} \vDash \varphi[e]$$

holds for quantifier-free formulas.

© The Author(s), under exclusive license to Springer Nature Switzerland AG 2022
L. Csirmaz and Z. Gyenis, *Mathematical Logic,* Problem Books
in Mathematics, https://doi.org/10.1007/978-3-030-79010-3_8

473. Give structures $\mathfrak{B} \subseteq \mathfrak{A}$ and a formula φ such that for all evaluations e over $\mathfrak{B}$ we get $\mathfrak{B} \models \exists x \varphi[e]$ but $\mathfrak{A} \not\models \exists x \varphi[e]$. Can φ be quantifier-free?

8.2 Definition (Elementary substructure). $\mathfrak{B}$ is an *elementary substructure* of $\mathfrak{A}$, written as $\mathfrak{B} \prec \mathfrak{A}$, if $\mathfrak{B}$ is a substructure of $\mathfrak{A}$ and for each evaluation e over the smaller structure $\mathfrak{B}$ we have

$$\mathfrak{B} \models \varphi[e] \quad \text{if and only if} \quad \mathfrak{A} \models \varphi[e].$$

474. Prove that $\mathfrak{B}$ is an elementary substructure of $\mathfrak{A}$ if and only if for all formulas $\varphi(\vec{x})$ with n free variables,

$$\{\vec{b} \in B^n : \mathfrak{A} \models \varphi[\vec{b}]\} = B^n \cap \{\vec{a} \in A^n : \mathfrak{A} \models \varphi[\vec{a}]\}.$$

475. Show that $\langle \mathbb{Z}, \le \rangle$ has no proper elementary submodels.

476. Let G be the infinite path (infinite in both directions). Find all elementary subgraphs of G.

477. Give two isomorphic structures $\mathfrak{A}$ and $\mathfrak{B}$ so that $\mathfrak{A}$ is a substructure of $\mathfrak{B}$, but $\mathfrak{A}$ is not an elementary substructure of $\mathfrak{B}$.

Notation. φ is a $\exists_n$-*formula* ($\forall_n$-formula) if φ has a sequence of n alternating blocks of quantifiers before a quantifier-free formula starting with a block of the quantifier $\exists$ (the quantifier $\forall$).
A $\forall_1$-formula is *universal*, and a $\exists_1$-formula is *existential*.

478. Find $\mathfrak{B} \subseteq \mathfrak{A}$ such that $\mathfrak{B}$ is *not* an elementary substructure of $\mathfrak{A}$, while for all evaluations e over $\mathfrak{B}$ and (a) for all $\exists_1$ formulas; (b) for all $\forall_1$ formulas we have $\mathfrak{A} \models \varphi[e]$ if and only if $\mathfrak{B} \models \varphi[e]$.

479. Prove or disprove: if both $\mathfrak{A} \prec \mathfrak{C}$ and $\mathfrak{B} \prec \mathfrak{C}$, moreover $\mathfrak{B}$ is a substructure of $\mathfrak{A}$, then (a) $\mathfrak{A}$ and $\mathfrak{B}$ are elementarily equivalent, (b) $\mathfrak{B} \prec \mathfrak{A}$.

480. Is every normal subgroup of a group an elementary subgroup? What about infinite groups?

481. Let G be an infinite Abelian group and H be an infinite subgroup. Is it true that H is an elementary subgroup of G?

8.3 Theorem (Tarski–Vaught test). Suppose $\mathfrak{B} \subseteq \mathfrak{A}$. $\mathfrak{B}$ is an elementary substructure of $\mathfrak{A}$ if and only if for all $\vec{b} \in B$ and $\varphi(x, \vec{v}) \in F(\tau)$ we have

$$\text{if } \mathfrak{A} \models (\exists x \varphi)[\vec{b}], \text{ then there is a } c \in B \text{ such that } \mathfrak{A} \models \varphi[c, \vec{b}].$$

482. Prove Theorem 8.3.

Hint. Use induction on the complexity of φ.

483. Let $\mathfrak{A}$ be a structure, and B be a subset of A. Suppose that for all $\vec{b} \in B$ and $a \in A$ there is an automorphism of $\mathfrak{A}$ which leaves all b_i in place and moves a into B. Show that in this case B is the ground set of a substructure $\mathfrak{B}$, and $\mathfrak{B}$ is an elementary submodel of $\mathfrak{A}$.

484. Let $a \leq b < c \leq d$ be real numbers; (b, c) is the open interval of the reals with endpoints b and c. Show that $\mathfrak{A} = \langle (b, c), \leq \rangle$ is an elementary submodel of $\mathfrak{B} = \langle (a, d), \leq \rangle$.

485. Prove that $\langle \mathbb{Q}, \leq \rangle$ is an elementary submodel of $\langle \mathbb{R}, \leq \rangle$.

486. Let $\mathfrak{A}$ be a countable dense order without endpoints, $a \in A$ and $B = A - \{a\}$. Show that $\mathfrak{B} \prec \mathfrak{A}$.

487. The graph G is the *broom* if it has a node which is the endpoint of an infinite path—the handle—and countably many edges are joined to it – the brush. Describe all elementary substructures of G.

488. The graph G consists of a vertex v to which continuum many disjoint paths of length two are joined. Describe all elementary subgraphs of G.

489. Give structures $\mathfrak{A}_i$ for $i < \omega$ such that $\mathfrak{A}_j \prec \mathfrak{A}_i$ whenever $i < j$, and

(a) $\bigcap A_i = \varnothing$.

(b) $\bigcap A_i \neq \varnothing$, but $\bigcap \mathfrak{A}_i$ is not elementarily equivalent to any $\mathfrak{A}_i$.

490. Let $\mathfrak{A}$ be a structure and B be a non-empty subset of A. Suppose B is closed for all Skolem functions, see Definition 7.6. Show that B is a ground set of a substructure $\mathfrak{B}$, and $\mathfrak{B} \prec \mathfrak{A}$.

> **8.4 Definition.** The theory $\Gamma \subset F(\tau)$ has *built-in Skolem functions* if for every formula $\psi(\vec{x})$ of the form $\exists y \varphi(y, \vec{x})$ there is a τ-term $t_\psi(\vec{x})$ such that $\Gamma \vDash \exists y \varphi(y, \vec{x}) \rightarrow \varphi(t_\psi(\vec{x}), \vec{x})$.

491. Suppose Γ has built-in Skolem functions and $\mathfrak{A} \vDash \Gamma$. Show that every substructure of $\mathfrak{A}$ is an elementary substructure.

> **8.5 Theorem** (Downward Löwenheim–Skolem). Let $\mathfrak{A}$ be a τ-structure and $X \subseteq A$. There is an elementary substructure $\mathfrak{B} \prec \mathfrak{A}$ such that $X \subseteq B$ and $|B| \leq \max(|X|, |\tau|, \omega)$.

492. Prove Theorem 8.5.

493. (a) Show that if Γ has an infinite model, then it has a model of cardinality κ for every $\kappa \geq \max(|\tau|, \omega)$.

(b) Suppose $\mathfrak{A}$ is infinite. Show that for each infinite $\kappa \geq |\tau|$ there is a structure $\mathfrak{B}$ of cardinality κ which is elementarily equivalent to $\mathfrak{A}$.

Remark. A countable structure has an elementarily equivalent structure of cardinality 2^ω, independently of $|\tau|$ (Problem 495). On the other hand there is a countable structure such that an elementarily equivalent structure is either isomorphic to it, or has cardinality at least 2^ω (Problem 583).

494. Let K be an arbitrary set of natural numbers. Give a theory Γ which has a model of size $i \in \omega$ iff $i \in K$.

495. Suppose Γ has a model of cardinality $\kappa \geq \omega$. Show that Γ has a model of size λ for every $\lambda \geq 2^\kappa$ independently of the cardinality of the type.

8.6 Definition (Complete theory). The theory $\Gamma \subset F(\tau)$ is *complete* if either $\Gamma \vDash \varphi$ or $\Gamma \vDash \neg\varphi$ for each closed formula $\varphi \in F(\tau)$.

496. Show that a consistent Γ is complete $\Leftrightarrow$ (i) $\Leftrightarrow$ (ii), where

(i) there is a structure $\mathfrak{A}$ such that $\mathrm{Th}(\mathfrak{A}) = \{\varphi : \Gamma \vDash \varphi\}$,

(ii) any two models of Γ are elementarily equivalent.

Notation. The theory $\Gamma \subset F(\tau)$ is κ-*categorical* if, up to isomorphism, Γ has a unique model of cardinality κ.

8.7 Theorem (Łoś–Vaught). If $\Gamma \subset F(\tau)$ has no finite models and Γ is κ-categorical for some infinite $\kappa \geq |\tau|$, then it is complete.

497. Prove Theorem 8.7.

498. (a) Give a counterexample for Theorem 8.7 without the assumption that Γ has no finite models.

(b) Find a set of formulas Γ which has exactly one model, up to isomorphism, for each infinite cardinal κ, but not all of its models are elementarily equivalent.

499. Do there exist elementarily equivalent graphs with different infinite chromatic numbers?

500. Let Γ be the theory of cycle-free graphs where each vertex has degree 2. Prove that Γ is complete.

501. Let $\tau = \langle 0, S \rangle$ where S is a unary function symbol. Show that the following theory is complete:

$$S(x) = S(y) \to x = y, \quad 0 \neq S(x), \quad x \neq 0 \to \exists y (x = S(y)),$$
$$x \neq S(x), \quad x \neq SS(x), \quad x \neq SSS(x), \quad \text{etc.}$$

502. Show that the theory of dense linear order without endpoints is $\aleph_0$-categorical, thus complete.

503. Prove that the theory of torsion-free divisible Abelian groups is κ-categorical for all $\kappa > \aleph_0$. Is it $\aleph_0$-categorical?

504. The theory of algebraically closed fields of characteristic 0 is κ-categorical for all $\kappa > \aleph_0$.

505. Let $\Gamma = \mathrm{Th}(\langle \mathbb{R}, \leq, Z \rangle)$, where $\mathbb{R}$ is the set of real numbers, and the interpretation of the unary relation symbol Z is the set of the integers. Show that Γ is not $\aleph_0$-categorical.

> **Notation.** An *embedding* of $\mathfrak{A}$ into $\mathfrak{B}$ is an isomorphism between $\mathfrak{A}$ and a substructure of $\mathfrak{B}$. It is an *elementary embedding* if the image of $\mathfrak{A}$ is an elementary substructure of $\mathfrak{B}$.

506. Find elementarily equivalent structures $\mathfrak{A}$ and $\mathfrak{B}$ such that there is no embedding from either one to the other.

> **8.8 Definition** (Diagram). Choose new constant symbols c_a for each element a in the ground set of the structure $\mathfrak{A}$. The *diagram* of $\mathfrak{A}$ is
>
> $$\Delta_{\mathfrak{A}} = \{\varphi[x_1/c_{a_1}, \ldots, x_n/c_{a_n}] : \varphi(\vec{x}) \in F(\tau) \ \text{ and } \ \mathfrak{A} \vDash \varphi[a_1, \ldots, a_n]\}.$$
>
> The *atomic diagram* of $\mathfrak{A}$, denoted by $\Delta_{\mathfrak{A}}^0$, contains the variable-free formulas of $\Delta_{\mathfrak{A}}$.

507. Let $\mathfrak{A}$ be a τ-type structure. Prove that

(a) if $\mathfrak{B} \vDash \Delta_{\mathfrak{A}}^0$, then $\mathfrak{A}$ can be embedded into $\mathfrak{B} \restriction \tau$ as a substructure.

(b) if $\mathfrak{B} \vDash \Delta_{\mathfrak{A}}$, then $\mathfrak{A}$ can be elementarily embedded into $\mathfrak{B} \restriction \tau$.

508. Show that every infinite structure has a proper elementary extension.

509. Let $\Sigma \subset F(\tau)$ and suppose that every finitely generated substructure of $\mathfrak{A}$ can be extended to a model of Σ. Show that $\mathfrak{A}$ can be extended to a model of Σ.

510. Let $\mathfrak{A}$ and $\mathfrak{B}$ be elementarily equivalent.

(a) Prove that $\Delta_{\mathfrak{A}} \cup \Delta_{\mathfrak{B}}$ is consistent.

(b) Show that there is a structure $\mathfrak{C}$ into which both $\mathfrak{A}$ and $\mathfrak{B}$ can be embedded elementarily.

511. Suppose $\mathfrak{A}_i$ are elementarily equivalent structures for $i \in I$. Prove that there exists $\mathfrak{C}$ into which all the $\mathfrak{A}_i$'s can be elementarily embedded.

512. Let τ_1 and τ_2 be compatible types. Show that the following statement is equivalent to Robinson's consistency theorem 7.11:

> Let $\mathfrak{A}_i$ be a τ_i-type structure. $\mathfrak{A}_1$ and $\mathfrak{A}_2$ can be elementarily embedded into a $\tau_1 \cup \tau_2$-type structure $\mathfrak{B}$ iff the τ-type reducts of $\mathfrak{A}_1$ and $\mathfrak{A}_2$ are elementarily equivalent.

513. Prove or disprove the following claim (compare to Problem 511):

Let τ_i be compatible types for $i \in I$, and $\mathfrak{A}_i$ be a τ_i-structure. For each $i, j \in I$ the structures $\mathfrak{A}_i$ and $\mathfrak{A}_j$ can be embedded into a structure of type $\tau_i \cup \tau_j$. Then there is a structure of type $\bigcup_i \tau_i$ into which every $\mathfrak{A}_i$ embeds elementarily.

514. Let $\Gamma \subset F(\tau)$, and let Δ be the set of universal consequences of Γ, that is, $\Delta = \{\varphi \in \forall_1(\tau) : \Gamma \vDash \varphi\}$. Show that every model of Δ can be embedded into some model of Γ.

515. * Suppose that there is no quantifier-free formula $\delta(\vec{x})$ such that $\Gamma \vDash \varphi(\vec{x}) \leftrightarrow \delta(\vec{x})$. Show that Γ has models $\mathfrak{A}$ and $\mathfrak{A}'$ such that $\mathfrak{B}$ is a common substructure of both models, and there is some $\vec{b} \in B$ such that $\mathfrak{A} \vDash \varphi[\vec{b}]$ and $\mathfrak{A}' \vDash \neg\varphi[\vec{b}]$.

Hint. Let $\mathfrak{A} \vDash \varphi[\vec{b}]$, $\mathfrak{B}$ be the substructure generated by $\vec{b}$, and $\mathfrak{A}'$ be a model of $\Delta_{\mathfrak{B}}^0 \cup \{\neg\varphi[\vec{b}]\}$.

8.9 Definition (Collection principle). Suppose $\in$ is a binary relation symbol in τ. The *collection principle* is the set of all formulas

$$(\forall x \in a)\,(\exists y)\,\varphi \;\rightarrow\; (\exists b)\,(\forall x \in a)\,(\exists y \in b)\,\varphi,$$

where b is not free in the formula φ.

516. Give examples for structures that satisfy the collection principle.

8.10 Definition (End extension). The structure $\mathfrak{B}$ is an *end extension* of its substructure $\mathfrak{A}$, or $\mathfrak{A}$ is an *initial segment* of $\mathfrak{B}$ with respect to $\in$, if $a \in^{\mathfrak{B}} b$ for all $a \in A$ and $b \in B \smallsetminus A$.

517. * Let $\mathfrak{A}$ be a countable structure, and $\in^{\mathfrak{A}}$ be a linear order without a maximal element which satisfies the collection principle. Prove that $\mathfrak{A}$ has a proper elementary end extension.

Hint. Use the omitting types theorem for the theory $\Delta_{\mathfrak{A}} \cup \{c_a \in c : a \in A\}$ and the types $T_a = \{x \in c_a \wedge x \neq c_b : b \in^{\mathfrak{A}} a\}$ for all $a \in A$.

8.2   EHRENFEUCHT–FRAÏSSÉ GAME

8.11 Definition. A *partial isomorphism* is a partial map $j : \mathfrak{A} \to \mathfrak{B}$ between τ-structures which preserves the truth of unnested atomic formulas of the form

$$x = y, \quad x = c, \quad y = f(\vec{x}), \quad r(\vec{x})$$

where all variables are evaluated to values in $\mathrm{dom}(j)$.

8.12 Definition. For an integer $N \geq 1$ the N-round *Ehrenfeucht–Fraïssé game*, denoted by $\mathrm{EF}(\mathfrak{A}, \mathfrak{B}, N)$, is played by players I and II as follows. In round $1 \leq i \leq N$ player I chooses one of the structures $\mathfrak{A}$ or $\mathfrak{B}$ and picks an element $a_i \in A$ ($b_i \in B$, respectively). II answers by choosing an element $b_i \in B$ ($a_i \in A$, respectively) from the other structure. After the N rounds they formed the partial function $a_i \mapsto b_i$. II wins if this function is a partial isomorphism, and I wins if it is not.

518. The graph G_1 is a circle (i.e., a finite connected graph where all vertices have degree 2) of length 2^N, and the graph G_2 is the infinite path (connected and all vertices have degree 2). Show that player II can win the N-round game.

519. Let $\varphi(x, y)$ be a quantifier-free formula with two free variables. Suppose $\mathfrak{A} \vDash \forall x \exists y \, \varphi(x, y)$, while $\mathfrak{B} \vDash \neg \forall x \exists y \, \varphi(x, y)$.

(a) Show that for large enough N, I wins the game.
(b) For each N construct structures $\mathfrak{A}$, $\mathfrak{B}$ and a formula $\varphi(x, y)$ as above such that II wins the N-round game.

520. Show that if for all N II wins the N-round game then $\mathfrak{A}$ and $\mathfrak{B}$ are elementarily equivalent.

521. (a) Let $\mathfrak{A}$ and $\mathfrak{B}$ be two discrete linear orderings with initial points and without endpoints. Show that II wins the N-round game for all N, thus $\mathfrak{A}$ and $\mathfrak{B}$ are elementarily equivalent.

(b) Show that if N is not fixed in advance, then I could win.

522. In two linear orders every point has either an immediate predecessor or an immediate successor but not both. In which cases can II win the N-round EF game for all N?

523. The type τ contains n unary relation symbols $R_1, \ldots, R_n$, and the τ-type structures $\mathfrak{A}$ and $\mathfrak{B}$ satisfy the following condition: the number of elements in every Boolean combination of the relations is either equal, or both numbers are larger than N (possibly infinite). Show that II wins the N-round EF game.

524. Let $\mathfrak{A} = \langle \omega, 0, \leq, S \rangle$ where S is the successor function, and $\mathfrak{B} = \langle \omega + \mathbb{Z}, 0, \leq, S \rangle$ where the set of integers $\mathbb{Z}$ is appended after ω. Show that $\mathfrak{A}$ and $\mathfrak{B}$ are elementarily equivalent.

525. Suppose there are neither constant nor function symbols in the type τ. Assume moreover that II wins the n-round Ehrenfeucht–Fraïssé game. Show that $\mathfrak{A}$ and $\mathfrak{B}$ are indistinguishable by closed formulas of the form $\forall x_1 \exists x_2 \ldots \delta(\vec{x})$ where δ is quantifier-free.

526. Take an infinite path P on vertices $\{v_i : i \in \mathbb{Z}\}$ where v_i is connected to v_{i-1} and v_{i+1}. Build a graph $\mathfrak{A}$ by dropping finite paths from vertices of P

so that it has the following property. For each finite sequence $\langle i_1, \ldots, i_k \rangle$ of natural numbers there are k consecutive vertices on P such that the length of paths dropped from these vertices are $i_1, \ldots, i_k$ in this order. Let $\mathfrak{B}$ be the same graph as $\mathfrak{A}$ with the only difference that the path dropped from v_0 is replaced by an infinite path. Show that II can win the EF game for every N.

527. Give elementarily equivalent structures $\mathfrak{A}$ and $\mathfrak{B}$ such that player I wins the N-round EF game for every N.

> **8.13 Theorem** (Fraïssé). Suppose τ is a finite type and $\mathfrak{A}$, $\mathfrak{B}$ are τ-structures. Then $\mathfrak{A}$ and $\mathfrak{B}$ are elementarily equivalent iff for all N II wins the N-round game $\mathrm{EF}(\mathfrak{A}, \mathfrak{B}, N)$.

> **8.14 Definition.** A set I of partial isomorphisms between $\mathfrak{A}$ and $\mathfrak{B}$ is a *back-and-forth system* if $I \neq \emptyset$ and the following stipulations hold:
>
> (Forth) For all $f \in I$ and $a \in A$ there is $g \in I$ such that $f \subseteq g$ and $a \in \mathrm{dom}(g)$.
>
> (Back) For all $f \in I$ and $b \in B$ there is $g \in I$ such that $f \subseteq g$ and $b \in \mathrm{ran}(g)$.
>
> $I : \mathfrak{A} \rightleftarrows \mathfrak{B}$ denotes that I is a back-and-forth system between $\mathfrak{A}$ and $\mathfrak{B}$, and $\mathfrak{A} \rightleftarrows \mathfrak{B}$ means that there is an I such that $I : \mathfrak{A} \rightleftarrows \mathfrak{B}$. In this case we say that $\mathfrak{A}$ and $\mathfrak{B}$ are *back-and-forth equivalent*.

528. If $\mathfrak{A}$ and $\mathfrak{B}$ are countable, then $\mathfrak{A} \cong \mathfrak{B}$ if and only if $\mathfrak{A} \rightleftarrows \mathfrak{B}$.

> What can be said about uncountable structures? Below $L_{\infty,\omega}$ denotes the language which is similar to a first-order language but in which arbitrary conjunctions with a bounded number of variables are allowed. We write $\mathfrak{A} \equiv_{\infty,\omega} \mathfrak{B}$ if for all closed $\varphi \in L_{\infty,\omega}$ we have $\mathfrak{A} \vDash \varphi$ if and only if $\mathfrak{B} \vDash \varphi$.

> **8.15 Theorem** (Karp). $\mathfrak{A} \equiv_{\infty,\omega} \mathfrak{B}$ if and only if $\mathfrak{A} \rightleftarrows \mathfrak{B}$.

529. Prove Theorem 8.15.

> **Hint.** The set
>
> $$I = \{f : A \to B : f \text{ is finite and preserves all } L_{\infty,\omega} - \text{formulas}\}$$
>
> is a back-and-forth system.

530. Give $\mathfrak{A}$ and $\mathfrak{B}$ such that $\mathfrak{A} \equiv \mathfrak{B}$ while $\mathfrak{A} \not\rightleftarrows \mathfrak{B}$.

531. Prove that countable models of the following theories are back-and-forth equivalent (and therefore these theories are $\aleph_0$-categorical, hence complete).

(a) Dense linear orderings without endpoints.

(b) Dense linear orderings with left (right) endpoint.

(c) Atomless Boolean algebras.

532. Construct two non-isomorphic atomless Boolean algebras of the same cardinality.

8.3   QUANTIFIER ELIMINATION

> **8.16 Definition.** The theory $\Gamma \subset F(\tau)$ has *quantifier elimination* if for every formula $\varphi(\bar{x}) \in F(\tau)$ there is a quantifier-free formula $\psi(\bar{x})$ such that $\Gamma \vDash \varphi(\bar{x}) \leftrightarrow \psi(\bar{x})$.

533. Let $\tau = \langle P \rangle$ and $\Gamma = \{\forall x P(x) \vee \forall x \neg P(x)\}$. Show that Γ does not admit quantifier elimination. Add a constant symbol to τ and extend Γ so that the extended theory eliminates quantifiers.

534. Let $\mathfrak{A} = \langle \mathbb{R}, +, \cdot, 0, 1, \leq \rangle$ and consider the formula $\Phi(a, b, c) \equiv \exists x (ax^2 + bx + c = 0)$. Is this formula equivalent in $\mathfrak{A}$ to a quantifier-free one?

535. Assume that for every formula φ of the form $\exists x (\ell_1 \wedge \cdots \wedge \ell_n)$ where each ℓ_i is a literal (that is, either an atomic formula or the negation of an atomic formula), there is a quantifier-free formula δ such that $\Gamma \vDash \varphi \leftrightarrow \delta$. Show that Γ admits quantifier elimination.

536. Prove that the theory of dense linear ordering without endpoints has quantifier elimination.

> **Hint.** Use strict inequality and observe that every literal can be assumed to be not negated.

537. Suppose Γ has built-in Skolem functions (Definition 8.4). Show that Γ has quantifier elimination. Is the reverse implication true?

538. (a) Suppose Γ has built-in Skolem functions, $\mathfrak{A} \vDash \Gamma$, and $\mathfrak{B}$ is a substructure of $\mathfrak{A}$. Show that $\mathfrak{B}$ is an elementary submodel of $\mathfrak{A}$.

(b) Suppose Γ has quantifier elimination, $\mathfrak{A} \vDash \Gamma$, $\mathfrak{B}$ is a substructure of $\mathfrak{A}$, and $\mathfrak{B} \vDash \Gamma$. Show that $\mathfrak{B}$ is an elementary submodel of $\mathfrak{A}$.

> **8.17 Definition** (Conservative extension). Let $\Gamma \subset F(\tau)$, $\tau \subset \tau'$ and $\Gamma' \subset F(\tau')$. Γ' is a *conservative extension of* Γ, if $\Gamma \subset \Gamma'$ and for each $\varphi \in F(\tau)$ we have $\Gamma \vDash \varphi$ if and only if $\Gamma' \vDash \varphi$.

539. Let $\Gamma \subset F(\tau)$ be a theory. Prove that Γ has a conservative extension which has quantifier elimination.

540. Assume $\Gamma \subset F(\tau)$ has quantifier elimination. Does it follow that Γ is complete?

541. Let τ be a finite similarity type and let $\mathfrak{A}$ be a τ-structure. Assume that every finite partial isomorphism of $\mathfrak{A}$ can be extended to an automorphism of $\mathfrak{A}$.

(a) Prove that for each formula φ there is a quantifier-free formula δ such that $\mathfrak{A} \vDash \varphi \leftrightarrow \delta$.

(b) Show that if $\mathfrak{A} \vDash \Gamma$ and Γ is complete, then Γ has quantifier elimination.

> **Hint.** The assumption that τ is finite implies that there is a formula which tells whether there is a partial isomorphism between two tuples.

542. Let $\Gamma \subset F(\tau)$ be a theory and $\varphi(\vec{x}) \in F(\tau)$ be a formula. Prove that the following two statements are equivalent:

(i) There is a quantifier-free formula $\delta(\vec{x})$ such that $\Gamma \vDash \varphi(\vec{x}) \leftrightarrow \delta(\vec{x})$.

(ii) Whenever $\mathfrak{A}, \mathfrak{A}' \vDash \Gamma$ and $\mathfrak{B}$ is a common substructure of $\mathfrak{A}$ and $\mathfrak{A}'$, then for all $\vec{b} \in B$ we have $\mathfrak{A} \vDash \varphi[\vec{b}]$ if and only if $\mathfrak{A}' \vDash \varphi[\vec{b}]$.

543. Let $\Gamma \subset F(\tau)$ be a theory. Suppose that for all quantifier-free formulas $\delta(x, \vec{y}) \in F(\tau)$, if $\mathfrak{A}, \mathfrak{A}' \vDash \Gamma$ and $\mathfrak{B}$ is a common substructure of $\mathfrak{A}$ and $\mathfrak{A}'$, $\vec{b} \in B$, and if there is an $a \in A$ such that $\mathfrak{A} \vDash \delta[a, \vec{b}]$, then there is an $a' \in A'$ such that $\mathfrak{A}' \vDash \delta[a', \vec{b}]$. Prove that Γ has quantifier elimination.

> **8.18 Definition.** A theory Γ is *model complete* if every embedding of models of Γ is an elementary embedding.

544. Prove that any theory with quantifier elimination is model complete.

545. Let $\mathfrak{A}$ be a dense linear order without endpoints and $a \in A$. Show that the substructure with ground set $B = A - \{a\}$ is an elementary submodel.

546. Let Γ be the theory asserting that the equivalence relation E has infinitely many classes and each class is infinite, see Problem 361. Prove that Γ is model complete.

547. Demonstrate that $\mathrm{Th}(\langle \omega, S \rangle)$, where S is the successor function, is not model complete.

548. Let $\Gamma \subset F(\tau)$ and let $\mathfrak{B}$ be a τ-structure. Assume that for every universal formula φ whenever $\Gamma \vDash \varphi$ then $\mathfrak{B} \vDash \varphi$. Prove that $\mathfrak{B}$ can be embedded into some model of Γ.

549. Show that the following statements are equivalent:

(i) Γ is model complete.

(ii) For any formula $\varphi(\vec{x})$ there is an *existential* formula $\psi(\vec{x})$ such that

$$\Gamma \vDash \forall \vec{x} \, (\varphi(\vec{x}) \leftrightarrow \psi(\vec{x})).$$

(iii) For any formula $\varphi(\vec{x})$ there is a *universal* formula $\psi(\vec{x})$ such that

$$\Gamma \vDash \forall \vec{x} \, (\varphi(\vec{x}) \leftrightarrow \psi(\vec{x})).$$

550. Suppose $\Gamma \subset F(\tau)$ has a conservative extension $\Gamma' \subset F(\tau')$ such that (i) Γ' admits elimination of quantifiers, and (ii) for each closed quantifier-free formula $\delta \in F(\tau')$ either $\Gamma' \vDash \delta$ or $\Gamma' \vDash \neg\delta$. Prove that Γ is complete.

551. Show that the theory of $\langle\omega, S\rangle$ does not have quantifier elimination while that of $\langle\mathbb{Z}, S\rangle$ does.

552. (a) Let $\tau = \langle 0, S\rangle$ and $\Gamma \subset F(\tau)$. Show that the following formula set from Problem 501 admits quantifier elimination:

$$S(x) = S(y) \to x = y, \quad 0 \neq S(x), \quad x \neq 0 \to \exists y(x = S(y)),$$
$$x \neq S(x), \quad x \neq SS(x), \quad x \neq SSS(x), \text{ etc.}$$

(b) Show that it is the theory of $\langle\omega, 0, S\rangle$.

(c) Show that every parametrically definable subset in the structure $\langle\omega, 0, S\rangle$ is either finite or co-finite.

553. Let Γ be the theory of discrete linear orderings without endpoints. Prove the following:

(a) Γ does not have quantifier elimination,

(b) Γ is not κ-categorical for any infinite cardinal κ,

(c) Γ is complete.

554. Let Γ be the theory of discrete linear ordering with an initial element.

(a) Show that Γ does not have quantifier elimination, while a conservative extension of type $\langle\leq, 0, S\rangle$ does.

(b) Show that Γ is complete.

555. The similarity type τ consists of the binary relation symbol $\leq$ and the constants c_n for $n \in \omega$. Γ is a set of axioms expressing that $\leq$ is a dense linear ordering without endpoints and the sequence c_n is strictly increasing. Prove that Γ is complete. What is the number of non-isomorphic countable models of Γ?

556. Prove that some conservative extension of Γ admits quantifier elimination where Γ is the theory of the models of an equivalence relation E

(a) with infinitely many classes all of which are infinite (see Problem 361),

(b) with infinitely many classes all of size 2,

(c) with infinitely many 2- and 3-element classes and every class has either 2 or 3 elements,

(d) with one equivalence class of size n for each $n \in \omega$,

(e) with infinitely many classes all of which are finite.

In each case determine the number of non-isomorphic countable models.

557. The following theories have quantifier elimination and are complete:

(a) infinite Abelian groups in which each non-unit element has order p,

(b) divisible torsion-free Abelian groups,

(c) algebraically closed fields of characteristics 0 or p,

(d) real closed fields,

(e) atomless Boolean algebras.

558. Let $\mathfrak{F}$ be an algebraically closed field and $\mathfrak{F}'$ its proper subfield which also is algebraically closed. Is it true that $\mathfrak{F}'$ is an elementary subfield of $\mathfrak{F}$?

> **Notation.** The *Presburger arithmetic* is the theory of the structure $\langle \omega, 0, 1, +, \leq \rangle$ of the natural numbers without multiplication.

> **8.19 Theorem** (M. Presburger). Extend the type $\langle 0, 1, +, \leq \rangle$ with the unary predicate symbols R_n for $n \geq 2$ with the interpretation that $R_n(x)$ holds if x is divisible by n. With this conservative extension Presburger arithmetic admits elimination of quantifiers.

8.4 EXAMPLES

559. Suppose that the infinite τ-structure $\mathfrak{A}$ has cardinality at least $|\tau|$. Prove that $\mathfrak{A}$ has a proper elementary extension of the same cardinality.

560. Work in the language $\langle +, 0 \rangle$ of groups and show that $\mathbb{Z} \oplus \mathbb{Z}$ is not elementarily equivalent to $\mathbb{Z}$.

561. Does there exist a non-Archimedean field elementarily equivalent to $\mathbb{R}$?

> **Hint.** Use compactness.

562. Let $\mathfrak{A}$ be a proper elementary extension of the ordered field $\mathbb{R}$. Show that $\mathfrak{A}$ is non-Archimedean.

563. Using that no free group contains infinitely divisible elements prove that the class of free groups is not closed under elementary equivalence.

> **Notation.** A graph G is *locally finite* if every vertex has finite degree.

564. Let G be a connected, locally finite graph. Show that G has no proper elementary subgraphs.

565. Let G_1 and G_2 be two locally finite graphs. For each vertex $v \in G_1$ and natural number $n \in \omega$ let $G_1(v, n)$ be the spanned subgraph of G_1 that consists of vertices of distance at most n from v. Assume

$$\{G_1(v, n) : v \in G_1, n \in \omega\} = \{G_2(v, n) : v \in G_2, n \in \omega\}.$$

Prove that G_1 and G_2 are elementarily equivalent.

> **Hint.** Use the Ehrenfeucht–Fraïssé game.

566. Construct two connected, countable, elementarily equivalent graphs which are not isomorphic. Can these graphs be locally finite?

567. For each pair of infinite cardinals κ and λ construct two elementarily equivalent graphs G and H such that $|G| = |H|$, $\chi(G) = \kappa$ while $\chi(H) = \lambda$.

568. Give two elementarily equivalent orderings, exactly one of them is well ordered.

569. Let $\mathfrak{A}$ be a dense linear ordering without endpoints. Prove that the truth of a formula depends only on the order of its free variables (and whether they are equal).

570. Let $\mathfrak{A}$ and $\mathfrak{B}$ be two countable dense linear orderings without endpoints and $f : A \to B$ a finite partial isomorphism. Prove that f extends to an isomorphism.

571. Construct two dense linear orderings $\mathfrak{A}$ and $\mathfrak{B}$ without endpoints of the same cardinality and a finite partial isomorphism $f : A \to B$ which cannot be extended to an isomorphism.

572. Let $\mathfrak{A}$ be a dense ordering without endpoints and let $B \subseteq A$ be such that the ordering restricted to B is dense without endpoints. Prove that the structure $\mathfrak{B}$ with ground set B is an elementary submodel of $\mathfrak{A}$.

573. Let $\mathfrak{A}$ be a discrete linear order with initial element 0. B is a subset of A which contains 0 and with each $b \in B$ it also contains the successor of b, and for $b \neq 0$ the predecessor of b. Show that $\mathfrak{B}$ is an elementary submodel of $\mathfrak{A}$.

574. Find a structure $\mathfrak{A}$ elementary equivalent to $\langle \omega, \leq \rangle$ such that $\langle \mathbb{R}, \leq \rangle$ can be embedded into $\mathfrak{A}$.

575. Find a countable structure $\mathfrak{A}$ which can be properly elementary embedded into itself and

(a) $\mathfrak{A}$ is a dense ordering without endpoints.

(b) $\mathfrak{A}$ is a discrete ordering without endpoints.

(c) $\mathfrak{A}$ is a discrete ordering with initial point but without right endpoint.

576. Give a structure $\mathfrak{A}$ and $X \subset A$ such that $\mathfrak{A}$ has no minimal elementary substructure containing X.

577. Give two elementarily equivalent but not isomorphic structures such that each is isomorphic to a substructure of the other.

> **Hint.** Any two discrete linear orderings without endpoints are elementarily equivalent (see Problem 553).

578. Let τ be a countable type and $\mathfrak{A}$ be a τ-type structure whose universe is just ω_1. Prove that there exists a countable ordinal $\alpha < \omega_1$ such that α is the universe of a (a) substructure (b) elementary substructure of $\mathfrak{A}$.

579. The ordinal α is regarded as an ordered structure with ordering $<$.

(a) Show that ω is not an elementary substructure of $\omega + 4$.

(b) Show that $\omega + \omega$ is not an elementary substructure of $\omega \cdot \omega$.

(c) Show that $\omega \cdot \omega$ is not an elementary substructure of ω_1.

580. Let $\mathfrak{A}$ be a countable ordinal with its usual ordering. Can $\mathfrak{A}$ have proper elementary submodels?

581. Find a complete theory on finite or countable language which has (a) one; (b) countably many; (c) continuum many countable models up to isomorphism.

582. Find a complete theory on a finite or countable language which has (a) three, (b) four, (c) n for some $n > 4$ pairwise non-isomorphic countable models.

583.* Give a theory Γ which has a countable model, but every other model of Γ has size at least 2^ω.

> **Hint.** Consider an almost disjoint family of subsets of ω (Problem 1) and the usual ordering on ω.

584. Let $\Gamma \subseteq F(\tau)$ be a theory. Prove that Γ has a conservative extension which has built-in Skolem functions.

585. $\mathfrak{A}$ is an infinite structure in which there are two orderings: $\leq_1$ and $\leq_2$. Show that there is an infinite subset $B \subset A$ where $\leq_1$ and $\leq_2$ either coincide, or one is the reverse of the other.

586. Suppose $\Gamma \subset F(\tau)$ has an infinite model. Show that Γ has a model $\mathfrak{A}$ in which there are distinct elements $\{a_i : i \in \omega\}$ in A such that

(a) $\mathfrak{A} \vDash \varphi[a_{i_1}] \leftrightarrow \varphi[a_{i_2}]$ for every $i_1, i_2 \in \omega$ and $\varphi(x) \in F(\tau)$;

(b) $\mathfrak{A} \vDash \varphi[a_{i_1}, a_{j_1}] \leftrightarrow \varphi[a_{i_2}, a_{j_2}]$ for every $i_1 < j_1$, $i_2 < j_2$, and $\varphi(x, y) \in F(\tau)$.

> **8.20 Definition** (Indiscernibles). Let $\mathfrak{A}$ be a τ-structure, $H \subset A$. $(H, \ll)$ is *indiscernible in* $\mathfrak{A}$ if $\ll$ is an ordering on H, and for every pair of $\ll$-increasing sequences $\vec{a}, \vec{b} \in H$ of the same length, $\mathfrak{A} \vDash \varphi[\vec{a}] \leftrightarrow \varphi[\vec{b}]$ for every formula $\varphi(\vec{x}) \in F(\tau)$.

587. Suppose the binary symbol $<$ is in the type τ, and the interpretation $<^{\mathfrak{A}}$ is an ordering in $\mathfrak{A}$. Let $(H, \ll)$ be indiscernible in $\mathfrak{A}$. Show that $\ll$ and $<$ on H are either the same or the reverse of each other.

588. Let $\mathfrak{A} = \langle A, < \rangle$ be a dense linear ordering without endpoints. Pick a strictly decreasing sequence $\langle a_i : i \in \omega \rangle$. Show that $H = \{a_i : i \in \omega\}$ with the ordering $a_i \ll a_j$ iff $i < j$ is indiscernible.

589. Suppose τ contains function symbols only. Let $\mathfrak{A}$ be a free algebra generated by the free set of generators X. Show that $(X, \ll)$ is indiscernible in $\mathfrak{A}$ for every ordering $\ll$.

590. Let $\mathfrak{A}$ be a discrete linear order with initial point. By Problem 368 it has the structure $\omega + \mathbb{Z} \times M$ for some ordered set M. Suppose $H \subset A$ contains a single element from each copy of $\mathbb{Z}$. Show that H is indiscernible with the inherited ordering.

591. Suppose $(H, \ll)$ is indiscernible in $\mathfrak{A}$ where $\ll$ is a well-ordering. Show that at most $|F(\tau)|$ different types can be realized (Definition 7.18) in the substructure generated by H.

> **8.21 Theorem.** Let $\mathfrak{A}$ be infinite and $(H, \ll)$ be an arbitrary ordering. There is an elementary extension $\mathfrak{B}$ of $\mathfrak{A}$ such that $H \subset B$ and $(H, \ll)$ is indiscernible in $\mathfrak{B}$.

592. Prove Theorem 8.21.

593. For each infinite κ some linear ordering on κ has 2^κ many automorphisms.

594. Suppose $(H, \ll)$ is indiscernible in $\mathfrak{A}$, $B \subseteq H$, and $\pi : B \to H$ is a $\ll$-preserving injection. Show that the substructures of $\mathfrak{A}$ generated by B and generated by $\pi[B]$ are isomorphic.

595. Suppose $\Gamma \subset F(\tau)$ has infinite models, and $\kappa \geq |\tau|$ is infinite. Show that Γ has a model of cardinality κ which has 2^κ many automorphisms.

596. Suppose τ is countable, and Γ has infinite models. Show that Γ has arbitrary large models in which only countably many different types are realized.

ULTRAPRODUCTS

9

> **Notation.** In this chapter the i-th coordinate of an element a from the product $\prod_{i\in I} A_i$ will be denoted by $a(i)$. For a vector $\vec{a} = \langle a_1,\ldots,a_n\rangle$ formed from elements of the product, $\vec{a}(i)$ denotes the vector of the i-th elements: $\vec{a}(i) = \langle a_1(i),\ldots,a_n(i)\rangle \in A_i^n$.

> **9.1 Definition** (Product of structures). The *direct product* of the τ-structures $\{\mathfrak{A}_i : i \in I\}$, denoted by $\prod_{i\in I}\mathfrak{A}_i$, has universe $\prod_{i\in I} A_i$, and the interpretation of symbols is done coordinatewise:
>
> $$
> \begin{aligned}
> c^{\prod\mathfrak{A}} &= \langle c^{\mathfrak{A}_i} : i \in I\rangle, \\
> f^{\prod\mathfrak{A}}(\vec{a}) &= \langle f^{\mathfrak{A}_i}(\vec{a}(i)) : i \in I\rangle, \ \text{ and} \\
> \vec{a} \in r^{\prod\mathfrak{A}} &\Leftrightarrow \vec{a}(i) \in r^{\mathfrak{A}_i} \ \text{ for all } i \in I.
> \end{aligned}
> $$

597. Let $e\colon X \to \prod_i A_i$ be an evaluation over $\prod_i\mathfrak{A}_i$, then $e(i)$ is an evaluation over $\mathfrak{A}_i$. Show that terms are evaluated coordinatewise, that is, the i-th coordinate of $t^{\prod\mathfrak{A}}[e]$ is $t^{\mathfrak{A}_i}[e(i)]$ for each term $t \in E(\tau)$.

598. Find a structure $\mathfrak{A}$ and a closed formula φ such that $\mathfrak{A} \vDash \varphi$, while $\mathfrak{A} \times \mathfrak{A} \nvDash \varphi$. Can φ be an atomic formula?

599. Let $\varphi \in F(t)$ be a closed $\exists_1$ formula, i.e., it is of the form $\exists \vec{x}\, \delta(\vec{x})$ where δ is quantifier-free. Is it true that if $\mathfrak{A}_i \vDash \varphi$ for all $i \in I$, then $\prod_i\mathfrak{A}_i \vDash \varphi$?

600. Let $R_1(\vec{x})$ and $R_2(\vec{x})$ be two n-variable relation symbols and ϑ be the formula $Q_1 x_1 \cdots Q_n x_n\,(R_1(\vec{x}) \to R_2(\vec{x}))$, where each Q_i is a quantifier. Suppose $\mathfrak{A}_i \vDash \vartheta$ for all $i \in I$. Show that $\prod_i\mathfrak{A}_i \vDash \vartheta$.

> **Notation.** For a family $\mathcal{F} \subset \wp(I)$ of indices and $a \in \prod_i A_i$ define
>
> $$
> a/\mathcal{F} = \Big\{ b \in \prod_i A_i : \{i \in I : a(i) = b(i)\} \in \mathcal{F} \Big\}.
> $$
>
> The *factor* $\prod_i A_i / \mathcal{F}$ is the collection $\{a/\mathcal{F} : a \in \prod_i A_i\}$.

601. Suppose $\mathcal{F}$ is a filter. (a) Show that $a \in a/\mathcal{F}$, and $a/\mathcal{F}$ and $b/\mathcal{F}$ are either disjoint or identical. (b) Show that in this case the partition $\prod_i A_i / \mathcal{F}$ of $\prod_i A_i$ is compatible with the interpretation of the function symbols in $\prod_i\mathfrak{A}_i$.

© The Author(s), under exclusive license to Springer Nature Switzerland AG 2022
L. Csirmaz and Z. Gyenis, *Mathematical Logic,* Problem Books
in Mathematics, https://doi.org/10.1007/978-3-030-79010-3_9

9.2 Definition (Ultraproduct). Let $\mathcal{U}$ be an ultrafilter on I. The *ultraproduct* $\prod_{i\in I}\mathfrak{A}_i/\mathcal{U}$ of the τ-structures $\mathfrak{A}_i$ has ground set $\prod_i A_i/\mathcal{U}$, and the interpretation of the symbols are

$$c^{\prod\mathfrak{A}/\mathcal{U}} = c^{\prod\mathfrak{A}}/\mathcal{U},$$
$$f^{\prod\mathfrak{A}/\mathcal{U}}(\bar{a}/\mathcal{U}) = f^{\prod\mathfrak{A}}(\bar{a})/\mathcal{U},$$
$$\bar{a}/\mathcal{U} \in r^{\prod\mathfrak{A}/\mathcal{U}} \Leftrightarrow \{i \in I : \bar{a}(i) \in r^{\mathfrak{A}_i}\} \in \mathcal{U}.$$

If $\mathcal{U}$ is a filter only then $\prod_i\mathfrak{A}_i/\mathcal{U}$ is called *reduced product*.

602. Show that Definition 9.2 is sound.

603. Show that the ultraproduct of finitely many structures is always isomorphic to one of the components.

604. Show that an ultraproduct of ultraproducts is isomorphic to an ultraproduct of the original structures.

605. Suppose all structures $\mathfrak{A}_i$ are finite, and there are finitely many non-isomorphic structures among $\{\mathfrak{A}_i : i \in I\}$. Show that $\prod_{i\in I}\mathfrak{A}_i/\mathcal{U}$ is isomorphic to one of the $\mathfrak{A}_i$'s.

9.3 Theorem (Łoś lemma—fundamental theorem of ultraproducts). Let $\mathfrak{A}_i$ be τ-structures for $i \in I$, $\mathcal{U}$ be an ultrafilter on I and $e: X \to \prod_i A_i$ be an evaluation. Then the following statements hold:

- for a term $t \in E(\tau)$,

$$t^{\prod\mathfrak{A}/\mathcal{U}}[e/\mathcal{U}] = t^{\prod\mathfrak{A}}[e]/\mathcal{U},$$

- for a formula $\varphi \in F(\tau)$,

$$\prod_i\mathfrak{A}_i/\mathcal{U} \vDash \varphi[e/\mathcal{U}] \quad \text{if and only if} \quad \{i \in I : \mathfrak{A}_i \vDash \varphi[e(i)]\} \in \mathcal{U}.$$

606. Show that $\prod_i\mathfrak{A}_i/\mathcal{U} \vDash \varphi$ if and only if $\{i \in I : \mathfrak{A}_i \vDash \varphi\} \in \mathcal{U}$.

Notation. If all structures $\mathfrak{A}_i$ are isomorphic to $\mathfrak{A}$, then the index set I is identified with the cardinal $\kappa = |I|$, the ultraproduct $\prod_{i\in\kappa}\mathfrak{A}_i/\mathcal{U}$ is called *ultrapower*, and is denoted by $^{\kappa}\mathfrak{A}/\mathcal{U}$.

607. Show that the ultrapower of $^{\kappa}\mathfrak{A}/\mathcal{U}$ is elementarily equivalent to $\mathfrak{A}$.

608. Let $\mathfrak{A}_i$ be a set of structures for $i \in I$ and let $\mathfrak{A} = \prod_i\mathfrak{A}_\alpha/\mathcal{U}$.

(a) Suppose $|\mathfrak{A}_i| \leq n$ for each $i \in I$. Prove that $|\mathfrak{A}| \leq n$.

(b) Suppose $X = \{|\mathfrak{A}_i| : i \in I\} \subset \omega$ is a finite set of natural numbers. Prove that $|\mathfrak{A}| \in X$.

(c) Suppose X is a finite set of natural numbers such that $\{i \in I : |\mathfrak{A}_i| \in X\} \in \mathcal{U}$. Prove that $|\mathfrak{A}| \in X$.

609. Let $\mathcal{U}$ be a non-trivial ultrafilter on ω, and $\mathfrak{A}$ be a countable structure. Show that $^{\omega}\mathfrak{A}/\mathcal{U}$ has cardinality 2^{ω}.

> **Hint.** Use almost disjoint sets.

610. Let $\mathfrak{A}_i$ be a finite structure with n_i elements and $\mathcal{U}$ a non-trivial ultrafilter on ω. Show that if $\lim n_i = \infty$ then the ultraproduct $\Pi_i \mathfrak{A}_i / \mathcal{U}$ has continuum many elements. What happens if we suppose $\limsup n_i = \infty$?

611. Let $\delta : \mathfrak{A} \to {}^{\kappa}\mathfrak{A}/\mathcal{U}$ be the diagonal mapping $\delta(a) = \langle a : i \in \kappa \rangle / \mathcal{U}$. Prove that δ is an elementary embedding of $\mathfrak{A}$ into $^{\kappa}\mathfrak{A}/\mathcal{U}$, i.e., $\delta[\mathfrak{A}]$ is isomorphic to $\mathfrak{A}$ and is an elementary substructure of $^{\kappa}\mathfrak{A}/\mathcal{U}$.

612. Let $\mathfrak{A}$ be infinite, and $\mathcal{U}$ be a non-trivial ultrafilter over ω. Show that $\mathfrak{A}$ has a proper elementary embedding to $^{\omega}\mathfrak{A}/\mathcal{U}$.

613. Every countable structure has an elementary extension of cardinality continuum, independently of the cardinality of the type. (See also Problems 495 and 583.)

614. Let κ be an infinite cardinal, $|I| = \kappa$, and suppose that the structures $\mathfrak{A}_i$ are infinite for $i \in I$. Show that for some ultrafilter $\mathcal{U}$ over I the cardinality of $\prod_i \mathfrak{A}_i / \mathcal{U}$ is at least 2^{κ}.

615. Suppose $\mathfrak{A}$ is infinite and has cardinality $< 2^{\kappa}$. Show that $\mathfrak{A}$ has an elementary extension of cardinality 2^{κ} independently of the cardinality of the type.

616. Construct an infinite structure $\mathfrak{A}$ so that $\mathfrak{A}$ and all ultrapowers $^{\omega}\mathfrak{A}/\mathcal{U}$ are isomorphic whenever $\mathcal{U}$ is an ultrafilter on ω. Compare it to Problem 612 which says that an infinite structure can be properly embedded into its ultrapowers.

617. Show that the ω-ultrapower of the complex numbers $\mathbb{C}$, as a field, is isomorphic to itself.

618. Give two sequence of structures $\mathfrak{A}_i$ and $\mathfrak{B}_i$ for $i \in \omega$ such that $\mathfrak{A}_i \not\equiv \mathfrak{B}_j$ for all $i, j \in \omega$, but $\Pi_i \mathfrak{A}_i / \mathcal{U} \cong \Pi_i \mathfrak{B}_i / \mathcal{U}$.

619. The structures $\mathfrak{A}_i$ can be classified into finitely many classes regarding elementary equivalence. Show that $\Pi_i \mathfrak{A}_i / \mathcal{U}$ belongs to one of these classes, as well.

620. Give two element structures $\mathfrak{A}_i$ such that the non-trivial ultraproduct $\Pi_{i<\omega} \mathfrak{A}_i / \mathcal{U}$ is elementary equivalent with none of the $\mathfrak{A}_i$'s.

621. Let $\mathcal{U} \subseteq \wp(I)$ be an ultrafilter, $J \in \mathcal{U}$ and $\mathcal{V}$ be the trace of $\mathcal{U}$ on $J \subseteq I$ (see Problem 37). Show that $\Pi_{i \in I} \mathfrak{A}_i / \mathcal{U}$ and $\Pi_{j \in J} \mathfrak{A}_j / \mathcal{V}$ are isomorphic.

9.4 Theorem (Compactness theorem). Γ has a model if and only if each finite subset of Γ has a model.

622. Prove Theorem 9.4 using ultraproducts (see Problem 418).

623. Prove that if Γ has arbitrarily large finite models, then it has an infinite model. (Cf. Problem 419)

624. Let $\mathcal{U}$ be an ultrafilter on I, $\mathfrak{A}_\xi$ be structures for $\xi \in I$, and $E \subset \mathcal{U}$. For each $e \in E$ we have a formula φ_e. Suppose that for each $\xi \in I$ there is an evaluation f_ξ over $\mathfrak{A}_\xi$ such that $\mathfrak{A}_\xi \vDash \varphi_e[f_\xi]$ for every $e \in E$ with $\xi \in e$. Show that $\Pi_\xi \mathfrak{A}_\xi / \mathcal{U} \vDash \varphi_e[\langle f_\xi \rangle / \mathcal{U}]$ for each $e \in E$.

625. Let $\mathcal{U}$ be a regular ultrafilter on I (see Definition 1.17) witnessed by $E \subset \mathcal{U}$, and $\mathcal{K}$ be a collection of structures. Suppose that every finite subset of $\Gamma = \{\varphi_e : e \in E\}$ has a model in $\mathcal{K}$. Show that there are structures $\mathfrak{A}_\xi \in \mathcal{K}$ for $\xi \in I$ such that $\Pi_\xi \mathfrak{A}_\xi / \mathcal{U} \vDash \Gamma$ (Cf. Problem 622).

626. Show that every structure can be embedded into an appropriate ultraproduct of its finitely generated substructures.

627. Let $T(x) = \{\varphi_i(x) : i \in \omega\}$ be a set of countably many formulas with one free variable x, and suppose that any finite subset of $T(x)$ can be realized (Definition 7.18) in $\Pi_{i \in \omega} \mathfrak{A}_i / \mathcal{U}$ where $\mathcal{U}$ is a non-trivial ultrafilter on ω. Prove that $T(x)$ can also be realized in $\Pi_i \mathfrak{A}_i / \mathcal{U}$.

An easy consequence of the Łoś lemme is that if two structures have isomorphic ultrapowers, then they are elementarily equivalent (see Problem 607). The converse is the celebrated theorem of H. J. Keisler and S. Shelah.

9.5 Theorem (Keisler–Shelah). $\mathfrak{A}$ is elementarily equivalent to $\mathfrak{B}$ if and only if they have isomorphic ultrapowers $^\kappa\mathfrak{A}/\mathcal{U}$ and $^\kappa\mathfrak{B}/\mathcal{U}$.

9.1 WHAT ULTRAPRODUCTS LOOK LIKE

628. Suppose the structure $\mathfrak{A}$ is the disjoint union of the structures $\mathfrak{A}_1$ and $\mathfrak{A}_2$. Show that the ultrapower $^\kappa\mathfrak{A}/\mathcal{U}$ is the disjoint union of the ultrapowers $^\kappa\mathfrak{A}_1/\mathcal{U}$ and $^\kappa\mathfrak{A}_2/\mathcal{U}$.

629. Let G_n be the cyclic graph on n vertices. Describe, up to isomorphism, the product $\Pi_{i \in \omega} G_i / \mathcal{U}$ for non-trivial $\mathcal{U}$.

630. Denote by G_n the path on n vertices. Describe, up to isomorphism, all ultraproducts $\Pi_{n \in \omega} G_n / \mathcal{U}$, where $\mathcal{U}$ is non-trivial.

631. Show that if the non-trivial ultrapower $^\omega G / \mathcal{U}$ of a graph G is connected, then the diameter of G is finite. (See also Problem 423.)

632. Let G_1 and G_2 be countable, 3-regular, cycle-free graphs. Is it true that they are isomorphic? Describe, up to isomorphism, all graphs $^\omega G_1/\mathcal{U}$ where $\mathcal{U}$ is a non-trivial ultrafilter on ω.

What about countable, 4-regular, cycle-free graphs?

633. Let H be the complete, infinite binary tree. Describe, up to isomorphism, the graph $^\omega H/\mathcal{U}$.

634. Let G be the countable broom (see Problem 487): a vertex v which is the endpoint of an infinite path, and countably many edges are attached to it. Describe, up to isomorphism, all non-trivial ultrapowers $^\omega G/\mathcal{U}$.

635. Prove that if the chromatic number of the ultraproduct $\Pi_{i\in\omega}G_i/\mathcal{U}$ is $k \in \omega$, then there is a finite k-chromatic graph which can be embedded into $\mathcal{U}$-almost every G_i.

636. Let $\mathfrak{A} = \langle \omega, S \rangle$ where S is the successor function. Describe, up to isomorphism, all non-trivial ultrapowers $^\omega\mathfrak{A}/\mathcal{U}$.

637. The similarity type τ contains the constant symbol 0 and the unary function symbol S. Give a set $\Gamma \subset F(\tau)$ of formulas expressing that 0 is not a successor of anything, non-zero elements are successors, and different elements have different successors.

Describe, up to isomorphism, all non-trivial ultrapowers $^\omega\mathfrak{A}/\mathcal{U}$ of some countable model $\mathfrak{A} \vDash \Gamma$.

638. Let $\mathfrak{A} = \langle \omega, < \rangle$ be the ordering of the natural numbers. In a non-trivial ultrapower $^\omega\mathfrak{A}/\mathcal{U}$ each element has an immediate successor. For $a_1, a_1 \in {}^\omega\mathfrak{A}/\mathcal{U}$ and $a_1 < a_2$ we say that the distance from a_1 to a_2 is finite if a_2 can be reached from a_1 using finitely many successor steps. Show that if the distance from a_1 to a_2 is infinite, then there is an $a_1 < c < a_2$ such that both distances from a_1 to c, and from c to a_2 are infinite.

639. Let $\mathfrak{A} = \langle \mathbb{Z}, < \rangle$ be the discrete ordering of the integers. By Problem 553 the order in the ultraproduct $^\omega\mathfrak{A}/\mathcal{U}$ is isomorphic to $\mathbb{Z} \times M$ for some linear ordering $\langle M, < \rangle$. Show that M is a dense linear ordering without endpoints.

640. Let $\mathfrak{A} = \langle \omega, < \rangle$. Show that any countable set in a non-trivial ultrapower $^\omega\mathfrak{A}/\mathcal{U}$ is bounded. Find a strictly increasing sequence of length ω_1 in $^\omega\mathfrak{A}/\mathcal{U}$.

641. Construct an uncountable ordering with the following property: the intersection of countably many strictly decreasing intervals $I_1 \supset I_2 \supset \ldots$ always contains a whole interval.

642. Does there exist a dense ordering without endpoints with the following property: if A and B are two countable subsets so that all elements of A are below the elements of B, then there is a whole interval between A and B?

643. Let $\mathfrak{B}$ be the usual ordering of the reals and $\mathfrak{A} = {}^\omega\mathfrak{B}/\mathcal{U}$ with a non-trivial $\mathcal{U}$. Write $0^\mathfrak{A}$ for the equivalence class of the all-zero sequence.

(a) Show that if $a_0 > a_1 > \cdots > a_n > \cdots > 0^{\mathfrak{A}}$ is a decreasing sequence, then one can always find a $b > 0$ which separates all the a_n's from $0^{\mathfrak{A}}$, that is, $0^{\mathfrak{A}} < b < a_n$ for all $n < \omega$.

(b) Prove that the ordering of $^{\omega}\mathfrak{A}/\mathcal{U}$ is not complete: there is a bounded set that has no least upper bound.

644. The partial order E is *well founded* if each non-empty subset contains an E-minimal element, see Problem 422. Let $\mathfrak{A}$ be a structure and suppose $E^{\mathfrak{A}}$ is well founded. Give sufficient and necessary condition for E to be well founded in a non-trivial ultrapower $^{\omega}\mathfrak{A}/\mathcal{U}$.

645. The similarity type of $\mathfrak{A}$ contains countably many unary relation symbols R_i. Suppose that each element of $\mathfrak{A}$ satisfies finitely many of the R_i's only; and for all $i_1, \ldots, i_n$ there exists $a \in A$ such that a satisfies $R_{i_1}, \ldots, R_{i_n}$ only. Let $X \subseteq \omega$ be infinite and $\mathcal{U}$ be a non-trivial ultrafilter on ω.

(a) Prove that $^{\omega}\mathfrak{A}/\mathcal{U}$ contains an element a such that a satisfies R_i if and only if $i \in X$.

(b) Show that $^{\omega}\mathfrak{A}/\mathcal{U}$ contains infinitely many such elements.

(c) Show that $^{\omega}\mathfrak{A}/\mathcal{U}$ contains continuum many such elements.

9.2 APPLICATIONS

646. A group is a *torsion group* if all elements are of finite order. Give torsion groups G_n such that $\Pi_{n<\omega} G_n/\mathcal{U}$ for non-trivial $\mathcal{U}$ is torsion-free.

647. Let $\mathbb{Z}$ be free the group generated by a single element and $\mathcal{U}$ be a non-trivial ultrafilter on ω. Show that $^{\omega}\mathbb{Z}/\mathcal{U}$ contains infinitely divisible elements.

Prove that if G is a free group, then G contains no infinitely divisible elements but $^{\omega}G/\mathcal{U}$ does contain such elements.

648. Let C_i be the cyclic group of order n and $\mathcal{U}$ be a non-trivial ultrafilter on ω. Which of the following statements are correct?

(a) $\Pi_{i<\omega} C_i/\mathcal{U}$ is infinite.

(b) $\Pi_{i<\omega} C_i/\mathcal{U}$ is torsion-free.

(c) $\Pi_{i<\omega} C_i/\mathcal{U}$ can be torsion-free.

(d) $\Pi_{i<\omega} C_i/\mathcal{U}$ can be a torsion group.

649. Let $\mathfrak{F}_n$ be the n-th finite field. Show that there is a non-trivial ultrafilter $\mathcal{U}$ on ω such that $\Pi_n \mathfrak{F}_n/\mathcal{U}$ has (a) characteristic 2, (b) characteristic zero.

650. Prove that there exists no ultrafilter $\mathcal{U}$ such that $\Pi_{p\text{ prime}} \mathbb{F}_p/\mathcal{U}$ is isomorphic to the field of complex numbers.

651. Prove that for any prime p there is a non-trivial ultraproduct of all finite fields of characteristics p which contains the algebraic closure of the field $\mathbb{F}_p$.

652. Prove that for any prime p, an ultraproduct of all finite fields of characteristic p is not algebraically closed.

653. Let φ be a closed formula in the language of fields and suppose φ is true on all fields of characteristics zero. Show that φ is true in all fields of characteristics p for every large enough p.

654. * Find not algebraically closed fields $\mathfrak{F}_n$ of characteristics p (zero or prime) such that each non-trivial ultraproduct $\Pi_n \mathfrak{F}_n / \mathcal{U}$ is algebraically closed.

655. Let $\mathbb{F}$ be a finite field, and $\{p_e : e \in E\}$ be a collection of polynomials with variables $x_1, x_2, \ldots, x_n, \ldots$ over $\mathbb{F}$. Suppose that for all finite $J \subseteq E$ the system of equations $\{p_j = 0 : j \in J\}$ has a solution in $\mathbb{F}$. Show that in this case the collection $\{p_e = 0 : e \in E\}$ has a solution in $\mathbb{F}$, too. (Cf. Problem 259)

> **Hint.** Let $\mathcal{U}$ be a regular ultrafilter witnessed by $E \subset \mathcal{U}$ (Definition 1.17), and f_ξ be the solution for the subsystem $\{p_e = 0 : \xi \in e\}$. Then use Problem 624.

656. Let $\{p_e : e \in E\}$ be a *countable* collection of polynomials over the complex numbers $\mathbb{C}$. Suppose each finite subcollection has a solution in $\mathbb{C}$. Show that in this case the whole collection has a solution as well.

> **Hint.** The ultrapower $^\omega \mathbb{C} / \mathcal{U}$ is isomorphic to $\mathbb{C}$ by Problem 617.

657. Let $\mathfrak{F}$ be a field and $\mathcal{U}$ be a non-trivial ultrafilter on κ. Identify $\mathfrak{F}$ with its diagonal image in $\mathfrak{B} = {}^\kappa \mathfrak{F} / \mathcal{U}$. This way $\mathfrak{B}$ can be considered a field extension of $\mathfrak{F}$. Prove that all the elements in $B \smallsetminus F$ are transcendental over $\mathfrak{F}$.

658. An *ordered field* is a field with an ordering $\leq$ which is translation invariant: $x \leq y$ iff $x + u \leq y + u$, and the product of non-negative elements is non-negative: $(x \geq 0 \land y \geq 0) \rightarrow xy \geq 0$. The field $\mathfrak{F}$ is *orderable* if an order can be defined on it which makes it an ordered field.

(a) Prove that there is a unique field ordering of the reals.

(b) Show that finite fields cannot be ordered.

(c) Show that an ultraproduct of finite fields cannot be ordered.

(d) Prove that if a field is orderable, then it has characteristics 0.

(e) Give a field which admits (at least) two different orderings.

(f) Prove that $\mathfrak{F}$ is orderable if and only if -1 is not a sum of squares.

9.3 ADVANCED EXERCISES

659. Let $|\mathfrak{A}| = \lambda$, where λ is an infinite cardinal. Show that if $\mathcal{U}$ is a regular ultrafilter on κ (Definition 1.17), then $^\kappa \mathfrak{A} / \mathcal{U}$ has cardinality λ^κ.

660. Show that one can find an infinite cardinal κ, a structure $\mathfrak{A}$ of cardinality κ, and two non-trivial ultrafilters $\mathcal{U}$ and V on κ such that $^\kappa \mathfrak{A} / \mathcal{U}$ and $^\kappa \mathfrak{A} / V$ have different cardinalities.

661. Without assuming the continuum hypothesis give two structures $\mathfrak{A}$ and $\mathfrak{B}$ such that $\mathfrak{A} \equiv \mathfrak{B}$, $|A| = |B| = \aleph_2$, and there are no ultrafilters $\mathcal{U}, V$ on ω with $^\omega \mathfrak{A} / \mathcal{U}$ and $^\omega \mathfrak{B} / V$ isomorphic.

Hint. Use total orders with cofinality ω_2 and ω_1.

662. Without assuming the continuum hypothesis give two non-isomorphic structures $\mathfrak{A}$ and $\mathfrak{B}$ such that $|\mathfrak{A}| = |\mathfrak{B}| = \aleph_2$, and that $^\omega\mathfrak{A}/\mathcal{U}$ and $^\omega\mathfrak{B}/\mathcal{V}$ are isomorphic for all non-trivial ultrafilters on ω.

663. Give an example of a countable structure $\mathfrak{A}$ which has 2^{2^ω} many non-isomorphic ultrapowers $^\omega\mathfrak{A}/\mathcal{U}$.

Hint. Use an uncountable language: for each subset of ω introduce a unary relation.

664. Let $\mathcal{U}$ be a regular ultrafilter on κ. Show that $\langle\kappa^+,<\rangle$ can be embedded into $^\kappa\langle\omega,<\rangle/\mathcal{U}$.

665. For a graph $\mathfrak{A}$ let $\mathrm{Fin}(\mathfrak{A})$ be the set of finite subgraphs of $\mathfrak{A}$. Show that if $\mathrm{Fin}(\mathfrak{A}) \subseteq \mathrm{Fin}(\mathfrak{B})$, then $\mathfrak{A}$ can be embedded into some ultrapower of $\mathfrak{B}$.

666. Let $\mathfrak{A}$ be $\aleph_0$-categorical and $\mathfrak{B}$ be countable. Suppose every finitely generated substructure of $\mathfrak{B}$ can be embedded into $\mathfrak{A}$. Prove that in this case $\mathfrak{B}$ can be embedded into $\mathfrak{A}$.

667. Prove that on each torsion-free Abelian group there is a shift-invariant total order, i.e., $a \le b$ iff $a + c \le b + c$.

Hint. A finitely generated substructure is isomorphic to $\mathbb{Z}^n$ for some $n \in \omega$.

668. $\mathrm{GL}(n,\mathfrak{F})$ is the multiplicative group of non-singular $n \times n$ matrices over the field $\mathfrak{F}$. The group G is *n-representable* if it is isomorphic to a subgroup of $\mathrm{GL}(n,\mathfrak{F})$ for some field $\mathfrak{F}$.

Prove that G is n-representable if all of its finitely generated subgroups are n-representable.

9.6 Definition (Types). Let $\mathfrak{A}$ be a τ-structure and $X \subseteq A$. $\tau_X \supset \tau$ denotes the similarity type which contains the additional constant symbols c_a for each $a \in X$ with interpretation $c_a^{\mathfrak{A}} = a$. A subset $T(\vec{x}) \subset F(\tau_X)$ (with free variables $\vec{x}$) is a *type in $\mathfrak{A}$ over X* if $T(\vec{x})$ is finitely satisfiable in $\mathfrak{A}$.

9.7 Definition (Saturated and compact structures). The τ-structure $\mathfrak{A}$ is *κ-saturated* if all types $T(\vec{x}) \subset F(\tau_X)$ with $|X| < \kappa$ can be satisfied in $\mathfrak{A}$. $\mathfrak{A}$ is saturated if it is $|A|$-saturated, and σ-saturated if it is ω_1-saturated.

$\mathfrak{A}$ is *κ-compact* if all types $T(\vec{x})$ with $|T(\vec{x})| < \kappa$ can be satisfied in $\mathfrak{A}$. $\mathfrak{A}$ is *compact* if it is $|A|$-compact, and σ-compact if it is ω_1-compact.

669. Prove that $\langle\mathbb{Q},<\rangle$ is saturated.

670. Show that $\langle\mathbb{Q},<,\cdot,+,0,1\rangle$ is not saturated.

Hint. Try to describe an irrational number via types.

671. Prove that $\langle \mathbb{R}, < \rangle$ is $\aleph_0$-saturated but not $\aleph_1$-saturated.

672. No infinite model $\mathfrak{A}$ can be $|A|^+$-saturated.

673. Show that a structure $\mathfrak{A}$ is κ-saturated for all cardinals κ if and only if $\mathfrak{A}$ is finite.

674. Every κ-saturated model is κ-compact. If $\kappa > |F(\tau)|$, then a κ-compact structure is also κ-saturated.

675. $T(\vec{x}) \subseteq F(\tau_X)$ is an n-type if $\vec{x} = \langle x_1, \ldots, x_n \rangle$, and it is an ω-type if countably many different free variables might occur in T. Show that if $\mathfrak{A}$ realizes all 1-types $T(x) \subset F(\tau_X)$ over every $|X| < \kappa$, then it also realizes all ω-types.

676. Let $\{R_i : i < \omega\}$ be a countable set of unary relation symbols, and let Σ be the set of formulas

$$\exists v (R_{i_0}(v) \wedge \cdots \wedge R_{i_n}(v) \wedge \neg R_{j_0}(v) \wedge \cdots \wedge \neg R_{j_m}(v))$$

for all $\{i_0, \ldots, i_n\} \cap \{j_0, \ldots, j_m\} = \varnothing$. Prove that Σ has a saturated model.

677. (a) Prove that $^\omega \mathfrak{A}/\mathcal{U}$ is σ-compact, i.e., all countable types can be realized in $^\omega \mathfrak{A}/\mathcal{U}$.

(b) Show that each structure of countable similarity type has a σ-saturated elementary extension.

> **9.8 Theorem.** Elementarily equivalent, saturated structures of the same cardinality are isomorphic.

678. Prove Theorem 9.8.

679. Assume $2^{\aleph_0} = \aleph_1$ and let $\mathfrak{A}$ and $\mathfrak{B}$ be elementarily equivalent, countable structures of a countable similarity type. Prove that $^\omega \mathfrak{A}/\mathcal{U}$ and $^\omega \mathfrak{B}/\mathcal{U}$ are isomorphic for any non-trivial ultrafilter $\mathcal{U}$.

680. (a) Show that the set of complex numbers, as an algebraic structure, is σ-saturated.

(b) Let F be a countable collection of multivariate polynomial equations $p_i(z_{i_1}, \ldots, z_{i_k}) = 0$ over the complex numbers. Show that if every finite subset of F has a solution, then F has a solution.

681. Prove that in any σ-saturated, dense, linear ordering without endpoints, the intersection of countably many intervals $I_1 \supseteq I_2 \supseteq \ldots$ always contains a whole interval.

682. Let $\mathcal{U}$ be a regular ultrafilter over the infinite κ. Prove that $^\kappa \mathfrak{A}/\mathcal{U}$ is κ^+-compact over $\mathfrak{A}$ (more precisely, over the diagonal image of $\mathfrak{A}$), i.e., all types over A that have cardinality at most κ can be realized in $^\omega \mathfrak{A}/\mathcal{U}$.

Remark. It follows that if the similarity type of $\mathfrak{A}$ contains at most κ symbols, then $^{\kappa}\mathfrak{A}/\mathcal{U}$ is κ^{+}-saturated over (the diagonal image of) $\mathfrak{A}$.

9.4 AXIOMATIZABILITY

9.9 Definition. A non-empty class $\mathcal{K}$ of τ-type structures is *axiomatizable* or *elementary* if for some set of formulas $\Gamma \subseteq F(\tau)$ we have

$$\mathcal{K} = \{\mathfrak{A} : \mathfrak{A} \vDash \Gamma\},$$

in which case $\mathcal{K}$ is *axiomatized by* Γ. $\mathcal{K}$ is *finitely axiomatizable* if there is a finite Γ which axiomatizes it.

683. Show that if $\mathcal{K}$ is axiomatizable then it contains a structure of cardinality at most $|\tau| \cdot \omega$.

684. Prove that the class of finite τ-structures is not axiomatizable, while the class of infinite τ-structures is axiomatizable.

685. Prove that a class consisting of isomorphic structures is axiomatizable if and only if these structures are finite.

686. Let $\mathfrak{A}$ be a fixed infinite model and set $\mathcal{K}_1 = \{\mathfrak{B} : \mathfrak{A} \prec \mathfrak{B}\}$, $\mathcal{K}_2 = \{\mathfrak{B} : \mathfrak{A}$ can be elementarily embedded into $\mathfrak{B}\}$. Which one of the following cases can occur?

(a) Both $\mathcal{K}_1$ and $\mathcal{K}_2$ are axiomatizable.
(b) $\mathcal{K}_1$ is axiomatizable but $\mathcal{K}_2$ is not.
(c) $\mathcal{K}_2$ is axiomatizable but $\mathcal{K}_1$ is not.
(d) Neither $\mathcal{K}_1$ nor $\mathcal{K}_2$ are axiomatizable.

687.[*] The formula set $\Gamma \subset F(\tau)$ is *independent* if $\Gamma \smallsetminus \{\varphi\} \nvdash \varphi$ for every $\varphi \in \Gamma$. Suppose $|\tau| \leq \omega$ and $\mathcal{K}$ is an axiomatizable class of τ-type structures. Show that $\mathcal{K}$ can be axiomatized by an independent set of formulas.

Remark (Reznikoff). Every axiomatizable class is axiomatizable by an independent set of formulas.

688. Let $\mathcal{K}$ be a class of τ-type structures, and let $\Gamma = \{\varphi : \mathfrak{A} \vDash \varphi$ for each $\mathfrak{A} \in \mathcal{K}\}$. Show that a structure is elementarily equivalent to an ultraproduct of structures from $\mathcal{K}$ if and only if it is a model of Γ.

689. Suppose that the class $\mathcal{K}$ is axiomatizable but not finitely axiomatizable. Is it true that (a) at least one structure from $\mathcal{K}$, (b) all structures from $\mathcal{K}$ are elementarily equivalent to an ultraproduct of structures *not* in $\mathcal{K}$?

9.10 Theorem (Keisler). A class $\mathcal{K}$ of τ-type structures is axiomatizable iff $\mathcal{K}$ is closed under elementary equivalence and ultraproducts.

> $\mathcal{K}$ is finitely axiomatizable if and only if both $\mathcal{K}$ and its complement are axiomatizable.

690. Prove Theorem 9.10.

691. Give an example of a class $\mathcal{K}$ of τ-type structures, if possible, such that

(a) $\mathcal{K}$ is closed under ultraproducts but not under elementary equivalence.

(b) $\mathcal{K}$ is closed under elementary equivalence but not under ultraproducts.

(c) $\mathcal{K}$ and its complement are closed under ultraproducts and isomorphism but not under elementary equivalence.

> **Hint.** Use the Keisler–Shelah isomorphism theorem 9.5.

692. Let $\Gamma_1, \Gamma_2 \subseteq F(\tau)$ and suppose that for every τ-type structure $\mathfrak{A}$ we have either $\mathfrak{A} \vDash \Gamma_1$ or $\mathfrak{A} \vDash \Gamma_2$ but not both. Prove that the class of models of Γ_1 is finitely axiomatizable.

693. Suppose $\mathcal{K}$ is axiomatized by Γ and $\mathcal{K}$ is finitely axiomatizable. Show that some finite subset of Γ axiomatizes $\mathcal{K}$.

694. Prove that the class $\mathcal{K}$ of models of Γ is axiomatizable by

(a) closed $\forall_1$ formulas if and only if $\mathcal{K}$ is closed under substructures,

(b) closed $\exists_1$ formulas if and only if $\mathcal{K}$ is closed under extensions.

695. Suppose $\mathcal{K}$ is axiomatizable by $\forall\exists$-formulas. Show that $\mathcal{K}$ is closed under unions of chains.

696. Show that the theory of $\mathfrak{A} = \langle \omega, 0, S, < \rangle$ is not axiomatizable by universal formulas. Find a conservative extension which has such an axiomatization.

9.4.1 NON-AXIOMATIZABILITY

697. Show that the following structure classes are not axiomatizable:

(a) cyclic graphs,

(b) connected graphs,

(c) finite groups,

(d) torsion groups (each element has finite order),

(e) free groups,

(f) simple groups,

(g) fields of positive characteristics,

(h) Archimedean fields,

(i) rings isomorphic to $F[x]$ for some field F of zero characteristics,

(j) well-founded binary relations,

(k) well-orders.

698. Prove that the class of linear orderings can be axiomatized by universal formulas, while the class of dense linear orderings without endpoints cannot.

699. A partition $F = A \cup^* B$ for two non-empty sets of a totally ordered field $\mathfrak{F}$ (Problem 658) is a *Dedekind-cut*, if each element of A is smaller than all elements of B. $\mathfrak{F}$ is *Dedekind-complete* if for each Dedekind-cut (A, B) either A has a largest element or B has a smallest element.

Show that the class of totally ordered Dedekind-complete fields is not axiomatizable.

700. Let $\mathcal{G}$ be the class of all simple graphs and $\chi(G)$ be the chromatic number of the graph G. Prove that for each $k \geq 3$ none of the classes

$$\mathcal{K}_{=k} = \{G \in \mathcal{G} : \chi(G) = k\},$$
$$\mathcal{K}_{>k} = \{G \in \mathcal{G} : \chi(G) > k\},$$
$$\mathcal{K}_{\neq k} = \{G \in \mathcal{G} : \chi(G) \neq k\}$$

are axiomatizable, while $\mathcal{K}_2 = \{G \in \mathcal{G} : \chi(G) = 2\}$ is axiomatizable.

> **Hint.** Use a theorem of Erdős stating that for every pair $g, k \geq 3$ of natural numbers there is a graph $G(g, k)$ of girth g and chromatic number k.

701. Prove that the class of planar graphs is not axiomatizable.

9.4.2 NOT FINITE AXIOMATIZABILITY

702. Prove that the theory of planar graphs is axiomatizable, but not finitely axiomatizable.

703. Let $\mathcal{K}$ consists of those graphs whose vertices have infinite degree. Show that $\mathcal{K}$ is not finitely axiomatizable.

704.* Give a finitely axiomatizable (non-empty) class of graphs in which each vertex of each graph has infinite degree.

> **Hint.** Using (the finitely axiomatizable) ordering, the formula $\forall x \exists y (x < y)$ guarantees that the structure is infinite. To embed the ordering to the graph use Problem 449.

705. Show that the following classes are axiomatizable but not finitely:

(a) infinite groups,

(b) torsion-free Abelian groups,

(c) divisible torsion-free Abelian groups,

(d) fields of characteristics 0,

(e) algebraically closed fields,

(f) real closed fields,

(g) cycle-free graphs,

(h) 3-regular cycle-free graphs.

706. The similarity type τ contains the constant 0 and the unary function symbol S. Γ is the set of axioms expressing: 0 is not a successor of anything; non-zero elements are successors; different elements have different successors; finally, for all $n \geq 1$ no element is the n-th successor of itself (see Problems 501 and 552). Let $\mathcal{K}$ be the class of models of Γ. Show that $\mathcal{K}$ is not finitely axiomatizable.

10

ARITHMETIC

10.1 ROBINSON'S AXIOM SYSTEM

Notation. The *standard model of arithmetic* is the structure $\mathfrak{N} = \langle \omega, 0, 1, +, \cdot \rangle$ of type $\tau_N = \langle 0, 1, +, \cdot \rangle$ with the natural interpretation. Next to the symbols in τ_N, $x \leq y$ will be used as an abbreviation for $\exists z (z + x = y)$, and $x < y$ for $(x \leq y \land x \neq y)$.

The axiom system Q, first set out by *R. M. Robinson*, describes some basic features of $\mathfrak{N}$ and plays an essential role in establishing undecidability.

10.1 Definition (Robinson's axioms). The system Q consists of the τ_N-type formulas Q1–Q7 below:

Q1 $x + 1 \neq 0$,
Q2 $x + 1 = y + 1 \rightarrow x = y$,
Q3 $x + 0 = x$,
Q4 $x + (y + 1) = (x + y) + 1$,
Q5 $x \cdot 0 = 0$,
Q6 $x \cdot (y + 1) = (x \cdot y) + x$,
Q7 $y \neq 0 \rightarrow \exists x (y = x + 1)$.

707. Check that $\mathfrak{N} \vDash Q$.

708. Show that $Q \vdash 0 \neq 1$.

709. Show that $Q \vdash 0 + x = 0 \rightarrow x = 0$, and $Q \vdash 0 + 1 = 1$.

Notation. For natural numbers n the τ_N-term π_n is defined inductively as follows: $\pi_0 = 0$, $\pi_1 = 0 + 1$, and, in general, $\pi_{n+1} = \pi_n + 1$. Clearly, in the standard model $\mathfrak{N}$, $\pi_n^{\mathfrak{N}} = n$ for every $n \in \omega$.

710. Prove that for natural numbers n, p the following statements are theorems of Q: (a) $\pi_n + \pi_p = \pi_{n+p}$, (b) $\pi_n \cdot \pi_p = \pi_{n \cdot p}$.

© The Author(s), under exclusive license to Springer Nature Switzerland AG 2022
L. Csirmaz and Z. Gyenis, *Mathematical Logic,* Problem Books
in Mathematics, https://doi.org/10.1007/978-3-030-79010-3_10

711. Show that $Q \vdash (x \leq \pi_n) \vee (\pi_n \leq x)$.

712. Prove that in each model of Q we have

$$(x \leq \pi_n) \rightarrow (x = \pi_0 \vee \cdots \vee x = \pi_n).$$

713. Show that $Q \vdash \pi_n \neq \pi_p$ whenever $n \neq p$. Show that $Q \vdash \pi_n \leq \pi_p$ if $n \leq p$, and $Q \vdash \neg(\pi_n \leq \pi_p)$ otherwise.

714. Let $\mathfrak{A} \vDash Q$ and assume $a + b = \pi_n$ for two elements $a, b \in A$. Show that both a and b are equal to π_i for some i.

715. Find a model of Q where $x + 1 \neq x$ fails.

716. Find a model of Q where the addition is not commutative.

717. In each model of Q the axioms imply $a + 0 = a$. Is it true that $0 + a = a$ always holds?

718. Let D be a partially ordered set of "infinite numbers" where every pair has a least upper bound. The ground set of $\mathfrak{A}$ is the disjoint union $\omega \cup D$. The result of addition or multiplication of two infinite numbers is their least upper bound. Complete the interpretation so that $\mathfrak{A}$ becomes a model of Q.

719. Find models of Q where

(a) $x + x = x$ holds for some $x \neq 0$,

(b) $\leq$ is not an ordering,

(c) there are different elements such that neither $a \leq b$ nor $b \leq a$,

(d) there are different elements such that both $a \leq b$ and $b \leq a$.

720.* Find a model of Q in which $(x < y) \rightarrow (x + 1 \leq y)$ does not hold.

> **Notation.** *Bounded quantifiers* are of the form $(\forall x < t)$ and $(\exists x < t)$ where the term $t \in E(\tau_N)$ does not contain x. Their meaning is
>
> $$(\forall x < t)\,\varphi \quad \text{means} \quad \forall x(x < t \rightarrow \varphi),$$
> $$(\exists x < t)\,\varphi \quad \text{means} \quad \exists x(x < t \wedge \varphi).$$

721. Show that every model $\mathfrak{A}$ of Q contains an isomorphic copy of $\mathfrak{N}$ as an initial segment (see Definition 8.10) with respect to the relation $\leq^{\mathfrak{A}}$.

722. Suppose that the closed $\varphi \in F(\tau_N)$ contains bounded quantifiers only. Show that $\mathfrak{N} \vDash \varphi$ implies $Q \vdash \varphi$.

723. Find a formula $\varphi(x)$ such that $Q \vdash \varphi(\pi_n)$ for all $n \in \omega$, while $Q \nvdash \forall x \varphi(x)$.

10.2   UNDECIDABILITY

In this section τ is a fixed finite type, and α is the coding of τ-type terms and formulas as defined in Section 6.5.

10.2 Definition. The theory $\Gamma \subseteq F(\tau)$ is *recursive* (recursively enumerable) if the set of codes $\{\alpha(\varphi) : \varphi \in \Gamma\}$ is recursive (recursively enumerable, respectively). Γ is *decidable* if the set of its consequences, namely, $\{\varphi \in F(\tau) : \Gamma \vdash \varphi\}$ is recursive; otherwise Γ is *undecidable*.

724. If Γ is recursively enumerable, then so is $\{\varphi : \Gamma \vdash \varphi\}$.

725. Show that every undecidable set of formulas can be extended to a decidable one.

10.2.1   DECIDABLE THEORIES

726. Suppose that Γ has no infinite models. Prove that Γ is decidable.

727. (a)   The empty set of formulas in the empty similarity type is decidable.
(b)   Let τ consist of k unary relation symbols. Then $\emptyset \subset F(\tau)$ is decidable.
(c)   The theory of dense linear ordering without endpoints is decidable.

728. Create a recursive set $\Gamma \subset F(\emptyset)$ which is undecidable.

Hint. See Problem 307.

729 (Craig).   Suppose $\Gamma \subseteq F(\tau)$ is recursively enumerable. Find a recursive set Σ with the same consequences, that is, $\{\varphi : \Sigma \vdash \varphi\} = \{\varphi : \Gamma \vdash \varphi\}$.

10.3 Theorem. Γ is decidable if it is recursive and complete (as in Definition 8.6).

730. Prove Theorem 10.3.

731. Let Γ be recursive and κ-categorical for some infinite cardinal κ. Show that Γ is decidable if, in addition, it has no finite models. Give a counterexample when Γ has finite models.

732. Show that these theories are decidable:
(a)   dense linear orderings without endpoints,
(b)   discrete linear orderings without endpoint,
(c)   infinite Abelian groups in which each non-unit element has order p,
(d)   divisible torsion-free Abelian groups,
(e)   algebraically closed fields of characteristic 0,
(f)   real closed fields,
(g)   atomless Boolean algebras.

Remark. Decidability of Abelian groups was proved by Wanda Szmielew.

10.2.2 UNDECIDABLE THEORIES

733. (a) Suppose $\Gamma, \Delta \subset F(\tau)$, Δ is finite, and $\Gamma \cup \Delta$ is undecidable. Prove that Γ is undecidable.

(b) Suppose there is a finite undecidable theory $\Delta \subseteq F(\tau)$. Show that the empty theory of type τ is undecidable.

734. Let c be a new constant symbol not in τ. Suppose $\Gamma \subset F(\tau)$ is undecidable in the type $\tau \cup \{c\}$. Show that it is undecidable in type τ.

Notation. The theory $\Delta \subset F(\tau)$ is *essentially undecidable*, if it is finite, and every consistent τ-type extension of Δ is undecidable.

735. Let $\Delta \subset F(\tau)$ be essentially undecidable, and τ' be another finite type. Suppose that a model of Δ can be defined semantically (see Definition 7.16) in a model of $\Gamma \subseteq F(\tau')$.

(a) Show that Γ is undecidable.

(b) Show that there is an essentially undecidable $\Delta' \subset F(\tau')$ such that $\Delta' \cup \Gamma$ is consistent.

736. In Problem 735 "essentially" is essential: give an undecidable theory Δ and a model of Δ which can be semantically defined in a model of a decidable theory.

The topic of the next few problems is the representation of $\omega^n \to \omega$ functions by formulas. To this end the type τ is assumed to extend τ_N so that natural numbers can be identified with the τ-terms π_0, π_1, etc. It is also assumed tacitly that all considered theories prove $\pi_k \neq \pi_\ell$ for distinct k and ℓ.

10.4 Definition. Let $\tau_N \subseteq \tau$. The theory $\Gamma \subset F(\tau)$ *represents the function* $f : \omega^n \to \omega$ if there is a formula $\varphi(\vec{x}, y) \in F(\tau)$ such that for all $\vec{a} \in \omega^n$,

$$\Gamma \vdash \forall y (\varphi(\pi_{\vec{a}}, y) \leftrightarrow y = \pi_{f(\vec{a})}).$$

Γ *represents the relation* $R \subseteq \omega^n$ if there is a formula $\varphi(\vec{x}) \in F(\tau)$ such that $\vec{a} \in R$ implies $\Gamma \vdash \varphi(\pi_{\vec{a}})$, and $\vec{a} \notin R$ implies $\Gamma \vdash \neg\varphi(\pi_{\vec{a}})$.

737. (a) An inconsistent theory represents every function and relation.

(b) If Γ represents f and $\Gamma \subseteq \Gamma' \subseteq F(\tau)$, then Γ' represents f as well.

(c) A consistent theory can represent at most countably many functions and relations.

(d) For every countable set of unary functions there is a consistent theory Γ over a finite type $\tau \supseteq \tau_N$ which represents all of them.

(e) If a consistent, recursive Γ represents f, then f is recursive.

738. Prove that Robinson's axiom system Q represents all recursive functions.

> **Hint.** Use induction on the definition of recursive functions.

739. Show that Γ represents the relation R if and only if it represents its characteristic function χ_R.

740. Give an example for a relation $R \subseteq \omega$ and a theory Γ so that $n \in R$ if $\Gamma \vdash \varphi(\pi_n)$, and $n \notin R$ if $\Gamma \nvdash \varphi(\pi_n)$, while R is not representable in Γ.

741. Find a consistent Γ that represents recursive functions such that for any pair of different infinite recursive sets A and B there are representations $\varphi(x)$ and $\psi(x)$, respectively, of A and B such that $\Gamma \vdash \exists x (\neg \varphi(x) \wedge \psi(x))$.

742. Suppose Γ is decidable. Show that the set $\{\langle \alpha(\varphi), n \rangle : \Gamma \nvdash \varphi(\pi_n)\} \subset \omega^2$ is recursive.

> **10.5 Theorem** (Church). If Γ is consistent and represents all recursive functions, then Γ is undecidable.

743. Prove Theorem 10.5.

> **Hint.** Use the diagonal of the relation from Problem 742.

744. Show that both consistency and representability of recursive functions are necessary in Church's theorem 10.5.

745. Show that Q is undecidable. Show that every consistent extension of Q is undecidable.

746. Let $\Gamma \subset F(\tau_N)$ be such that it has a model $\mathfrak{A} \vDash \Gamma$ in which Q holds. Prove that Γ is undecidable.

747. Suppose $\Gamma \subseteq F(\tau)$ has a model in which the standard arithmetic $\mathfrak{N} = \langle \omega, 0, 1, +, \cdot \rangle$ can be semantically defined. Show that Γ is undecidable.

748. Show that these theories are undecidable:

(a) theory of the natural numbers: $\mathrm{Th}(\mathfrak{N})$,
(b) set theory,
(c) theory of rings,
(d) theory of graphs,
(e) theory of lattices,
(f) theory of groups,
(g) theory of fields.

749. Suppose τ contains a binary relation symbol or a binary function symbol. Show that it is undecidable whether a τ-type formula is true in every τ-structure.

750. Suppose τ contains two unary function symbols. Show that the empty τ-type theory is undecidable.

Remark. If τ contains only unary relation symbols and at most one unary function symbol, then the τ-type empty theory is decidable, see also Problem 727(b).

10.6 Theorem (Gödel, first incompleteness). If Γ is consistent, recursive, and represents recursive functions, then it is incomplete.

751. Prove Theorem 10.6.

752. Assume Γ is consistent, recursive, and represents recursive functions. Construct a formula $\varphi(x)$ such that $\Gamma \vdash \varphi(0) \vee \varphi(1)$ but neither $\Gamma \vdash \varphi(0)$ nor $\Gamma \vdash \varphi(1)$.

753. Let Γ be a consistent, recursive theory that represents recursive functions. Construct a formula $\varphi(x,y)$ such that $\Gamma \vdash \forall x \exists! y\, \varphi(x,y)$, but for all natural numbers i and j, $\Gamma \nvdash \varphi(\pi_i, \pi_j)$.

754. Find, if possible, a consistent theory such that it is

(a) recursive, complete, decidable;

(b) recursive, complete, undecidable;

(c) recursive, incomplete, decidable;

(d) recursive, incomplete, undecidable;

(e) not recursive, complete, decidable;

(f) not recursive, complete, undecidable;

(g) not recursive, incomplete, decidable;

(h) not recursive, incomplete, undecidable.

10.3 DERIVABILITY

In this section the similarity type τ is finite, extends τ_N, all theories prove $\pi_k \neq \pi_\ell$ for different k and ℓ, and α is the coding of τ-type terms and formulas from Section 6.5. In addition, all theories are assumed to represent all recursive functions.

Notation. We write $\ulcorner \varphi \urcorner$ to denote the term $\pi_{\alpha(\varphi)}$.

755. Suppose Γ represents the (recursive) function $\langle \alpha(\varphi), n \rangle \mapsto \alpha(\varphi(\pi_n))$ by the formula $\chi(x_1, x_2, y) \in F(\tau)$. Let $\Psi(x) \equiv \forall y\, (\chi(x, x, y) \to \Phi(y))$. Show that

$$\Gamma \vdash \Psi(\pi_m) \leftrightarrow \Phi(\ulcorner \varphi(\pi_m) \urcorner)$$

for every $\varphi(x) \in F(\tau)$ where $m = \alpha(\varphi)$.

> **10.7 Theorem** (Fixed point theorem). Assume Γ represents recursive functions. For every formula $\Phi(x)$ with a single free variable x there there is a closed formula ν such that
>
> $$\Gamma \vdash \nu \leftrightarrow \Phi(\ulcorner \nu \urcorner).$$

756. Prove Theorem 10.7.

757. Suppose Γ is consistent. Show that there is no formula $\Phi(x)$ such that $\Gamma \vdash \varphi$ implies $\Gamma \vdash \Phi(\ulcorner \varphi \urcorner)$, and $\Gamma \nvdash \varphi$ implies $\Gamma \vdash \neg\Phi(\ulcorner \varphi \urcorner)$ for every closed formula φ.

758. Prove Church's theorem 10.5 using the fixed point theorem.

759. Suppose Γ is consistent. Show that there is no formula $\Phi(x)$ such that $\Gamma \vdash \varphi \leftrightarrow \Phi(\ulcorner \varphi \urcorner)$ holds for every closed φ.

760. Prove that every formula $\Phi(x)$ has infinitely many different fixed points.

761. Prove that for any pair of formulas $\Phi_1(x)$ and $\Phi_2(x)$ there are closed ν_1 and ν_2 such that

$$\Gamma \vdash \nu_1 \leftrightarrow \Phi_1(\ulcorner \nu_2 \urcorner),$$
$$\Gamma \vdash \nu_2 \leftrightarrow \Phi_2(\ulcorner \nu_1 \urcorner).$$

762. Prove that for any pair of formulas $\Phi_1(x, y)$ and $\Phi_2(x, y)$ there are closed ν_1 and ν_2 such that

$$\Gamma \vdash \nu_1 \leftrightarrow \Phi_1(\ulcorner \nu_1 \urcorner, \ulcorner \nu_2 \urcorner),$$
$$\Gamma \vdash \nu_2 \leftrightarrow \Phi_2(\ulcorner \nu_1 \urcorner, \ulcorner \nu_2 \urcorner).$$

763. Let Γ be consistent. Write $A = \{\alpha(\varphi) : \varphi \text{ is closed and } \Gamma \vdash \varphi\}$, and $B = \{\alpha(\varphi) : \varphi \text{ is closed and } \Gamma \vdash \neg\varphi\}$. Show that there is no recursive C that separates A and B, that is $A \subset C$ and $B \subset (\omega \smallsetminus C)$.

> **10.8 Definition** (Provability predicate). The code of the formula sequence $\varphi_1, \ldots, \varphi_n$ is $\langle \alpha(\varphi_1), \ldots, \alpha(\varphi_n) \rangle \in \omega$. Let $\Gamma \subset F(\tau)$ be a theory. The *provability predicate* is the binary relation
>
> $$\mathrm{PP}_\Gamma = \{\langle u, \alpha(\varphi) \rangle \in \omega^2 : u \text{ is the code of a derivation of } \varphi \text{ from } \Gamma\},$$
>
> which is recursive when Γ is recursive.

> **Notation.** For a recursive $\Gamma \subseteq F(\tau)$ let $\mathrm{Prov}_\Gamma(x, y) \in F(\tau)$ be an arbitrary Γ-representation of the provability predicate PP_Γ, and define $\mathrm{Pr}_\Gamma(x) \equiv \exists u\, \mathrm{Prov}_\Gamma(u, x)$. If there is no danger of confusion, the index Γ is omitted.

764. (a) If $\Gamma \vdash \varphi$, then $\Gamma \vdash \mathsf{Pr}(\ulcorner \varphi \urcorner)$.

(b) If $\Gamma \nvdash \varphi$, then $\Gamma \vdash \neg \mathsf{Prov}(\pi_n, \ulcorner \varphi \urcorner)$ for every $n \in \omega$.

765. Suppose $\Gamma \subset F(\tau)$ is recursive and consistent. For any total recursive function f there is a formula $\varphi \in F(\tau)$ such that $\Gamma \vdash \varphi$, but φ has no derivation whose code would be smaller than $f(\alpha(\varphi))$.

766. (a) Suppose $\Gamma \vdash \forall x \varphi(x)$. Prove that there is a recursive function f such that for every $i \in \omega$, $\Gamma \vdash \varphi(\pi_i)$ has a derivation with code less than $f(i)$.

(b) Find a recursive Γ and a formula $\varphi(x)$ such that for all $i \in \omega$ we have $\Gamma \vdash \varphi(\pi_i)$ but there is no recursive function f such that some derivation of $\Gamma \vdash \varphi(\pi_i)$ has code less than $f(i)$.

767. Let v be a fixed point of the formula $\neg \mathsf{Pr}(x)$, that is, $\Gamma \vdash v \leftrightarrow \neg \mathsf{Pr}(\ulcorner v \urcorner)$. Show that if Γ is consistent, then $\Gamma \nvdash v$.

> **Remark.** This v is the famous "I am not derivable" formula created first by K. Gödel. While it expresses the truth (it is not derivable), without further assumptions it might happen that $\Gamma \vdash \neg v$, see Problems 772 and 814.

768. Suppose Γ proves that "what is provable is true," namely, $\Gamma \vdash \mathsf{Pr}(\ulcorner \varphi \urcorner) \rightarrow \varphi$ for every closed φ. Show that Γ is inconsistent.

769. Γ is recursive and consistent. Show that Γ is not ω-consistent, that is, there is a formula $\varphi(x)$ such that $\Gamma \vdash \varphi(\pi_n)$ for all $n \in \omega$, while $\Gamma \nvdash \forall x \varphi(x)$.

> **Notation** (J. B. Rosser). Denote by $n(x)$ the representation of the recursive function that gives back the code of the negation of the universal closure of the formula that has code x. Take any representation formula Prov_Γ of the provability predicate and write
> $$\mathsf{Prov}_\Gamma^*(u, i) \equiv \mathsf{Prov}_\Gamma(u, i) \wedge (\forall v \leq u) \neg \mathsf{Prov}_\Gamma(v, n(i)).$$
> Let moreover $\mathsf{Pr}_\Gamma^*(x) \equiv \exists u \, \mathsf{Prov}_\Gamma^*(u, x)$, and $\mathsf{Con}_\Gamma^* \equiv \neg \mathsf{Pr}_\Gamma^*(\ulcorner \bot \urcorner)$.

770. Suppose $Q \subseteq \Gamma$. Show that Prov_Γ^* also represents the provability predicate.

771. Suppose $Q \subseteq \Gamma$, φ is closed, and $\Gamma \vdash \neg \varphi$. Show that $\Gamma \vdash \neg \mathsf{Pr}^*(\ulcorner \varphi \urcorner)$.

772. Suppose $Q \subseteq \Gamma$ is consistent. Let v^* be the fixed point of $\neg \mathsf{Pr}^*(x)$. Show that $\Gamma \nvdash v^*$ and $\Gamma \nvdash \neg v^*$.

773. If $Q \subseteq \Gamma$ is recursive, then $\Gamma \vdash \mathsf{Con}_\Gamma^*$.

774. Suppose $Q \subseteq \Gamma$ is recursive. Show that there is a fixed point v of the formula $\mathsf{Pr}^*(x)$ such that (a) $\Gamma \vdash v$, (b) $\Gamma \vdash \neg v$.

775. Suppose $Q \subseteq \Gamma$, and A, B are disjoint recursively enumerable sets.

(a) Find a formula $\varphi(x)$ so that if $n \in A$ then $\Gamma \vdash \varphi(\pi_n)$, and if $n \in B$ then $\Gamma \vdash \neg\varphi(\pi_n)$.

(b) Suppose Γ is recursive and consistent. Show that it is not always possible to require in addition to (a) that for every $n \in \omega$ either $\Gamma \vdash \varphi(\pi_n)$ or $\Gamma \vdash \neg\varphi(\pi_n)$.

10.4   PEANO'S AXIOM SYSTEM

Peano's axiom system consists of infinitely many axioms. The first few ones describe the basic properties of the successor, addition and multiplications; while the most important one is the axiom of *full induction*. This latter one is, in fact, infinitely many formulas, and rather should be called an *axiom scheme*.

The main purpose of Peano's axioms was to give a faithful description of the standard model $\mathfrak{N}$ of the arithmetic.

10.9 Definition (Peano axioms). The similarity type is τ_N, the axioms system is also called *Peano arithmetic* and is denoted by PA. The first six axioms are the same as that of Q, and Q7 is replaced by the *induction axiom* (scheme):

PA1 $x + 1 \neq 0$;

PA2 $x + 1 = y + 1 \to x = y$;

PA3 $x + 0 = x$;

PA4 $x + (y + 1) = (x + y) + 1$;

PA5 $x \cdot 0 = 0$;

PA6 $x \cdot (y + 1) = (x \cdot y) + x$;

PA7 $\left[\varphi(0, \vec{p}) \wedge \forall x \big(\varphi(x, \vec{p}) \to \varphi(x + 1, \vec{p}) \big) \right] \to \forall x \varphi(x, \vec{p})$.

In PA7 the variables $\vec{p}$ are the *parameters* of the induction.
Similar to Q, the relations $\leq$ and $<$ are defined as abbreviations for $\exists z(z + x = y)$ and $(x \leq y \wedge x \neq y)$, respectively.

776. Show that every model of PA is also a model of Q.

777. Show that the first axiom PA1 does not follow from the other Peano axioms.

778. Show that the following formulas are consequences of PA:

(a) $(x + y) + z = x + (y + z)$,

(b) $x + z = y + z \to x = y$,

(c) $x + y = 0 \to x = 0 \wedge y = 0$,

(d) $x + y = y + x$,

(e) $x \cdot y = y \cdot x$.

779. Show that in every Peano model $\leq$ is a total discrete order with 0 as the smallest element, and $x + 1$ as the immediate successor of x.

780. $x \mid y$ is an abbreviation for $\exists z (z \cdot x = y)$. Show that

$$\mathrm{PA} \vdash (\forall a \geq 1) \, \forall b \, (\exists! r < a) \, (a \mid b + r).$$

781. The following scheme is usually dubbed as *strong induction*:

$$\forall x \big[(\forall y < x) \varphi(y, \vec{p}) \to \varphi(x, \vec{p}) \big] \to \forall x \varphi(x, \vec{p}). \qquad (\star)$$

Show that the strong induction follows from the Peano axioms.

782. Find a model $\mathfrak{A}$ of PA1–PA6 such that $\mathfrak{A} \vDash 0 + 1 \neq 1$, while the strong induction holds in $\mathfrak{A}$ for every formula $\varphi(x, \vec{p})$.

783. Let PA^- be the set of Peano axioms where the induction axiom PA7 can only be used without the parameters $\vec{p}$. Show that induction with parameters is a theorem of PA^-.

784. Show that the collection principle holds for the relation $\leq$ (see Definition 8.9) in every Peano model.

785. Suppose $\mathfrak{A} \vDash$ PA1–PA6, and $\leq^{\mathfrak{A}}$ is a total discrete order with 0 as the minimal element and $x + 1$ as the immediate successor of x. Show that $\mathfrak{A} \vDash$PA7 if and only if every subset of $\mathfrak{A}$ definable using parameters (see Definition 7.15) has a minimal element.

786. (a) Show that a subset $B \subset \omega$ is definable in $\mathfrak{N}$ with parameters (Definition 7.15) iff it is definable without parameters.
(b) Find a Peano model $\mathfrak{A}$ and a subset $X \subset A$ which can be defined with parameters but not without parameters.

787. Suppose $\mathfrak{A} \vDash$ PA1–PA6, and $\leq^{\mathfrak{A}}$ is a total discrete order compatible with the addition (see Problem 785). Show that if every set $X \subseteq A$ definable without parameters has a $\leq^A$-minimal element, then every set definable by parameters also has a minimal element.

788. Let $\mathfrak{A} \vDash$ PA, and $B \subseteq A$ be the set of definable elements in $\mathfrak{A}$. Show that $\mathfrak{B}$ is an elementary substructure of $\mathfrak{A}$ (in particular, $\mathfrak{B}$ is a PA model).

789. Show that every model $\mathfrak{A} \vDash$ PA is an end extension of the standard model $\mathfrak{N}$ for the total ordering $\leq^{\mathfrak{A}}$ (see Definition 8.10).

> **Notation.** In a Peano model an element is *finite* or *standard* if it is in the initial segment isomorphic to $\mathfrak{N}$, and *infinite* or *non-standard* otherwise.

790. Show that every non-standard model contains infinitely many infinite elements.

791. The standard part $\mathfrak{N}$ cannot be defined in any non-standard Peano model.

792 (Overspill principle). Let $\mathfrak{A}$ be a non-standard model and $\vec{p} \in A$ be parameters. Then $\mathfrak{A} \vDash \varphi(\pi_n, \vec{p})$ for all $n \in \omega$ iff there exists an infinite $a \in A$ such that $\mathfrak{A} \vDash (\forall x < a)\varphi(x, \vec{p})$.

793. Show that every countable Peano model has a proper elementary end extension.

> **Remark.** The MacDowell–Specker theorem states that not only countable, but every Peano model has a proper elementary end extension.

794. Prove that there are $2^{\aleph_0}$ non-isomorphic countable non-standard models.

> **Hint.** For a subset X of the prime numbers construct a countable model which realizes the type $T_X(c) = \{p \mid c : p \in X\} \cup \{p \nmid c : p \notin X\}$.

795. Assume we know the following two statements:

(i) If the closed formula $\varphi \in F(\tau_{PA})$ does not contain multiplication, then either $PA \vdash \varphi$, or $PA \vdash \neg\varphi$.

(ii) There is a closed formula $\psi \in F(\tau_{PA})$ such that neither $PA \vdash \psi$, nor $PA \vdash \neg\psi$.

Show that there are Peano models $\mathfrak{A}$ and $\mathfrak{B}$ such that their reducts to the symbols $\langle 0, 1, + \rangle$ are isomorphic, the multiplication, however, is different.

> **Remark.** (i) follows from the fact that the $\langle 0, 1, + \rangle$-reduct of Peano models—the Presburger arithmetic—has quantifier elimination, see Theorem 8.19. (ii) follows from Gödel's first incompleteness theorem 10.6.

796. The order type of a non-standard Peano model $\mathfrak{A}$ is $\omega + \mathbb{Z} \times M$ for some linear ordering $\langle M, \leq \rangle$ (Problems 779 and 554).

(a) Show that $\langle M, \leq \rangle$ is a dense linear ordering without endpoints.

(b) A countable non-standard model has order type $\omega + \mathbb{Z} \times \mathbb{Q}$.

(c) The order type cannot be $\omega + \mathbb{Z} \times \mathbb{R}$ where $\mathbb{R}$ is the set of reals.

> **10.10 Definition.** A function $f : \omega^n \to \omega$ is PA-*definable* if there is a formula $\varphi_f(\vec{x}, y)$ such that in every PA-model φ_f defines a function and $\mathfrak{N} \vDash \varphi_f[\vec{a}, b]$ iff $f(\vec{a}) = b$.

797. If the function f is PA-defined by φ_f, then $PA \vdash \forall \vec{x} \exists! y \varphi_f(\vec{x}, y)$ and φ_f represents f in PA in sense of Definition 10.4.

798. Give a function f and two PA-representations φ and ψ of f such that φ is a PA-definition of f while ψ is not.

799. Show that the following functions are PA-definable: $x \dotminus y$, $K_<(x, y)$, $[\sqrt{x}]$, $\mathrm{rem}(x, y)$ (remainder), the function which returns the smallest power of 2 which is bigger than x.

800. Suppose $f_1(x)$ and $f_2(x)$ are PA-definable functions, moreover $PA \vdash f_i(0) = 1$ and $PA \vdash \forall x(f_i(x+1) = 2 \cdot f_i(x))$ for $i = 1, 2$. Show that f_1 and f_2 are the same functions in every PA-model.

10.11 Theorem. There are PA-definable functions $\mathrm{Elem}(u, i)$ and $\mathrm{Len}(u)$ so that $\mathrm{PA} \vdash \mathrm{Len}(0) = 0$, and

$$\mathrm{PA} \vdash \forall u \forall z \exists v \big(\, \mathrm{Len}(v) = \mathrm{Len}(u) + 1 \, \wedge$$
$$\mathrm{Elem}(v, \mathrm{Len}(u)) = z \, \wedge$$
$$(\forall i < \mathrm{Len}(u)) \, \mathrm{Elem}(v, i) = \mathrm{Elem}(u, i) \big).$$

Remark. The coding functions based on Gödel's β-function (Problem 126) have this property. The proof requires establishing a significant part of elementary number theory in PA. Proving the same statement for the alternative coding outlined in Problems 129 and 130 is more straightforward.

801. Show that there is a PA-definable function f such that $\mathrm{PA} \vdash f(0) = 1 \wedge \forall x(f(x+1) = 2 \cdot f(x))$. Show that $\mathrm{PA} \vdash f(\pi_n) = \pi_{2^n}$ for all $n \in \omega$.

10.12 Definition. The *faithful representation* $\varphi_f(\vec{x}, y) \in F(\tau_N)$ of the primitive recursive functions f is defined by structural induction:

- for initial functions $\varphi_0(x, y) \equiv y = 0$ (zero), $\varphi_S(x, y) \equiv y = x + 1$ (successor), and $\varphi_{U_i^k}(\vec{x}, y) \equiv y = x_i$ (projections),
- for composition $f = \mathrm{Comp}(g, h_1, \ldots, h_\ell)$

$$\varphi_f(\vec{x}, y) \equiv \exists z_1 \ldots \exists z_\ell \big(\bigwedge_i \varphi_{g_i}(\vec{x}, z_i) \wedge \varphi_h(\vec{z}, y) \big),$$

- and for the primitive recursion operator $f = \mathrm{PrRec}(g, h)$

$$\varphi(\vec{p}, x, y) \equiv \exists u \big(\mathrm{Len}(u) = x + 1 \wedge \varphi_g(\vec{p}, 0, \mathrm{Elem}(u, 0)) \, \wedge$$
$$\wedge \, (\forall i < x) \varphi_h(\vec{p}, i, \mathrm{Elem}(u, i), \mathrm{Elem}(u, i+1)) \, \wedge$$
$$\wedge \, y = \mathrm{Elem}(u, x) \big).$$

802. Let f be primitive recursive, and φ_f be its faithful representation. Show that (a) $\mathrm{PA} \vdash \forall \vec{x} \exists! y \varphi_f(\vec{x}, y)$, (b) φ_f represents f in PA in sense of Definition 10.4.

Remark. It follows that every primitive recursive function is PA-defined by its faithful representation.

803.[*] Show that the Ackermann function (Definition 4.4) is PA-definable.

804. Let τ be a finite similarity type, and $\Gamma \subset F(\tau)$ be a primitive recursive theory. Show that the provability predicate

$$\mathrm{PP}_\Gamma = \{ \langle u, \alpha(\varphi) \rangle \in \omega^2 : u \text{ is the code of a derivation of } \varphi \text{ from } \Gamma \}$$

is primitive recursive.

Notation. For a finite type $\tau \supseteq \tau_N$ and a primitive recursive $\Gamma \subseteq F(\tau)$, $\mathrm{Prov}_\Gamma^\circ(u, x) \in F(\tau_N)$ is the faithful representation of the provability predicate PP_Γ. Let moreover $\mathrm{Pr}_\Gamma^\circ(x) \equiv \exists u \, \mathrm{Prov}_\Gamma^\circ(u, x)$.

805. Suppose $\Gamma, \Delta \subseteq F(\tau)$ are primitive recursive theories. Show that

(a) $\Gamma \vdash \varphi$ implies $\mathrm{PA} \vdash \mathrm{Pr}_\Gamma^\circ(\ulcorner\varphi\urcorner)$,

(b) $\mathrm{PA} \vdash \mathrm{Pr}_\Gamma^\circ(\ulcorner\varphi\urcorner) \to \mathrm{Pr}_{\Gamma \cup \Delta}^\circ(\ulcorner\varphi\urcorner)$,

(c) $\mathrm{PA} \vdash \big(\mathrm{Pr}_\Gamma^\circ(\ulcorner\varphi\urcorner) \wedge \mathrm{Pr}_\Gamma^\circ(\ulcorner\varphi \to \psi\urcorner)\big) \to \mathrm{Pr}_\Gamma^\circ(\ulcorner\psi\urcorner)$.

10.13 Theorem. For a primitive recursive theory Γ, $\mathrm{PA} \vdash \mathrm{Pr}_\Gamma^\circ(\ulcorner\varphi\urcorner) \to \mathrm{Pr}_{\mathrm{PA}}^\circ(\ulcorner\mathrm{Pr}_\Gamma^\circ(\ulcorner\varphi\urcorner)\urcorner)$.

Remark. The statement follows from the facts that $\mathrm{Pr}_\Gamma^\circ(x)$ is PA-equivalent to a Σ_1^*-formula, and $\mathrm{PA} \vdash \psi \to \mathrm{Pr}_{\mathrm{PA}}^\circ(\ulcorner\psi\urcorner)$ for every closed Σ_1^*-formula ψ.

In the rest of this section $\mathrm{PA} \subseteq \Gamma \subset F(\tau)$ is a fixed primitive recursive theory, and $\mathrm{Pr}(x)$ means $\mathrm{Pr}_\Gamma^\circ(x)$.

806. Prove that

(a) $\Gamma \vdash \big(\mathrm{Pr}(\ulcorner\varphi\urcorner) \wedge \mathrm{Pr}(\ulcorner\varphi \to \psi\urcorner)\big) \to \mathrm{Pr}(\ulcorner\psi\urcorner)$,

(b) $\Gamma \vdash \varphi \to \psi$ implies $\Gamma \vdash \mathrm{Pr}(\ulcorner\varphi\urcorner) \to \mathrm{Pr}(\ulcorner\psi\urcorner)$,

(c) $\Gamma \vdash \bigwedge_i \varphi_i \to \psi$ implies $\Gamma \vdash \bigwedge_i \mathrm{Pr}(\ulcorner\varphi_i\urcorner) \to \mathrm{Pr}(\ulcorner\psi\urcorner)$,

(d) $\Gamma \vdash \mathrm{Pr}(\ulcorner\varphi\urcorner) \to \mathrm{Pr}(\ulcorner\mathrm{Pr}(\ulcorner\varphi\urcorner)\urcorner)$.

807. Show that $\Gamma \vdash \big(\mathrm{Pr}(\ulcorner\varphi\urcorner) \wedge \mathrm{Pr}(\ulcorner\neg\varphi\urcorner)\big) \to \mathrm{Pr}(\ulcorner\bot\urcorner)$ where φ is closed.

808. Find a closed φ such that $\Gamma \vdash \varphi \leftrightarrow \mathrm{Pr}(\ulcorner\neg\varphi\urcorner)$.

809. Find a closed φ such that both $\Gamma \vdash \varphi \to \mathrm{Pr}(\ulcorner\varphi\urcorner)$ and $\Gamma \vdash \varphi \to \mathrm{Pr}(\ulcorner\neg\varphi\urcorner)$. Find φ in the form of $\mathrm{Pr}(\bullet)$.

810. Let ν be a fixed point of $\neg\mathrm{Pr}(x)$, that is, $\Gamma \vdash \nu \leftrightarrow \neg\mathrm{Pr}(\ulcorner\nu\urcorner)$. Show that $\Gamma \vdash \neg\nu \leftrightarrow \mathrm{Pr}(\ulcorner\bot\urcorner)$.

811. Show that $\Gamma \vdash \varphi \to \mathrm{Pr}(\ulcorner\varphi\urcorner)$ for all closed φ if and only if $\Gamma \vdash \mathrm{Pr}(\ulcorner\bot\urcorner)$.

10.14 Theorem (Gödel's second incompleteness). Suppose $\mathrm{PA} \subseteq \Gamma$ is consistent and primitive recursive. Let $\mathrm{Prov}^\circ(u, x)$ be the faithful representation of the provability predicate, and $\mathrm{Con}_\Gamma \equiv \neg\mathrm{Pr}_\Gamma^\circ(\ulcorner\bot\urcorner)$. Then $\Gamma \nvdash \mathrm{Con}_\Gamma$.

812. Prove Theorem 10.14

813. Show that $\mathrm{PA} \nvdash \neg\mathrm{Con}_{\mathrm{PA}}$. Create a consistent, primitive recursive theory $\Gamma \supset \mathrm{PA}$ such that $\Gamma \vdash \neg\mathrm{Con}_\Gamma$.

814. Show that there is a theory Γ and a fixed point $\Gamma \vdash \nu \leftrightarrow \neg\mathrm{Pr}(\ulcorner\nu\urcorner)$ such that $\Gamma \vdash \neg\nu$ (see Problem 767).

815. Find a (consistent) Γ such that both $\Gamma \vdash \mathrm{Pr}(\ulcorner\varphi\urcorner)$ and $\Gamma \vdash \mathrm{Pr}(\ulcorner\neg\varphi\urcorner)$ hold for some closed formula φ.

816. Does $\Gamma \vdash \mathrm{Pr}(\ulcorner\varphi\urcorner)$ imply $\Gamma \vdash \varphi$ for closed formulas φ?

817. Let φ be closed, and μ be a fixed point of $\mathrm{Pr}(x) \rightarrow \varphi$. Show that

(a) $\Gamma \vdash \mathrm{Pr}(\ulcorner\varphi\urcorner) \rightarrow \mathrm{Pr}(\ulcorner\mu\urcorner)$,

(b) $\Gamma \vdash \mathrm{Pr}(\ulcorner\mu\urcorner) \rightarrow \mathrm{Pr}(\ulcorner\varphi\urcorner)$,

(c) $\Gamma \vdash \mathrm{Pr}(\ulcorner\mathrm{Pr}(\ulcorner\varphi\urcorner) \rightarrow \varphi\urcorner) \rightarrow \mathrm{Pr}(\ulcorner\varphi\urcorner)$.

> **10.15 Theorem** (Löb). If $\Gamma \vdash \mathrm{Pr}(\ulcorner\varphi\urcorner) \rightarrow \varphi$ for some closed φ, then $\Gamma \vdash \varphi$.

818. Prove Theorem 10.15.

819. Show that formulas which say "I am derivable" are derivable.

820. Show that all fixed points of (a) $\mathrm{Pr}(x)$, (b) $\neg\mathrm{Pr}(x)$ are Γ-equivalent.

821. Let μ be a fixed point of $\mathrm{Pr}(x) \rightarrow \varphi$. Show that $\Gamma \vdash \mu \leftrightarrow (\mathrm{Pr}(\ulcorner\varphi\urcorner) \rightarrow \varphi)$.

10.5 ARITHMETICAL HIERARCHY

> **10.16 Definition** (Σ_n, Π_n, Δ_n formulas). Δ_0 is the set of all τ_N-type formulas in which all quantifiers are bounded.
> Let $\Sigma_0 = \Pi_0 = \Delta_0$. Σ_n, Π_n and Δ_n are defined inductively as
> - $\psi \in \Sigma_{n+1}$ if there is a $\varphi \in \Pi_n$ such that $\mathfrak{N} \vDash \psi(\vec{y}) \leftrightarrow \exists x \varphi(x, \vec{y})$
> - $\psi \in \Pi_{n+1}$ if there is a $\varphi \in \Sigma_n$ such that $\mathfrak{N} \vDash \psi(\vec{y}) \leftrightarrow \forall x \varphi(x, \vec{y})$
> - $\Delta_{n+1} = \Sigma_{n+1} \cap \Pi_{n+1}$.

> **10.17 Definition** (Σ_n, Π_n, Δ_n sets). The subset $A \subseteq \omega^n$ is Δ_k (Σ_k, Π_k) if there exists a formula $\varphi(\vec{x}) \in \Delta_k$ (Σ_k, Π_k, respectively) so that $\vec{a} \in A$ iff $\mathfrak{N} \vDash \varphi[\vec{a}]$. A subset of ω^n is *arithmetical* if it belongs to $\bigcup_k(\Sigma_k \cup \Pi_k)$.

822. Show that the set of prime numbers is Δ_0.

823. Find, for each Δ_0-formula $\psi(x, y)$, another Δ_0-formula $\varphi(y, z)$ such that $\mathfrak{N} \vDash (\forall x < z) \exists y \psi(x, y) \leftrightarrow \exists y \varphi(y, z)$.

824. Prove that the set of Σ_1-formulas is closed under $\vee$, $\wedge$, bounded quantifiers ($\exists x < t$), ($\forall x < t$), and the existential quantifier $\exists x$.

825. Show that the Σ_n and Π_n sets are closed under union and intersection, and Δ_n sets are closed under complementation.

826. Show that for each Σ_n formula $\varphi(\vec{x})$ there is a Δ_0 formula $\vartheta(\vec{x}, y_1, \ldots, y_n)$ such that $\mathfrak{N} \vDash \varphi(\vec{x}) \leftrightarrow \exists y_1 \forall y_2 \exists y_3 \ldots \vartheta(\vec{x}, \vec{y})$.

827. Let $\mathfrak{A}$ be a τ_N-type structure which is an end extension of $\mathfrak{N}$. Let e be an evaluation over $\mathfrak{N}$. Show that for every Δ_0 formula φ

(a) $\mathfrak{N} \vDash \varphi[e]$ iff $\mathfrak{A} \vDash \varphi[e]$,

(b) $\mathfrak{N} \vDash (\exists y \varphi)[e]$ implies $\mathfrak{A} \vDash (\exists y \varphi)[e]$,

(c) $\mathfrak{N} \vDash (\forall y \varphi)[e]$ provided $\mathfrak{A} \vDash (\forall y \varphi)[e]$.

828. Let $\varphi(\vec{x}, y) \in \Delta_0$. Prove that $\mathfrak{N} \vDash (\exists y \varphi)[\vec{a}]$ iff $Q \vdash (\exists y \varphi)[\pi_{\vec{a}}]$.

829. Find a recursive function $b(\cdot)$ such that for every closed Δ_0 formula φ, whenever i is the code of φ, then $b(i)$ is the code of one of the derivations $Q \vdash \varphi$ or $Q \vdash \neg \varphi$.

830. Show that for every Δ_1 subset $A \subseteq \omega^n$ there is a Σ_1 formula $\vartheta(\vec{x})$ such that $\vec{a} \in A$ implies $Q \vdash \vartheta(\pi_{\vec{a}})$, and $\vec{a} \notin A$ implies $Q \vdash \neg \vartheta(\pi_{\vec{a}})$.

> **Hint.** Use Rosser's trick as in Problem 775.

831. Show that if the graph of the function f is Σ_1, then it is also Δ_1.

832. (a) Show that the graph of Gödel's β-function (Definition 4.8) is Δ_0.

(b) The graphs of the functions $\mathsf{Len}(u)$, $\mathsf{Elem}(u, i)$, and $u \frown z$ are also Δ_0.

833. Construct Σ_1 formulas so that

(a) $\mathfrak{N} \vDash \varphi[x, y, z]$ iff $z = x^y$.

(b) $\mathfrak{N} \vDash \psi[x, y]$ iff $x = y!$ (factorial).

834. Prove that every Δ_0 set is primitive recursive.

835. Prove that if f is primitive recursive, then its graph is Σ_1.

836. (Kleene's T). For each $n \geq 1$ there is a Σ_1 set $T_n \subseteq \omega^{n+1}$ such that for every n-variable partial recursive function f there exists an index $e \in \omega$ such that so that $f(\vec{a}) = \downarrow$ if and only if $\langle e, \vec{a} \rangle \in T_n$.

837. Prove that a set $A \subseteq \omega^n$ is recursively enumerable if and only if it is Σ_1.

838. Prove that recursive sets and Δ_1-sets coincide.

839. (a) Prove that the Δ_0 subsets of ω are uniformly primitive recursive: there is a primitive recursive relation $U \subset \omega^2$ such that for every $\varphi(x) \in \Delta_0$,

$$\{n \in \omega : \mathfrak{N} \vDash \varphi[n]\} = \{n \in \omega : \langle \alpha(\varphi), n \rangle \in U\}.$$

(b) There is a primitive recursive set which is not Δ_0.

840. Suppose $\varphi(x, y)$ is a Σ_1 formula and $\mathfrak{N} \vDash \exists y \varphi(\pi_n, y)$ for every $n \in \omega$. Show that there is a recursive function f such that $\mathfrak{N} \vDash \varphi(\pi_n, \pi_{f(n)})$.

841. (a) Show that every recursive function f has a Σ_1 representation $\varphi(x, y)$, that is, $\mathfrak{N} \vDash \forall z (\varphi(\pi_n, z) \leftrightarrow z = \pi_{f(n)})$ for all $n \in \omega$.

(b) Show that if a function has such a Σ_1 representation, then it is recursive.

842.[*] Let $\Gamma \subset F(\tau_N)$ be a fixed primitive recursive theory extending PA such that $\mathfrak{N}$ is a model of Γ. The function f is *provably recursive* if it has a Σ_1 representation $\varphi(x, y) \equiv \exists z \psi(x, y, z)$ with $\psi \in \Delta_0$ such that $\Gamma \vdash \forall x \exists y \varphi(x, y)$. Construct a recursive function that grows faster than any provably recursive function (and thus it is not provably recursive).

843. (a) There are universal Σ_1 sets $U_k^1 \subset \omega^{k+1}$ such that for every Σ_1 set $A \subseteq \omega^k$ one can find an $e \in \omega$ such that $\vec{a} \in A$ iff $\langle e, \vec{a} \rangle \in U_k^1$.

(b) Prove that Σ_1 sets are not closed under taking complements.

(c) Conclude that $\Sigma_1 \neq \Pi_1$, and $\Delta_1 \neq \Sigma_1$, $\Delta_1 \neq \Pi_1$.

844. There is no universal recursive relation, i.e., a recursive $U \subset \omega^2$ so that for all recursive $A \subset \omega$ one can find $i \in \omega$ with $a \in A$ iff $\langle i, a \rangle \in U$.

845. (a) For every $n \geq 2$ there are universal Σ_n sets $U_k^n \subset \omega^{k+1}$ such that for every Σ_n set $A \subseteq \omega^k$ one can find an $e \in \omega$ such that $\vec{a} \in A$ iff $\langle e, \vec{a} \rangle \in U_k^n$.

(b) Σ_n sets are not closed under taking complements.

(c) Δ_n is a proper subset of both Σ_n, and Π_n.

846. A Σ_n^* formula has a sequence of n alternating quantifiers before a Δ_0 formula starting with an existential quantifiers. Show that for $n \geq 1$

$$\mathrm{Sat}_n = \{\alpha(\varphi) : \varphi \in \Sigma_n^* \text{ is closed and } \mathfrak{N} \vDash \varphi\} \in \Sigma_n.$$

10.18 Theorem (Tarski). Truth is not arithmetical, that is, the set of codes of the true arithmetical formulas $\{\alpha(\varphi) : \mathfrak{N} \vDash \varphi\}$ is not in any Σ_n.

847. Prove Theorem 10.18.

848. Find a Peano model $\mathfrak{A}$ such that the set $\{\alpha(\varphi) : \mathfrak{A} \vDash \varphi\}$ is arithmetical.

11

SELECTED APPLICATIONS

11.1 INDEPENDENT UNARY RELATIONS

In this section κ is an *infinite* cardinal. The similarity type $\tau = \langle U_i : i < \kappa \rangle$ contains unary relation symbols only, and $\Gamma \subset F(\tau)$ is the theory which says that any Boolean combination of finitely many of the U_i's contains at least one element.

849. Show that every Boolean combination of finitely many relations contain infinitely many elements.

850. Show that Γ is consistent.

851. Construct $\mathfrak{B}$ from $\mathfrak{A} \vDash \Gamma$ by adding a new element that realizes all the relations U_i for $i < \kappa$. Show that $\mathfrak{A}$ and $\mathfrak{B}$ are elementarily equivalent.

852. Suppose $\kappa = \omega$. Construct two non-isomorphic countable models of Γ.

853. Prove that $\{\mathfrak{A} : \mathfrak{A} \vDash \Gamma\}$ is not finitely axiomatizable.

854. Suppose $\kappa = \omega$ and let $\mathfrak{A}$ be a countable model of Γ. Prove that any Boolean combination of finitely many of the U_i's contains continuum many elements in $\mathfrak{B} = \mathfrak{A}^{\omega}/U$.

855. Prove that Γ admits quantifier elimination. Show that Γ is complete.

856. Find two elementarily equivalent structures $\mathfrak{A}$ and $\mathfrak{B}$ such that I wins the Ehrenfeucht–Fraïssé game for all N. (See Problems 520, 527).

857. For which cardinality λ is Γ λ-categorical?

11.2 UNIVERSAL GRAPHS

11.1 Definition. Let τ be the similarity type of graphs. For $n, m \in \omega$ let $\varphi_{n,m}$ be the formula expressing that whenever one picks disjoint finite sets A and B with $|A| = n$ and $|B| = m$, then there exists an element

© The Author(s), under exclusive license to Springer Nature Switzerland AG 2022
L. Csirmaz and Z. Gyenis, *Mathematical Logic,* Problem Books
in Mathematics, https://doi.org/10.1007/978-3-030-79010-3_11

which is connected to each element of A but is not connected to any element of B. Let $\Gamma \subset F(\tau)$ be the theory

$$\Gamma = \{\varphi_{n,m} : n, m < \omega\} \cup \{\forall x \neg E(x,x), \forall x \forall y (E(x,y) \rightarrow E(y,x))\}$$

Models of Γ are called *universal graphs*.

858. Prove that Γ is consistent.

859. Show that Γ has infinite models only.

860. In a universal graph every vertex has infinite degree.

861. Let $\mathfrak{G}$ be a countable universal graph. Show that each countable graph can be embedded into $\mathfrak{G}$.

862. Let $\mathfrak{A}$ and $\mathfrak{B}$ be countable universal graphs. Prove that each finite partial isomorphism $f : A \rightarrow B$ extends to an isomorphism. Particularly, $\mathfrak{A} \cong \mathfrak{B}$.

863. Show that Γ is complete, $\aleph_0$-categorical, and admits elimination of quantifiers.

864. Let $\mathfrak{G}$ be a universal graph. Show that the subgraph with one less vertex is an elementary substructure of $\mathfrak{G}$.

865. Let $\mathfrak{G}$ be a universal graph on ω_1. Show that it has a vertex of degree ω_1.

866. Let $\mathfrak{G}$ be a countable universal graph and U be a non-principal ultrafilter over ω. Show that ${}^{\omega}\mathfrak{G}/U$ contains a complete subgraph of cardinality $\aleph_1$.

867. Construct a universal graph $\mathfrak{G}$ of cardinality $\aleph_1$ such that

(a) $\mathfrak{G}$ contains a complete subgraph of cardinality $\aleph_1$.

(b) $\mathfrak{G}$ does not contain a complete subgraph of cardinality $\aleph_1$.

Hint. Consider a bijection $f : \kappa \rightarrow [\kappa]^{<\omega}$ and for $a, b \in \kappa$ draw an edge iff $a \in f(b)$ or $b \in f(a)$.

868.[*] Construct a universal graph of cardinality $\aleph_1$ which has at least one countable degree vertex.

11.3   UNIVERSAL TOURNAMENTS

11.2 Definition. A *tournament* is a directed graph with exactly one directed edge between every pair of points. Formally, it is a structure

$\mathfrak{T} = \langle V, E \rangle$ where V is a non-empty set of vertices and E is a binary relation such that $\mathfrak{T} \vDash \eta$, where η is the formula

$$\forall x \forall y \big(E(x, y) \rightarrow (x \neq y \wedge \neg E(y, x)) \wedge x \neq y \rightarrow (E(x, y) \vee E(y, x)) \big)$$

$E^{\mathfrak{T}}$ is the *dominance relation*, and we say that a dominates b if $\langle a, b \rangle \in E^{\mathfrak{T}}$.

869. For $k \in \omega$ the formula χ_k asserts "for each set of k vertices there is a vertex which dominates each of them."

(a) Fix $k \in \omega$. Is there a *finite* tournament that satisfies φ_k?

(b) Construct a tournament that satisfies χ_k for all $k \in \omega$.

11.3 Definition. For $n, m \in \omega$ let $\psi_{n,m}$ be the first-order formula expressing that whenever one picks two disjoint finite sets of vertices A and B with $|A| = n$ and $|B| = m$, then there is a vertex z which dominates each vertex in A and is dominated by every vertex in B. Write

$$\Gamma = \{ \psi_{n,m} : n, m < \omega \} \cup \{ \eta \}.$$

Models of Γ are called *universal tournaments*.

870. Show that Γ is consistent and Γ has infinite models only.

871. Let $\mathfrak{T}$ be a countable universal tournament. Prove that each countable tournament can be embedded into $\mathfrak{T}$.

872. Prove that Γ is complete and $\aleph_0$-categorical.

873. Call a tournament $\mathfrak{T} = \langle V, E \rangle$ *transitive* if E is transitive. Let $\mathfrak{T}$ be a countable universal tournament and U a non-principal ultrafilter over ω. Show that $^{\omega}\mathfrak{T}/U$ contains a transitive subtournament of cardinality $\aleph_1$.

874. Let $(X, <)$ be an arbitrary strict total ordering and let $\mathfrak{T}$ be a countable universal tournament. Show that for suitable κ and ultrafilter U, $(X, <)$ can be embedded into $^{\kappa}\mathfrak{T}/U$.

875. Construct, without reference to Problems 873 and **??**, a universal tournament $\mathfrak{T}$ of cardinality $\aleph_1$ such that $\mathfrak{T}$ contains a transitive subtournament of cardinality $\aleph_1$.

11.4 ZERO-ONE LAW

Notation. For a non-empty finite set G of structures and a set Σ of formulas $\text{Prob}(G \vDash \Sigma)$ is the probability that a randomly chosen (with uniform distribution) structure from G satisfies Σ:

$$\text{Prob}(G \vDash \Sigma) = \frac{|\{\mathfrak{A} \in G : \mathfrak{A} \vDash \Sigma\}|}{|G|}.$$

Let M be a class of structures closed under isomorphism and denote by M_n the set of structures in M that has universe n. Then $\text{Prob}(M_n \vDash \vartheta)$ is the fraction of all M-structures of size n where ϑ holds. If the limit

$$\lim_{n \to \infty} \text{Prob}(M_n \vDash \vartheta)$$

exists, then this limit is called the asymptotic probability of ϑ in M. If the limit is 1 (or 0), then ϑ is almost surely true (or false) in M.

876. Verify the statements below:

(a) ϑ is almost surely true iff $\neg\vartheta$ is almost surely false.

(b) $\vartheta \wedge \varphi$ is almost surely true iff both ϑ and φ are almost surely true.

(We tacitly assume that all limits in question exist.)

877. Let $\tau = \langle U, c \rangle$ where U is a unary relation symbol and c is a constant symbol. Determine the asymptotic probability of the formula $U(c)$ in the class of all τ-structures.

878. Let τ consists of a single unary function symbol f. Determine the asymptotic probability of the formula $\forall x(f(x) \neq x)$ in the class of all τ-structures.

879. Suppose $\Gamma \subseteq F(\tau)$ is complete and that every $\gamma \in \Gamma$ is almost surely true in M. Prove that every closed $\varphi \in F(\tau)$ is either almost surely true or almost surely false in M.

880. Prove that each $\varphi_{k,\ell}$ from Definition 11.1 is almost surely true in the class of simple graphs. Prove the analogous statement for $\psi_{k,\ell}$ from Definition 11.3 for tournaments.

881 (0–1 law for graphs and tournaments). Every closed formula in the language of graphs is either almost surely true or almost surely false in the class of simple graphs and in the class of tournaments.

882. Let G be a fixed finite simple graph. What is the asymptotic probability that a randomly chosen finite simple graph contains G as a subgraph?

883. Let $\tau = \langle E \rangle$ be the similarity type of graphs. Determine the asymptotic probabilities of the formulas $\forall x \neg E(x, x)$ and $\forall x \forall y (E(x, y) \to E(y, x))$.

The goal of the next few problems is to show that the zero-one law holds in the class of all structures of a given finite relational language.

884. Let τ be a finite similarity type that contains relation symbols only. Construct a set Γ_τ of formulas expressing the following property:

For any finite subset B of the domain, any k-ary relation $R \in \tau$, and any set $P_R \subseteq (B \cup \{y\})^k$ of k-tuples such that every $\vec{a} \in P_R$ contains y, there is $y \notin B$ such that $R(\vec{a})$ holds iff $\vec{a} \in P_R$.

885. Show that Γ_τ from the previous problem has a countable model.

886. Show that Γ_τ is $\aleph_0$-categorical and complete.

887. Prove that each formula in Γ_τ is almost surely true in the class of all τ-structures.

888 (R. Fagin). Show that each formula in a finite relational language is either almost surely true or almost surely false.

SOLUTIONS

12.1   SPECIAL SET SYSTEMS

1. Solution 1. $\mathbb{Q}$ is dense in $\mathbb{R}$, hence for all $\gamma \in \mathbb{R}$ there is an increasing sequence $\{q_i^\gamma\}_{i \in \omega} \subseteq \mathbb{Q}$ converging to γ. Clearly two sequences with different limits can have at most finitely many common members. Thus $\{\{q_i^\gamma\} : \gamma \in \mathbb{R}\}$ is an almost disjoint family of cardinality continuum on $\mathbb{Q}$.

Solution 2. Consider the infinite binary tree. Every branch of it can be coded as function $a : \omega \to 2$: at a node in the ith level we turn left if $a(i) = 1$ and turn right if $a(i) = 0$. Two branches can have at most a finite number of common nodes, therefore the family consisting of branches of the countable set of nodes of an infinite binary tree is an almost disjoint family of cardinality $2^{\aleph_0}$.

Solution 3. Let S_α be an infinite strip of width five on the plane in direction α centered at the origin. Any two of these strips have finite intersection but contain infinitely many lattice points.

Solution 4. Let $a_0 a_1 a_2 \ldots$ be an infinite 0–1 sequence with $a_0 = 1$, and define b_n to be the integer with binary form $a_0 a_1 \ldots a_n$. For each such sequence let B consist of the integers of the form $\prod_{i=0}^{n} p_i^{b_i}$, where p_i is the i-th prime number.

2. (a) Let $\mathcal{F}$ consist of X and all finite subsets of X. Then $\mathcal{F}$ is maximal, and if X is countable so is $\mathcal{F}$.

(b) "The complement of the union of finitely many elements from $\mathcal{F}$ is infinite." Indeed, if this is the case, then let $\mathcal{F} = \{A_i : i \in \omega\}$ and pick b_n from $X - \bigcup_{i<n} A_i$ different from previously chosen b_i's. Then $B \cap A_n \subseteq \{b_0, \ldots, b_n\}$, i.e. $\mathcal{F} \cup \{B\}$ is almost disjoint. Furthermore B is infinite and $B \cap A_n$ is finite, therefore $B \neq A_n$.

(c) We have to maintain the condition in (b). Let $\mathcal{F} = \{A_i : i \in \omega\}$ and pick $b_n \neq c_n$ from $X - \bigcup_{i<n} A_i$ different from previously chosen b_i's and c_i's. Let $B = \{b_i\}$, and $C = \{c_i\}$. Then, as before, $B \neq A_n$ as B is infinite and $B \cap A_n \subseteq \{b_0, \ldots, b_n\}$. Moreover $X - \bigcup_{i<n} A_i - B \supseteq \{c_n, c_{n+1}, \ldots\}$, i.e. this is also infinite.

This construction can be used to define the subsets A_α for $\alpha < \omega_1$ as follows. A_0 is an infinite subset of X whose complement is infinite. If A_β has been

L. Csirmaz and Z. Gyenis, *Mathematical Logic*, Problem Books
in Mathematics, https://doi.org/10.1007/978-3-030-79010-3_12

defined for $\beta < \alpha$ then let A_α be the subset B from the previous construction. $\{A_\alpha : \alpha < \omega_1\}$ will be almost disjoint.

> **Remark.** It is consistent that an almost disjoint family of cardinality ω_1 over a countable set is not maximal. Over a countable set there is an almost disjoint family of cardinality 2^ω, see Problem 1.

3. We may replace κ with $\kappa \times \kappa$. It is enough to construct a sequence of functions $f_\alpha : \kappa \to \kappa$ for $\alpha < \kappa^+$ such that any two differ from a certain point onward. We do this by transfinite recursion. Suppose $\langle f_\beta : \beta < \alpha \rangle$ is already given and enumerate α as $\alpha = \{\alpha_\xi : \xi < \kappa\}$. Now let $f_\alpha(\xi)$ be different from all $f_{\alpha_\zeta}(\xi)$, $\zeta < \xi$. This will be good, since if we have $\beta < \alpha < \kappa^+$, then $\beta = \alpha_\zeta$ for some $\zeta < \kappa$ and then $f_\beta(\xi) \neq f_\alpha(\xi)$ for $\xi > \zeta$.

4. Without loss of generality, we may assume that every set in $\mathcal{F}$ is infinite. Let X be the set of all countable subsets of κ; obviously $|X| = \kappa^\omega$. For each $A \in \mathcal{F}$ let $X_A = \{x \in X : x \subseteq A\}$. As $\mathcal{F}$ is A.D, for different $A, B \in \mathcal{F}$, X_A and X_B are disjoint. Consequently $|\mathcal{F}| \leq |X|$, as was claimed.

Consider the tree of height ω where each node splits into κ siblings. It has κ^ω many branches and κ nodes. Let $\mathcal{F}$ consist of the branches. As any two branches have a finite intersection, this is an almost disjoint family of cardinality κ^ω.

5. Solution 1. Let $A_r = \{p : p$ is a polynomial with rational coefficients and $p(r) > 0\}$ and $\mathcal{F} = \{A_r : r \in \mathbb{R}\}$. Then $\mathcal{F}$ is independent: If $r_1, \ldots, r_n$ are different real numbers and $\varepsilon_1, \ldots, \varepsilon_n$ are 0-1 numbers then there is a polynomial p with rational coefficients such that $p(r_i) > 0$ if and only if $\varepsilon_i = 1$. Such a polynomial can be constructed, e.g., by interpolation.

Solution 2. Let $\{A_\alpha : \alpha < 2^\omega\}$ be a family of infinite subsets of ω such that no one is covered by a finite union of the others. The families created in Problem 1 have this property. The independent family will be defined on the set of finite subsets of ω as follows: $B_\alpha = \{F \subseteq \omega : |F| < \omega, A_\alpha \cap F \neq \emptyset\}$. To show that this family is independent, let $\alpha_1, \ldots, \alpha_n$ and $\beta_1, \ldots, \beta_k$ be pairwise different indices. By assumption for each i there is an element in $A_{\alpha_i} \setminus \bigcup_j A_{\beta_j}$. The set of these elements is a finite $F \subset \omega$ which intersects each A_{α_i} (is an element of B_{α_i}), and disjoint from each A_{β_j} (not in B_{β_j}), as required.

6. It is enough to prove the statement for one specific set of cardinality κ. Suppose κ is infinite and let X be the set of functions that are defined on the powerset of a finite subset of κ and take 0 and 1 only:

$$X = \{f : \wp(D_f) \to 2 : D_f \in [\kappa]^{<\omega}\}.$$

Then the cardinality of X is equal to $\sum_{n \in \omega} \kappa^n \cdot 2^{2^n} = \omega \cdot \kappa = \kappa$. For each subset $A \subseteq \kappa$ put

$$F_A = \{f \in X : f(A \cap D_f) = 1\}.$$

We claim that $\mathcal{F} = \{F_A : A \subseteq \kappa\} \subseteq \wp(X)$ is independent.

Pick different subsets $A_0, \ldots, A_{n-1} \subseteq \kappa$ and 0-1 numbers $\varepsilon_0, \ldots, \varepsilon_{n-1}$. We shall prove that $F_{A_0}^{\varepsilon_0} \cap \cdots \cap F_{A_{n-1}}^{\varepsilon_{n-1}}$ is not empty. Observe that

$$F_{A_0}^{\varepsilon_0} \cap \cdots \cap F_{A_{n-1}}^{\varepsilon_{n-1}} = \{f \in X : f(A_i \cap D_f) = \varepsilon_i, \, i \in n\}.$$

For each different $i, j < n$ pick an arbitrary element $x_{i,j}$ from the symmetric difference of A_i and A_j

$$x_{i,j} \in (A_i \cup A_j) - (A_i \cap A_j),$$

and write $D_f = \{x_{i,j} : i \neq j < n\}$. As $x_{i,j}$ cannot be contained in both A_i and A_j, it is straightforward that $D_f \cap A_i \neq D_f \cap A_j$. Therefore one can define a function $f : \wp(D_f) \to 2$ in such a way that $f(A_i \cap D_f) = \varepsilon_i$ for $i < n$. Then, obviously, $f \in F_{A_0}^{\varepsilon_0} \cap \cdots \cap F_{A_{n-1}}^{\varepsilon_{n-1}}$, as desired.

7. $\{a_1, \ldots, a_n\} \in X_{a_1} \cap \ldots \cap X_{a_n}$.

8. Yes. Let X be the set of the finite subsets of κ, and, for $\alpha \in \kappa$, let $A_\alpha = \{a \in X : \alpha \in a\}$. Then $a \in A_\alpha$ if and only if $\alpha \in a$, i.e. each $a \in X$ is in finitely many A_α only. On the other hand $a = \{\alpha_1, \ldots, \alpha_n\}$ is in $A_{\alpha_1} \cap \cdots \cap A_{\alpha_n}$, therefore $\{A_\alpha\}$ has the finite intersection property (see Problem 7).

9. Solution 1. For a prime p let $A_p = \{p^n q^k : q \neq p \text{ is prime}, 0 < n, k\}$. Then $A_p \cap A_q$ is infinite, as $p^i q^j$ belongs to the intersection for all $i, j \geq 1$, while $A_p \cap A_q \cap A_r$ is empty.

Solution 2. Split X into infinitely many disjoint infinite parts $X_{i,j}$ indexed by the two-element subsets of ω. Let $A_i = \bigcup\{X_{i,j} : j \in \omega, j \neq i\}$. Then $X_{i,j} = A_i \cap A_j$ is infinite, and $A_i \cap A_j \cap A_k$ is empty.

10. Similarly to Problem 9, split X into infinitely many disjoint infinite parts X_a indexed by the n-element subsets of ω. Let $A_i = \bigcup\{X_a : i \in a\}$. Then $A_{i_1} \cap \cdots \cap A_{i_n} = X_{\{i_1, \ldots, i_n\}}$, which is infinite, while $\bigcap_{k=1}^{n+1} A_{i_k} = \emptyset$.

11. Let $X = \aleph_1$ and for $\mathcal{F} \subseteq \wp(X)$ let $\Phi(\mathcal{F})$ be the property that $|\mathcal{F}|$ is countable. Then there is no maximal $\mathcal{F}$ satisfying Φ.

12. (a) Solution 1. Let $0 < \alpha < 1$ be a real number and write it as an infinite sequence of digits: $\alpha = 0.a_1 a_2 \ldots$. The function $f_\alpha : \omega \to \omega$ is defined as follows. $f_\alpha(0) = 1$, and $f_\alpha(k) = 1 a_1 \ldots a_k$ where this sequence is a 10-base decimal number. If α and β differ, then f_α and f_β take different values from somewhere on.

Solution 2. Let $\mathcal{F}$ be an almost disjoint family on ω (see Problem 1). For $A \in \mathcal{F}$ let $f_A : \omega \to A$ be a bijection. Now if $A, B \in \mathcal{F}$ are different, then $|A \cap B|$ is finite, hence $|\text{ran}(f_A) \cap \text{ran}(f_B)|$ is also finite, thus $|\{i \in \omega : f_A(i) = f_B(i)\}| < \aleph_0$.

(b) For $0 < \alpha < 1$ let $f_\alpha(i) = [\alpha i]$, the integer part of αi.

13. We may assume $A_n \supseteq \{0, 1, \ldots, 2^n - 1\}$. For a sequence $s : \omega \to 2$ let f_s be the element of $\prod_{n<\omega} A_n$ such that $f_s(n)$ is the natural number whose binary expansion is given by the first n digits of s. For different 0–1 sequences s and t we have $|f_s \cap f_t|$ is finite.

14. For each $n \in \omega$ let G_n be a *finite* group of size at least 2^n. Let X be the disjoint union $X = \bigcup_n G_n$. X is countably infinite. By Problem 13 there is a set $F \subseteq \prod_{n<\omega} G_n$ of cardinality continuum such that $|f \cap g| < \omega$ whenever $f, g \in F$ are different. For a given $f \in F$ define a permutation π_f of X as follows: for $x \in G_n$ let $\pi_f(x) = x \cdot f(n)$ where $\cdot$ is the multiplication in the group G_n. For distinct $f, g \in F$ the permutations π_f and π_g coincide exactly on elements $x \in G_n$ where $f(n) = g(n)$. This follows from properties of groups as $x \cdot f(n) = x \cdot g(n)$ implies $f(n) = g(n)$. By the choice of F, π_f and π_g coincide on at most finitely many elements. Finally, take any bijection between X and ω to get permutations defined on ω.

> **Remark.** This construction can be modified to obtain a *subgroup* of permutations of cardinality continuum such that distinct members of this subgroup can agree on finitely many places only. Assume that $A_n \subseteq G_n$ contains 2^n elements which freely generate the finite group G_n. Modify the previous construction so that F is a subset of $\prod_{n<\omega} A_n$ instead of $\prod_{n<\omega} G_n$. The rest remains the same. Then the subgroup generated by $\{\pi_f : f \in F\}$ is as desired.

15. We define the functions f_α for $\alpha < \kappa^+$. If $\alpha < \kappa$ then let $f_\alpha(\xi) = \alpha$. If $\alpha \geq \kappa$ then fix a one-to-one function $g_\alpha : \alpha \to \kappa$. If f_β has been defined for all $\beta < \alpha$, then pick $f_\alpha(\xi) < \kappa$ so that it differs from any $f_\beta(\xi)$ where $g_\alpha(\beta) < \xi$. As there are less than κ many such β, one can find an appropriate value. Thus, if $\alpha > \beta$ and $f_\alpha(\xi) = f_\beta(\xi)$ then $g_\alpha(\beta) > \xi$. For fixed α and β there are less than κ many such ξ's.

16. We need 2^κ many functions, so our functions will be indexed by the subsets of κ. Also, the functions will be defined on a set X of cardinality κ rather than on κ itself.

The requirement that the family $\{D(f, g)\}$ has the FIP is equivalent to requiring that given finitely many functions, one can find an $x \in X$ where all functions take different values. Given finitely many subsets $A_1, \ldots, A_n$ of κ, one can find a finite subset D of κ such that $A_1 \cap D, \ldots, A_n \cap D$ are different subsets of D. Now let X be the collection of functions x which map all subsets of a finite subset D_x of κ into different elements of κ, and let $f_A(x) = x(A \cap D_x)$. Clearly X has cardinality κ and if $A_1 \cap D_x, \ldots, A_n \cap D_x$ are all different, then $f_{A_1}(x), \ldots, f_{A_n}(x)$ are different as well.

17. It is enough to construct a family of one-to-one functions f with $\mathrm{dom}(f) = X$, $\mathrm{ran}(f) = Y$ where $|X| = |Y| = \kappa$. We will have $X = \kappa \cup A$ where $|A| = \kappa$ is disjoint from κ, and $Y = \kappa \times \kappa$. Let $\mathcal{F}$ be the family of functions from Problem 16 and for each $f \in \mathcal{F}$ construct the function $g : (A \cup \kappa) \to \kappa \times \kappa$ as follows. When $x \in \kappa$ then $g(x) = \langle x, f(x) \rangle$, and if $x \in A$, then $g(x)$ covers the remaining part of $\kappa \times \kappa$. It is clear that g is one-to-one and preserves the required FIP.

18. For each $A \in \mathcal{D}$ pick a one-to-one function $g_A : \kappa \to A$. Then

$$\left| \{ \xi < \kappa : g_A(\xi) = g_B(\xi) \} \right| \leq |A \cap B| < \kappa.$$

19. Very similar to Problem 16. The functions will be indexed by subsets of κ. Given the subsets $A_1, \ldots, A_n$ one can find a finite subsets D such that $A_i \cap D$ are all different. X will be the set of all functions x whose domain is the power set of a finite subset D_x of κ, and the values $x(A)$ for $A \subseteq D_x$ are different natural numbers from 0 to $2^{|D_x|} - 1$. Trivially, X has cardinality κ. For $A \subseteq \kappa$ let $f_A(x) = x(A \cap D_x)$. Now, if $A_i \cap D_x$ are all different, then $f_{A_i}(x)$ are different, as well.

20. For $f, g \in {}^{\omega}\omega$ we say that $f <^* g$ (f is dominated by g) if there is $n \in \omega$ such that for all $n < m$ we have $f(m) < g(m)$. We shall define an $<^*$-increasing ω_1-sequence of functions. This we do by transfinite recursion. Let f_0 be arbitrary and suppose f_α has already been defined for $\alpha < \beta < \omega_1$. As α is countable there is an enumeration $\{g_n : n \in \omega\}$ of the sequence $\langle f_\alpha : \alpha < \beta \rangle$. Then the function

$$f_\beta(n) = \max\{g_i(n) : i \le n\} + 1,$$

continues the sequence.

> **Remark.** It is consistent that 2^ω is arbitrarily large, and there is an $<^*$-increasing sequence of length 2^ω. It is also consistent that there is no $<^*$-increasing sequence of length ω_2 while 2^ω is arbitrarily large.

21. The union of a chain of filters is a filter, as well, hence applying Zorn's lemma 1.3 to the set $Q = \{\mathcal{G} : \mathcal{F} \subseteq \mathcal{G}, \mathcal{G} \text{ is a filter on } P\}$ ordered by inclusion gives the solution.

22. Let m be a maximal element of P. Then $\mathcal{F} = \{m\}$ is a filter which avoids q. If we do not allow principal filters then $P = \{0, \ldots, n\}$ (ordered by the usual ordering) and $q = n - 1$ is a counterexample: any proper filter should contain q due to upward closedness of P.

23. P will consists of finite partial functions from ω to 2, $P = \{p : \omega \to 2 : |p| < \omega\}$, and for $p, q \in P$ we let $p < q$ if $q \subseteq p$, that is, p extends q as a function. For $n \in \omega$ and $h : \omega \to 2$ let $D_n = \{p \in P : n \in \text{dom}(p)\}$ and $E_h = \{p \in P : \exists n \in \text{dom}(p)(p(n) \ne h(n))\}$. A moment of thought shows that D_n and E_h are dense in P. Let $\mathcal{F}$ be a filter on P and put $f = \bigcup \mathcal{F}$. It can be checked that f is a function. If $\mathcal{F}$ intersects each D_n, then f has as domain all of ω. Similarly, if $\mathcal{F}$ intersects each E_h, then f differs from every function from ω to 2, which is impossible. Therefore $\mathcal{F}$ must avoid some of the E_h's.

24. Using that D_{n+1} is dense, define, by induction on n, a sequence $p_0 \ge p_1 \ge \ldots$ so that $p_0 \in P$ is arbitrary and $p_{n+1} \in D_{n+1}$ is an extension of p_n. Write $\mathcal{F} = \{q \in P : \exists n(q \ge p_n)\}$. Then $\mathcal{F}$ is a filter (generated by the p_n's) which intersects each D_n.

25. For a normal filter $\mathcal{F}$ let $P = \{\mathcal{G} : \mathcal{F} \subseteq \mathcal{G}, \mathcal{G} \text{ is a normal filter on } G\}$ be ordered by inclusion. The union of an increasing chain of P belongs to P hence applying Zorn's lemma we get a maximal normal filter extending $\mathcal{F}$.

26. It is false. Let $G = S_3$ be the symmetric group of degree 3 and let $H_1 = \{(1), (12)\}$, $H_2 = \{(1), (13)\}$ be two conjugate subgroups. If a normal filter $\mathcal{F}$ contains H_1 then it must contain its conjugate H_2 as well. The intersection $H_1 \cap H_2 = \{(1)\}$. But no normal filter can contain the one-element subgroup, hence no normal filter may contain H_1.

27. $\mathcal{F}$ is closed for the intersection since $H_A \cap H_B = H_{A \cup B}$. Also, if $H \in \mathcal{F}$ then $gHg^{-1} \in \mathcal{F}$ as $gH_A g^{-1}$ is the pointwise stabilizer of gA. Finally the one-element subgroup (consisting of the identity permutation) is not in $\mathcal{F}$ as U is infinite.

28. Let $\mathcal{U} \subseteq \wp(X)$ be a set system which is maximal with respect to FIP. We have to check the properties of an ultrafilter. $\emptyset \notin \mathcal{U}$ is evident. If $A \in \mathcal{U}$ and $A \subseteq B$, then $\emptyset \neq A \cap A_0 \cap \ldots \cap A_{n-1} \subseteq B \cap A_0 \cap \ldots \cap A_{n-1}$. It follows that $\mathcal{U} \cup \{B\}$ still has FIP, therefore by maximality $B \in \mathcal{U}$, thus $\mathcal{U}$ is closed upwards. For $A, B \in \mathcal{U}$ we have $A \cap B \neq \emptyset$, consequently $\mathcal{U} \cup \{A \cap B\}$ still has FIP, so by maximality $A \cap B \in \mathcal{U}$. Finally, suppose $X \supseteq A \notin \mathcal{U}$. Then there exists $B \in \mathcal{U}$ such that $A \cap B = \emptyset$. But clearly $(X - A) \cap A = \emptyset$, hence $B \subseteq X - A$ and by upward closedness we have $X - A \in \mathcal{U}$.

29. Let $\mathcal{F} \subseteq \wp(X)$ be a set system with FIP, and let $P = \{\mathcal{H} \subseteq \wp(X) : \mathcal{F} \subseteq \mathcal{H}, \mathcal{H}$ has FIP $\}$. Then $(P, \subseteq)$ is a partially ordered set. If $\mathcal{C}$ is a chain in P, then $\bigcup \mathcal{C}$ still has FIP and it is an upper bound for the chain. Hence, by Zorn's lemma, we might conclude that there is a maximal set system with respect to FIP. By Problem 28 it is an ultrafilter.

30. Let $X = \{x_i : i < n\}$ be finite and let $\mathcal{U}$ be an ultrafilter on X. We show that $\{x_i\} \in \mathcal{U}$ for some $i < n$. For if not, then by maximality, for all $i < n$ we have $X - \{x_i\} \in \mathcal{U}$ and hence $\bigcap_{i<n}(X - \{x_i\}) \in \mathcal{U}$. But this latter intersection is empty; a contradiction.

Next, we show that over any infinite set there is a non-trivial ultrafilter. To this end, let X be infinite and let $\mathcal{F}$ consist of co-finite subsets of X: $\mathcal{F} = \{A \subseteq X : X - A$ is finite$\}$. Clearly $\mathcal{F}$ has the FIP and hence by Problems 29 and 28 it extends to an ultrafilter $\mathcal{U}$. Since for all $x \in X$ we have $X - \{x\} \in \mathcal{U}$, $\mathcal{U}$ is non-trivial.

Finally, suppose $\mathcal{U}$ is an ultrafilter on X and $A = \{a_0, \ldots, a_{n-1}\}$ is a finite subset of X. We show that $A \in \mathcal{U}$ implies that $\{a_i\} \in \mathcal{U}$ for some $i \in n$. For if not, then $X - \{a_i\} \in \mathcal{U}$ for all $i \in n$ and then

$$\bigcap_{i \in n}(X - \{a_i\}) = X - \bigcup_{i \in n}\{a_i\} = X - A$$

belongs to $\mathcal{U}$ which contradicts $A \in \mathcal{U}$.

31. Let $X = \bigcup_{i \in n}^{*} X_i$ be a partition and assume $X_i, X_j \in \mathcal{U}$ for two different i, j. Then by FIP we get $X_i \cap X_j = \emptyset \in \mathcal{U}$ which cannot be the case.

Next, we show that $X_i \in \mathcal{U}$ for some $i \in n$. For if not, then by maximality, for all $i \in n$ we have $X - X_i \in \mathcal{U}$ and by FIP we get $\bigcap_{i \in n}(X - X_i) \in \mathcal{U}$. But $\bigcap_{i \in n}(X - X_i) = X - \bigcup_{i \in n} X_i = X - X = \emptyset$, which contradicts $\emptyset \notin \mathcal{U}$.

For infinite partitions no similar statement holds in general. For example, no member of the partition $\bigcup_{n\in\omega}\{n\}$ is contained in any proper ultrafilter on ω.

32. It does follow. We shall show that $\mathcal{F}$ is closed upwards and closed under intersection. Let $A, B \in \mathcal{F}$ and $A \subseteq C \subseteq X$ be arbitrary.

If C is not a member of $\mathcal{F}$ then, as $C \cup^* (X - C)$ is a partition of X, we get that $X - C$ is in $\mathcal{F}$. Then the partition $(X - C) \cup^* A \cup^* (C - A) = X$ violates our assumption that exactly one member of a partition may belong to $\mathcal{F}$. Therefore $\mathcal{F}$ is closed upwards.

Now suppose $A \cap B \notin \mathcal{F}$. Using upward closedness, $A \cup B \in \mathcal{F}$ and thus $X - (A \cup B) \notin \mathcal{F}$. Considering the partition

$$X = \big(X - (A \cup B)\big) \cup^* (A - B) \cup^* (B - A) \cup^* (A \cap B),$$

we get that exactly one of $A - B$ or $B - A$ is in $\mathcal{F}$, say $A - B$. Then X can be partitioned into $(B - A) \cup^* (X - B) \cup^* (A \cap B)$; no member of this partition belongs to $\mathcal{F}$: a contradiction.

33. Let $\{A_\alpha : \alpha < 2^\kappa\}$ be an independent family of subsets of κ (see Problem 6). For a function $\varepsilon : 2^\kappa \to 2$ let $\mathcal{F}_\varepsilon$ be the following set

$$\mathcal{F}_\varepsilon = \big\{A_\alpha^{\varepsilon(\alpha)} : \alpha < 2^\kappa\big\},$$

which has the FIP, by the independence of the A_α's. Using Problems 28 and 29, $\mathcal{F}_\varepsilon$ can be extended to an ultrafilter $\mathcal{U}_\varepsilon$. For different ε's we get different $\mathcal{U}_\varepsilon$'s, consequently, the number of ultrafilters on κ is at least 2^{2^κ}. Since each ultrafilter on κ is an element of $\wp(\wp(\kappa))$, there are at most 2^{2^κ} many.

34. Let $\mathcal{U}$ be an ultrafilter on ω and pick an arbitrary $A \notin \mathcal{U}$. Let π be any non-trivial permutation of ω that is identical on $\omega - A$. We claim that $\mathcal{U} = \pi\mathcal{U}$.

Choose an arbitrary $B \in \mathcal{U}$ and consider the set $B \cap (\omega - \pi B)$. Since π is identical on $B - A$, it follows that this intersection is contained in A which, by upward closedness of $\mathcal{U}$, implies $B \cap (\omega - \pi B) \notin \mathcal{U}$. Therefore $\omega - \pi B \notin \mathcal{U}$ and hence $\pi B \in \mathcal{U}$.

Applying the same argument to π^{-1} we get that $B \in \mathcal{U}$ if and only if $\pi B \in \mathcal{U}$.

35. Yes. There are 2^{2^ω} many ultrafilters and only 2^ω many permutations.

36. Let $\omega = X_0 \cup \cdots \cup X_n$ be partitioned into $n + 1$ infinite distinct sets. For all $\mathcal{U}_i$ there is some n_i such that $X_{n_i} \in \mathcal{U}_i$. If X is the union of the X_{n_i}'s, then $X \in \mathcal{U}_1 \cap \cdots \cap \mathcal{U}_n$ and X is co-infinite since it is disjoint from the X_j for which $j \notin \{n_1, \ldots, n_n\}$.

37. Let $\mathcal{U}$ be an ultrafilter on X, let $Y \subseteq X$ and denote the trace of $\mathcal{U}$ on Y by $\mathcal{T}$. We claim that

$$\mathcal{T} = \begin{cases} \wp(Y) & \text{if } Y \notin \mathcal{U} \\ \text{an ultrafilter} & \text{if } Y \in \mathcal{U}. \end{cases}$$

Given any subset $A \subseteq Y$ we have $A = ((X - Y) \cup A) \cap Y$, which belongs to $\mathcal{T}$ provided $Y \notin \mathcal{U}$ as in this case $(X - Y) \cup A \in \mathcal{U}$.

Let us assume that $Y \in \mathcal{U}$ and let $A, B \in \mathcal{U}$ be arbitrary. Clearly $A \cap Y \neq \emptyset \notin \mathcal{T}$ and $(A \cap Y) \cap (B \cap Y) = (A \cap B) \cap Y \in \mathcal{T}$ therefore $\mathcal{T}$ is non-trivial and closed under intersection. If $A \cap Y \subseteq B \subseteq Y$ then $B \cap Y = B$ which proves that $\mathcal{T}$ is upward closed. Finally, if $Z \subseteq Y$ does not belong to $\mathcal{T}$ then Z cannot be an element of $\mathcal{U}$ and hence $(X - Z) \cap Y = Y - Z$ in is $\mathcal{T}$.

From the claim it follows that $\mathcal{V} = \{Y \subseteq X : Y \in \mathcal{U}\} = \mathcal{U}$.

38. We claim that $\mathcal{V}$ is an ultrafilter (over dom f) if and only if ran $f \in \mathcal{U}$.

Solution 1. Observe that f is a bijection between dom f and ran f and hence $\mathcal{V}$ is "isomorphic" to the trace of $\mathcal{U}$ on ran f, i.e. $B \subseteq$ dom f is in $\mathcal{V}$ if and only if $f[B]$ belongs to the trace of $\mathcal{U}$ on ran f. Using Problem 37 we get that $\mathcal{V}$ is an ultrafilter (on dom f) if and only if ran $f \in \mathcal{U}$, otherwise $\mathcal{V} = \wp(\text{dom } f)$.

Solution 2 (Without reference to Problem 37). $\mathcal{V}$ is closed under intersection because $f^{-1}(A) \cap f^{-1}(B) = f^{-1}(A \cap B)$ and is upwards closed: If $f^{-1}(A) \subseteq C \subseteq$ dom f for some $A \in \mathcal{U}$, then $A \subseteq f[C]$ hence $f[C] \in \mathcal{U}$ thus, by injectivity of f, we get $f^{-1}(f[C]) = C \in \mathcal{V}$. Maximality follows from the fact that

$$B \in \mathcal{V} \Leftrightarrow B = f^{-1}(A) \text{ for some } A \in \mathcal{U} \text{ such that } f[B] = A \cap \text{ran } f.$$

Similarly, $\emptyset \notin \mathcal{V}$ if and only if ran f intersects each member of $\mathcal{U}$, which is equivalent to ran $f \in \mathcal{U}$.

39. Each homomorphism $h: \mathbf{B} \to 2$ gives rise to an ultrafilter $\mathcal{U}_h = h^{-1}(1) = \{b \in \mathbf{B} : h(b) = 1\}$ and conversely using the ultrafilter $\mathcal{U}$ one can define the map

$$h_{\mathcal{U}}(x) = \begin{cases} 1 & \text{if } x \in \mathcal{U} \\ 0 & \text{otherwise.} \end{cases}$$

which is a homomorphism. Not surprisingly $h_{\mathcal{U}_h} = h$.

40. The map $f: \mathbf{B} \to \wp(\mathbf{B}^*)$, $f(x) = N_x$ is an isomorphism: $N_0 = \emptyset$, $N_{x \wedge y} = N_x \cap N_y$, $N_{x \vee y} = N_x \cup N_y$, $\mathbf{B}^* - N_x = N_{-x}$.

41. $\{N_x : x \in \mathbf{B}\}$ is a basis for a topology because it is closed under intersection: $N_x \cap N_y = N_{x \wedge y}$ and $N_0 = \emptyset$. Open sets are of the form $\bigcup_{x \in \Gamma} N_x$ for some $\Gamma \subseteq \mathbf{B}$. Since $\mathbf{B}^* - N_x = N_{-x}$, this basis consists of closed-open sets, thus $\mathbf{B}^*$ is zero dimensional.

Suppose $\mathcal{U}, \mathcal{V} \in \mathbf{B}^*$ are different. Then there is some $x \in \mathbf{B}$ such that $x \in \mathcal{U}$ and $-x \in \mathcal{V}$ which means $\mathcal{U} \in N_x$ and $\mathcal{V} \in N_{-x}$. Hence the disjoint open sets N_x and N_{-x} separate $\mathcal{U}$ and $\mathcal{V}$, therefore $\mathbf{B}^*$ is Hausdorff.

Next, we prove that $\mathbf{B}^*$ is compact. Let F be a family of closed sets having the FIP. We have to show that $\bigcap F$ is not empty. As $\mathbf{B}^* - N_x = N_{-x}$, each closed set is an intersection of basic open sets, hence, by replacing F with the family of those basic open sets that are contained in some elements of F, without loss of generality we may assume that F consists of basic open sets $F = \{N_x : x \in \Gamma\}$, where $\Gamma \subseteq \mathbf{B}$. Then Γ has the FIP, and thus there is an ultrafilter $\mathcal{U}$ containing Γ. Now, for any $N_x \in F$ we have $x \in \Gamma \subseteq \mathcal{U}$, equivalently $\mathcal{U} \in N_x$, consequently $\mathcal{U} \in \bigcap F$.

42. If $\mathcal{U}$ is trivial then $\mathcal{U} = \mathcal{U}_a$ for an atom $a \in \mathbf{B}$, but the only element of N_a is $\mathcal{U}$, hence $\mathcal{U}$ is isolated. Conversely, if $\mathcal{U}$ is isolated, say $\mathcal{U}$ is the only element of N_x, then $\mathcal{U}$ is the unique ultrafilter which extends the filter $\mathcal{F}_x$ concentrated on x. This means $\mathcal{F}_x$ is an ultrafilter and thus $\mathcal{F}_x = \mathcal{U}$.

It follows that there is some non-trivial ultrafilter on $\mathbf{B}$ provided $\mathbf{B}^*$ is infinite. Namely, since $\mathbf{B}^*$ is compact (Problem 41) and infinite, it has an accumulation point. Such a point is not isolated.

43. (a) For any non-trivial ultrafilter $\mathcal{U}$ on X, the family $\mathcal{U} \cup \{\emptyset\}$ defines a topology on X in which $\mathcal{U}$ converges to every point $p \in X$.

(b) Let X be the countable discrete topology (in which every set is open). As no non-trivial ultrafilter can contain a finite set and each point has a finite neighbourhood, no non-trivial ultrafilter converges. Note that each trivial ultrafilter converges to the point it is concentrated on, regardless of topology.

44. (a) If both p and q are limit points of $\mathcal{U}$, then they cannot have disjoint neighborhoods as in this case $\emptyset$ would be in $\mathcal{U}$, therefore X is not Hausdorff.

Conversely, suppose p and q do not have disjoint neighborhoods. Then the family $N(p) \cup N(q)$ of neighbours of p and q has the FIP, and hence it extends to an ultrafilter (see Problems 28, 29). This ultrafilter converges to both p and q.

(b) Suppose X is compact and let $\mathcal{U}$ be an ultrafilter on X. Since each finite intersection of (closed) elements of $\mathcal{U}$ is not empty, by compactness of X, the intersection of all closed sets of $\mathcal{U}$ is not empty. Pick an element x from this intersection. Since x is in the closure of every element of $\mathcal{U}$, every neighbourhood N of x meets every member of $\mathcal{U}$. This means $N \in \mathcal{U}$ since else $X - N$ would be an element of $\mathcal{U}$ and clearly N does not meet $X - N$. (To put it another way $\{A \cap B : A \in \mathcal{U}, B \in N(x)\}$ is a proper filter on X, which is also maximal, hence this filter is $\mathcal{U}$ itself, thus $N(x) \subseteq \mathcal{U}$).

Conversely, let $\{U_\alpha : \alpha \in A\}$ be an open cover of X with no finite subcover. Then the family $\{X - U_\alpha : \alpha \in A\}$ has the FIP, hence extends to an ultrafilter $\mathcal{U}$ (see Problems 28, 29). As $\mathcal{U}$ does not contain any of the U_α's, it cannot have any limit points in $\bigcup_\alpha U_\alpha = X$.

45. (a) Denote by $X \oplus i$ the set where the containment of i is "swapped," namely, it is $X - \{i\}$ when $i \in X$, and $X \cup \{i\}$ when $i \notin X$. Observe that if $\{i\} \notin \mathcal{U}$ then $X \in \mathcal{U}$ and $X \oplus i \in \mathcal{U}$ are equivalent.

If $\mathcal{U}$ is non-trivial, then for every $X \subseteq \omega$ and finite F there is an $Y \in \mathcal{U}$ such that $X \cap F = Y \cap F$, namely, $Y = (\omega - F) \cup (X \cap F)$ will do. It follows that all subsets of ω are in the closure of $\mathcal{U}$. If $\mathcal{U}$ is generated by the one-element set $\{i\}$, then all subsets of ω containing i (and only these subsets) are in the closure of $\mathcal{U}$.

(b) Let $X \in \mathcal{U}$, F be a finite set, and $i \notin F$ be so that $\{i\} \notin \mathcal{U}$. Then $Y = X \oplus i \in \mathcal{U}$, and $X \cap F = Y \cap F$, i.e. $\mathcal{U}$ is dense in itself.

46. $\mathbb{R} = \bigcup_{n \in \omega} (-n, n)$.

47. Take any partition A_n $(n < \omega)$ of ω into infinite sets. Let

$$\mathcal{V} = \{B \subset \omega : |B \cap A_i| < \omega \text{ for all but finitely many } i \in \omega\}.$$

The complements of the sets in $\mathcal{V}$ have the FIP, thus there is an ultrafilter which does not contain any element from $\mathcal{V}$. Then each $A_k \notin \mathcal{U}$.

48. As the sequence is bounded, there are $a < b \in \mathbb{R}$ such that for all $n \in \omega$, $x_n \in [a, b]$. Suppose on the contrary that there is no ultralimit. Then for every $r \in [a, b]$ there is $\varepsilon_r > 0$ such that $\{n : |x_n - r| < \varepsilon_r\}$ is not in $\mathcal{U}$. As $[a, b]$ is compact, there are finitely many such r's so that $X_r = \{n : |x_n - r| < \varepsilon_r\}$ is a partition of ω. But any finite partition contains a set in $\mathcal{U}$, which yields a contradiction.

The ultralimit clearly is an accumulation point, thus there must be a subsequence converging to it.

Finally, take $\mathcal{U}$ and a partition A_n as in Problem 47. Let $x_n = 1/k$ if $n \in A_k$. As there is no $X \in \mathcal{U}$ that meets each A_k in a finite set, on each $X \in \mathcal{U}$ the subsequence $(x_n)_{n \in X}$ does not converge to 0.

49. A is not Lebesgue measurable. By way of reaching a contradiction suppose otherwise. As exactly one of X and $\omega \smallsetminus X$ is in $\mathcal{U}$, exactly one of a and $1 - a$ is in A with the exception of countably many dyadic numbers. If A is measurable so is $1 - A$ and has the same measure; they are almost disjoint, and their union is the unit interval. Consequently, A has measure $1/2$.

As $\mathcal{U}$ is non-trivial ultrafilter, the same reasoning shows that in any dyadic interval $[j/2^n, (j+1)/2^n]$ the relative measure of A is $1/2$. As every interval of length d contains a dyadic interval of length at most $d/4$, the relative measure of A in any interval is at most $7/8$. But this contradicts Lebesgue's density theorem which says that a positive measure set has density arbitrarily close to 1 in some interval.

50. No such group and ultrafilter exist. Let $a \in G$ arbitrary, and H be the subgroup generated by a. Pick one element from each equivalence class in G/H; let this set be X. The sets $\{g + X : g \in H\}$ are disjoint, and their union is G. If H is finite, then at least one member of this family is in the ultrafilter, but all of them cannot be there.

If H is infinite, then let $Y = \bigcup\{k \cdot a + X : k \text{ is an even integer}\}$. Now Y and $a + Y$ are disjoint, their union is G, thus exactly one of these sets is in the ultrafilter.

51. If $\mathcal{U}$ and $\mathcal{V}$ are ultrafilters on G then so is $\mathcal{U} + \mathcal{V}$:

$$\begin{aligned}
\emptyset \in \mathcal{U} + \mathcal{V} \quad &\Leftrightarrow \quad \{a : \emptyset - a \in \mathcal{U}\} \in \mathcal{V} \quad \Leftrightarrow \quad \emptyset \in \mathcal{V}. \\
A, B \in \mathcal{U} + \mathcal{V} \quad &\Leftrightarrow \quad \left. \begin{cases} \{a : A - a \in \mathcal{U}\} \in \mathcal{V} \\ \{a : B - a \in \mathcal{U}\} \in \mathcal{V} \end{cases} \right\} \\
&\Leftrightarrow \quad \{a : (A - a) \cap (B - a) \in \mathcal{U}\} \in \mathcal{V}
\end{aligned}$$

$$\Rightarrow \quad \{a : (A \cap B) - a \in \mathcal{U}\} \in \mathcal{V}$$

$$\Leftrightarrow \quad A \cap B \in \mathcal{U} + \mathcal{V}.$$

$$A \in \mathcal{U} + \mathcal{V}, A \subseteq B \quad \Rightarrow \quad \{a : A - a \in \mathcal{U}\} \in \mathcal{V}$$

$$\Rightarrow \quad \{a : A - a \subseteq B - a \in \mathcal{U}\} \in \mathcal{V}$$

$$\Rightarrow \quad B \in \mathcal{U} + \mathcal{V}.$$

$$A \notin \mathcal{U} + \mathcal{V} \quad \Leftrightarrow \quad \{a : A - a \in \mathcal{U}\} \notin \mathcal{V}$$

$$\Leftrightarrow \quad G \smallsetminus \{a : A - a \in \mathcal{U}\} \in \mathcal{V}$$

$$\Leftrightarrow \quad \{a : A - a \notin \mathcal{U}\} \in \mathcal{V}$$

$$\Leftrightarrow \quad \{a : (G \smallsetminus A) - a \in \mathcal{U}\} \in \mathcal{V}$$

$$\Leftrightarrow \quad G \smallsetminus A \in \mathcal{U} + \mathcal{V}.$$

We note that the same result holds when G is a semigroup.

52. (a) For $g \in G$ let $\mathcal{U}_g$ be the trivial ultrafilter concentrated on g and suppose $\mathcal{V}$ is non-trivial. Observe first that

$$A \in \mathcal{V} + \mathcal{U}_g \quad \Leftrightarrow \quad \{x \in G : A - x \in \mathcal{V}\} \in \mathcal{U}_g$$

$$\Leftrightarrow \quad g \in \{x \in G : A - x \in \mathcal{V}\}$$

$$\Leftrightarrow \quad A - g \in \mathcal{V}.$$

$$A \in \mathcal{U}_g + \mathcal{V} \quad \Leftrightarrow \quad \{x \in G : A - x \in \mathcal{U}_g\} \in \mathcal{V}$$

$$\Leftrightarrow \quad \{x \in G : g \in A - x\} \in \mathcal{V}$$

$$\Leftrightarrow \quad A - g \in \mathcal{V}.$$

Consequently $\mathcal{U}_g + \mathcal{V} = \mathcal{V} + \mathcal{U}_g$, in particular $\mathcal{U}_a + \mathcal{U}_b = \mathcal{U}_{a+b}$. However, $\mathcal{U} + \mathcal{V} = \mathcal{V} + \mathcal{U}$ does not, in general, hold.

In order to construct counterexamples $\mathcal{U}$ and $\mathcal{V}$, we have to guarantee the existence of disjoint sets E and F such that $E \in \mathcal{U} + \mathcal{V}$ and $F \in \mathcal{V} + \mathcal{U}$. Suppose G is countably infinite and pick two infinite subsets $X = \{x_0, x_1, \ldots\}$ and $Y = \{y_0, y_1, \ldots\}$ of G. Furthermore, let $X_i = \{x_i, x_{i+1}, \ldots\}$ and $Y_i = \{y_i, y_{i+1}, \ldots\}$ be the initial segments respectively of X and Y. The families

$$\{A \subseteq G : X - A \text{ is finite}\}, \quad \text{and} \quad \{A \subseteq G : Y - A \text{ is finite}\}$$

both have the F.I.P. and hence they can be extended to ultrafilters $\mathcal{U}$ and $\mathcal{V}$. Obviously $X_i \in \mathcal{U}$ and $Y_i \in \mathcal{V}$ for all $i \in \omega$. It is enough to find disjoint subsets E and F such that $E - y_i \in \mathcal{U}$ and $F - x_i \in \mathcal{V}$ as in this case

$$Y \subseteq \{y : E - y \in \mathcal{U}\} \in \mathcal{V}, \quad \text{and} \quad X \subseteq \{x : F - x \in \mathcal{V}\} \in \mathcal{U},$$

and therefore $E \in \mathcal{U} + \mathcal{V}$ and $F \in \mathcal{V} + \mathcal{U}$. By letting $E = \{x_i + y_j : i \geq j\}$ and $F = \{x_i + y_j : i < j\}$ we get that $E - y_i \supseteq X_i \in \mathcal{U}$ and $F - x_i \supseteq Y_{i+1} \in \mathcal{V}$.

So it remained to find a suitable group G and two infinite subsets X and Y of G such that

$$\{x_i + y_j : i \geq j\} \cap \{x_i + y_j : i < j\} = \emptyset,$$

and it is not hard to check that $G = (\mathbb{Z}, +)$, $X = \{2^{2n} : n > 0\}$ and $Y = \{2^{2n+1} : n > 1\}$ is suitable.

(b) Associativity always holds

$$A \in (\mathcal{U} + \mathcal{V}) + \mathcal{W} \quad \Leftrightarrow \quad \{s : A - s \in \mathcal{U} + \mathcal{V}\} \in \mathcal{W}$$
$$\Leftrightarrow \quad \{s : \{t : A - s - t \in \mathcal{U}\} \in \mathcal{V}\} \in \mathcal{W}$$
$$\Leftrightarrow \quad \{s : \{t : A - t \in \mathcal{U}\} - s \in \mathcal{V}\} \in \mathcal{W}$$
$$\Leftrightarrow \quad \{t : A - t \in \mathcal{U}\} \in \mathcal{V} + \mathcal{W}$$
$$\Leftrightarrow \quad A \in \mathcal{U} + (\mathcal{V} + \mathcal{W}).$$

53. Observe that if either $\mathcal{U}$ or $\mathcal{V}$ is non-trivial then so is $\mathcal{U} + \mathcal{V}$. For if $\{g\} \in \mathcal{U} + \mathcal{V}$, then $\{x : \{g - x\} \in \mathcal{U}\}$ would be either empty (if $\mathcal{U}$ is non-trivial) or a one-element set (if $\mathcal{U}$ is trivial). Consequently, for a non-trivial $\mathcal{U}$ and a trivial $\mathcal{V}$ there is no solution of the equations $\mathcal{U} + \mathcal{W} = \mathcal{V}$ and $\mathcal{W} + \mathcal{U} = \mathcal{V}$.

54. Proceed by induction on k. Let $f : [\omega]^k \to n$ be a coloring and let $\mathcal{U}$ be a non-trivial ultrafilter on ω. Using $\mathcal{U}$ we obtain a coloring $f' : [\omega]^{k-1} \to n$ which is defined as

$$f'(\{i_1, \ldots, i_{k-1}\}) = c \quad \Leftrightarrow \quad \{j \in \omega : f(\{i_1, \ldots, i_{k-1}, j\}) = c\} \in \mathcal{U}.$$

Note, that because n is finite, exactly one of the partitions belongs to $\mathcal{U}$, thus f' is well defined. Now, by our inductive hypothesis there exists an infinite $S \subseteq \omega$ for which f' is homogeneous for some color. Let us say $f'|_S = c$ for a fixed $c \in n$. We may assume $S \in \mathcal{U}$, since it is trivial in the case $k = 1$, and this property remains true in any step of the induction. Now $f'(\{i_1, \ldots, i_{k-1}\}) = c$ means that $\{j \in \omega : f(\{i_1, \ldots, i_{k-1}, j\}) = c\} = S_0 \in \mathcal{U}$, hence $S \cap S_0 \in \mathcal{U}$. Then $f|_{S \cap S_0} = c$, as desired.

55. It is enough to prove that pairs of $^\kappa 2$ can be colored by κ colors such that no triangle becomes homogeneous. Define the coloring $F : [^\kappa 2]^2 \to \kappa$ as follows

$$F(\{f, g\}) = \text{ the least } \alpha < \kappa \text{ such that } f(\alpha) \neq g(\alpha).$$

Then, for distinct $f, g, h \in {}^\kappa 2$, it is impossible to have $F(\{f, g\}) = F(\{f, h\}) = F(\{g, h\})$.

56. We claim first that there is no increasing or decreasing chain of reals of length ω_1. For if $\langle r_\alpha : \alpha < \omega_1 \rangle$ were an increasing (decreasing) chain then there would exist ω_1 many different rationals $r_\alpha < q_\alpha < r_{\alpha+1}$ which is clearly impossible.

If there were a homogenous set S of cardinality $\aleph_1$, then one could select an increasing (if S is red) or decreasing (if S is blue) sequence of reals from S of order type ω_1; a contradiction.

> **Remark.** A similar construction shows that the edges of the complete graph on 2^κ the vertices can be colored by two colors so that it has no homogeneous subsets of size κ^+.

57. Since κ is infinite, there exists a bijection $f : \kappa \to [\kappa]^{<\omega}$. For all $\alpha < \kappa$, write

$$e_\alpha = \{\beta \in \kappa : \alpha \in f(\beta)\}.$$

Then the family $E = \{e_\alpha : \alpha < \kappa\}$ has the FIP, thus it generates an ultrafilter $\mathcal{U}$ (see Problems 28, 29). Now, E witnesses that $\mathcal{U}$ is regular: for $\xi \in \kappa$ we have $\xi \in e_\alpha$ if and only if $\alpha \in f(\xi)$, hence $\{e \in E : \xi \in e\} = \{e_\alpha : \alpha \in f(\xi)\}$ is finite.

58. (a) Let $\mathcal{U}$ be regular with $E \subseteq \mathcal{U}$ such that $|E| = \kappa$ and $f(\xi) = \{e \in E : \xi \in e\}$ is finite for all $\xi < \kappa$. Suppose $A \subseteq \kappa$ and $|A| < \kappa$. This implies $|\bigcup_{a \in A} f(a)| < \kappa$, therefore there exists some $e \in E - \bigcup_{a \in A} f(a)$. Since $e = \{\xi \in \kappa : e \in f(\xi)\} \subseteq \kappa - A$, we get that $\kappa - A \in \mathcal{U}$, or equivalently $A \notin \mathcal{U}$.
  (b) $E = \{\{n, n+1, n+2, \ldots\} : n \in \omega\}$.

59. (a) For all cardinals $\omega \le \lambda < \kappa$ we construct an ultrafilter on κ which has an element of cardinality λ. Partition κ into λ infinite parts: $\kappa = \bigcup^*_{\alpha < \lambda} X_\alpha$ and let $\mathcal{V}$ be an arbitrary non-trivial ultrafilter on λ. Define the family W of subsets of κ as

$$W = \Big\{ \bigcup_{\alpha \in V} X_\alpha : V \in \mathcal{V} \Big\}.$$

Clearly, W has the FIP (in fact, closed under intersection; but not closed upwards). Choose arbitrary elements $x_\alpha \in X_\alpha$ and put $H = \{x_\alpha : \alpha < \lambda\}$. Then $W \cup \{H\}$ also has the FIP and hence it extends to an ultrafilter $\mathcal{U}$, which cannot be uniform as $H \in \mathcal{U}$. Finally, note that for each $\alpha \in \kappa$ there is $A \in W$ so that $\alpha \notin A$ (this follows from non-triviality of $\mathcal{V}$), therefore $\mathcal{U}$ cannot be trivial.
  (b) The ultrafilter constructed in (a) is non-regular by Problem 58 (a).

> **Remark.** The statements "every uniform ultrafilter is regular" and "there exist uniform non-regular ultrafilters over each successor of a regular cardinal" are both consistent relative to ZFC.

60. Fix a regular ultrafilter $\mathcal{U}$ with $E \subseteq \mathcal{U}$ such that $|E| = \kappa$ and $f(\xi) = \{e \in E : \xi \in e\}$ is finite for all $\xi < \kappa$. For a filter $\mathcal{F}$ (on κ) the set system $\{X \times Y : X \in \mathcal{U}$ and $Y \in \mathcal{F}\}$ has the FIP, thus extends to an ultrafilter $\mathcal{F}^+$ which is regular, witnessed by $F = \{e \times \kappa : e \in E\}$: for each $\xi, \zeta < \kappa$ the set $\{e \times \kappa \in F : \langle \xi, \zeta \rangle \in e \times \kappa\}$ has cardinality $f(\xi)$.

  Using Problem 33 and that $\mathcal{F}_1 \ne \mathcal{F}_2$ implies $\mathcal{F}_1^+ \ne \mathcal{F}_2^+$, we get 2^{2^κ} different regular ultrafilters (on $\kappa \times \kappa$).

12.2   GAMES AND VOTING

61. (a) Player I has a simple winning strategy: just pick the least odd number that has not already been picked.

(b) Again, player I can always win: he can pick twenty primes on his first move.

(c) Player II can always win by always picking the least even number available.

62. This is the so-called strategy stealing argument.

(a) Suppose, by contradiction, that II has a winning strategy S. The first player I will use S to win the game as follows. He pretends to be the second player; picks an arbitrary element $i_0 \in \omega$ as the first move of the opponent, consults S for the reply, and takes that element. Now player II answers by taking i_1. If i_0 and i_1 differ, then I consults S for the next move, otherwise takes another untaken element from ω and considers it as the opponent's move. During the game S never advises a number which has been taken, thus the set of elements taken by I belongs to the ultrafilter, i.e. I wins the game.

> **Remark.** The argument does not exclude the possibility that player II has a drawing strategy, i.e. a strategy which prevents I from winning the game.

(b) Suppose player I has a winning strategy S. Player II will use S to win the game. Player I pick $i_0 \in \omega$ and player II pretends that she is player I following S and she chose i_0; pick an element $i_1 \in \omega$ as the reply of the opponent, and consult S for the reply which she takes. Following this strategy her points together with i_0 is in U because S always wins. As U is non-principal, her points are in U, thus she wins the game. But this is impossible if player I has a winning strategy.

63. We show that if I plays by a strategy, II can win against it; the other case is similar. In each of the three games the first choice of I is $a_1 \in \omega$, fixed by the strategy. In the subsequent moves II chooses one of the games, takes a move there, and then I responds in that game by the strategy. II will ensure that she takes every element of $\omega \smallsetminus \{a_1\}$ in one of the games. As U is non-trivial, surely she will win one of the games.

She can do it. If at his next move I picks a number which he also has taken in another game, then II chooses this number in the third game (forced move). Otherwise she chooses the smallest number which has not been claimed by her in any of the games, and takes it. In n rounds she has at most $n/2$ forced moves, so she has plenty of moves to take every element of ω.

64. (a) Let $A = [0, 1)$ and both players pick $a_n = 9$ at each round n. After each round player II does not have a winning strategy, for if player I picks any other digit than 9, then II loses. Yet, player I will lose as $0.\bar{9} \notin A$.

(b) Suppose that I has no winning strategy. Then II can play such that after her steps I still has no winning strategy (this is called the *defensive strategy*). If A is closed, then this strategy of II is in fact a winning strategy. For, suppose the real number $r = 0.a_1 a_2 \ldots$ has been selected. We need to see that $r \notin A$. By way of contradiction, if $r \in A$, then by closedness of A there exists an

$\varepsilon > 0$ such that the interval $(r - \varepsilon, r + \varepsilon)$ is contained in A. Thus there is a natural number N such that each real number the digits of which starts with $0.a_1 a_2 \ldots a_N$ belongs to A. Thus, after the N-th step player I has the winning strategy "play any number whatsoever"—a contradiction. (See also Solution $65(d)$).

> **Remark** (Martin, 1975). For every Borel set A one of the players has a winning strategy in the game $G(A)$.

(c) Enumerate $A = \{r_i : i < \omega\}$. At the i-th round II plays any number which is different from the $(2i)$-th digit of r_i.

> **Remark.** Similar argument shows that if $|A| < 2^\omega$, then Player I cannot have a winning strategy.

(d) Enumerate all the possible strategies and by transfinite induction construct a set that none of the strategies are winning strategies for either player. Each strategy S can be identified with a function $S : 10^{<\omega} \to 10$, therefore there are only 2^ω strategies. Let $\{\sigma_\alpha : \alpha < 2^\omega\}$ and $\{\tau_\alpha : \alpha < 2^\omega\}$ be an enumeration of the strategies of I and II, respectively. For a strategy S of Player I (player II) let $P(S)$ denote the set of all possible infinite plays (i.e. the real numbers $0.a_1 a_2 \ldots$) in which I (player II) plays according to S. It is clear that $P(S)$ has cardinality continuum.

By transfinite recursion define the sets $A = \{a_\alpha : \alpha < 2^\omega\}$ and $B = \{b_\alpha : \alpha < 2^\omega\}$ as follows. For $\alpha = 0$ pick any $a_0 \in P(\tau_0)$. As $P(\sigma_0)$ contains more than one element, pick $b_0 \in P(\sigma_0)$ such that $b_0 \neq a_0$. In the inductive step suppose the elements a_β and b_β have already been chosen for $\beta < \alpha$. Then $\{b_\beta : \beta < \alpha\}$ has cardinality smaller than continuum, thus the set $P(\tau_\alpha) \smallsetminus \{b_\beta : \beta < \alpha\}$ is non-empty. Pick any element of this set and call it a_α. Similarly pick any b_α from the set $P(\sigma_\alpha) \smallsetminus (\{a_\beta : \beta < \alpha\} \cup \{a_\alpha\})$.

It is not hard to check that A and B are disjoint. We claim that in the game $G(A)$ no player has a winning strategy. Indeed, suppose σ is a winning strategy for I. Then $P(\sigma) \subseteq A$ and $\sigma = \sigma_\alpha$ for some $\alpha < 2^\omega$. At the α-th stage of the induction we picked $b_\alpha \in P(\sigma_\alpha)$ and this b_α cannot be in A as $A \cap B = \emptyset$.

Assuming II has a winning strategy τ we get $P(\tau) \cap A = \emptyset$. There is an α such that $\tau = \tau_\alpha$ but at stage α of the recursion we picked $a_\alpha \in P(\tau_\alpha)$.

65. (a) The argument is similar to $64(c)$. Every branch can be identified with a function $s : \omega \to 2$. Enumerate $A = \{s_\alpha : \alpha < \omega\}$. Before the i-th move of II the token sits on a node. If s_i meets that node, then II moves the token in the direction to avoid the branch s_i, otherwise moves the token arbitrarily.

(b) Let τ be a strategy for II. We define a countable set $\bar{A}$ of branches so that I can travel any branch not in $\bar{A}$ while playing against the strategy τ. To this end assign a branch $b(p)$ to each partial play p (played against the strategy τ where the next move belongs to I) as follows. Suppose the token is at node s. For the "no move" move of I τ answers by $a_0 \in \{0, 1\}$ (going left or right). For the $\bar{a}_0$ move (go to the opposite direction) of I, τ answers by a_1. For the $\bar{a}_0 \bar{a}_1$

move τ answers by a_2, etc. The assigned branch is $s\bar{a}_0\bar{a}_1\bar{a}_2\cdots$. $\bar{A}$ is the set of all branches $b(p)$ for all partial plays p. As there are countably many partial plays, this is a countable set.

Let b be a branch not in $\bar{A}$, and p_0 be the empty partial play. As $b \neq b(p_0)$, there is a one-round partial play p_1 ending in an initial segment of b. As $b \neq b(p_1)$, there is a two-round partial play p_2 ending in a longer initial segment of b, etc. That is, the token travels along branch b, as claimed.

(c) Suppose I has a winning strategy σ in the game $G(A)$. Let $P(\sigma)$ denote the set of all possible infinite plays (branches) in which I plays according to σ. As σ is a winning strategy we have $P(\sigma) \subseteq A$. Suppose I plays according to σ. At any step II can move the token in two directions. These different directions necessarily determine different branches, and when the game finishes all these branches must be in $P(\sigma)$. Therefore $|P(\sigma)| \geq 2^\omega$.

Remark. If the continuum hypothesis holds (that is $\aleph_1 = 2^{\aleph_0}$), then any infinite set A such that $|A| < 2^{\aleph_0}$ is countable, hence, by (a), Player II has a winning strategy in the game $G(A)$. Both players cannot have winning strategies for the same game, therefore I must not have a winning strategy.

(d) Similar to Solution 64(b). Suppose I has no winning strategy. Then II can play such that after her steps I still has no winning strategy. Call this defensive strategy of II τ. We claim that τ is in fact a winning strategy for II. Suppose the branch $x \in B$ is the result of the infinite play (II following strategy τ). We need to see $x \notin A$. If $x \in A$, then there is an initial segment s of x such that every (full) branch that extends x must be in A. But then player I would have a trivial winning strategy: after reaching s move the token in any direction whatsoever.

66. Suppose I wins against this strategy. There is a minimal k such that the sequence $b_1, \ldots, b_k$ chosen by I contains a finite subset in $\mathcal{F}$. Then for all $n \geq k$, $S_n(b_1, \ldots, b_{k-1}) = b_k$ (and also, there could be no other element $b \neq b_k$ so that $\{b_1, \ldots, b_{k-1}, b\}$ covers a winning set in $\mathcal{F}$) because according to the strategy S_n, player I cannot win before the n-th move. But then $I_{b_k} \in U$ (as the complement of I_{b_k} is finite), therefore after the $(k-1)$-st round in the game, player II has picked b_k, which yields a contradiction.

67. (a) Argument similar to Solution 65(d). The result of each infinite play (a full evaluation of the propositional variables) can be identified with a function $f : \omega \to 2$ (which can be considered as a branch of the complete infinite binary tree). Let A denote the set of evaluations that make Γ false. Then A has the property described in Problem 65(d): if $f \in A$ then there is $\gamma \in \Gamma$ such that $f(\gamma)$ is false. Let γ use the first n variables only. Then every evaluation g that agrees with f up to the first n values makes γ false. (In other words, every branch of the infinite binary tree that extends the initial segment $f \restriction n$ falsifies Γ). By Problem 65(d) one of the players has a winning strategy.

(b) Take $\Gamma = \{A_1, A_2\}$. Clearly Γ is satisfiable, nevertheless II can set either A_1 or A_2 be false.

68. (a) We make use of the following observation: if no vote goes for x, then the result cannot be x. More precisely, let $f : M \to V$ be a voting such that for all $m \in M$, $f(m) \neq x$. Then $v(f) \neq x$ (provided v is a fair voting). For, let $c_x(m) = x$ for all $m \in M$. Then $v(c_x) = x$ and since f and c_x differ everywhere, $v(f)$ cannot be x.

1. If $f : M \to V$ is a voting where none of the members voted for x, then $v(f) \neq x$, thus $\emptyset \notin \mathcal{F}$.

2. Suppose $A, B \in \mathcal{F}$, witnessed by the votings f_A and f_B: $v(f_A) = a$ and $f_A(i) = a$ iff $i \in A$; similarly $v(f_B) = b$ and $f_B(i) = b$ iff $i \in B$. Write

$$g_A(i) = \begin{cases} b & \text{if } i \in A \\ a & \text{otherwise,} \end{cases} \qquad g_B(i) = \begin{cases} a & \text{if } i \in B \\ b & \text{otherwise.} \end{cases}$$

Then f_A differs everywhere from g_A, thus $v(f_A) \neq v(g_A)$. By the observation above we obtain $v(g_A) = b$. Similarly, $v(g_B) = a$. Finally, pick $x \in V \smallsetminus \{a, b\}$ (such exists as $|V| \geq 3$) and write

$$h(i) = \begin{cases} a & \text{if } i \in A \smallsetminus B \\ x & \text{if } i \in A \cap B \\ b & \text{if } i \in B \smallsetminus A. \end{cases}$$

Then h differs from g_A and g_B everywhere, thus $v(h)$ must be x. This shows $A \cap B \in \mathcal{F}$.

3. Say $A \in \mathcal{F}$ is witnessed by $f : M \to V$ in that $f(i) = a$ iff $i \in A$ and $v(f) = a$. Take any $B \supseteq A$ and write

$$g(i) = \begin{cases} b & \text{if } i \in A \\ a & \text{otherwise} \end{cases} \qquad h(i) = \begin{cases} a & \text{if } i \in B \\ b & \text{otherwise} \end{cases}$$

As f and g differ everywhere, we get $v(g) = b$, and since g and h differs everywhere, we obtain $v(h) = a$. This latter shows $B \in \mathcal{F}$.

4. Let

$$f(i) = \begin{cases} a & \text{if } i \in A \\ b & \text{otherwise} \end{cases}$$

If $A \notin \mathcal{F}$, then $v(f) \neq a$, thus $v(f) = b$, hence $M \smallsetminus A \in \mathcal{F}$.

(b) If M is finite, then the ultrafilter $\mathcal{F}$ in (a) must be principal. Therefore there is $m \in M$ such that $A \in \mathcal{F}$ iff $m \in A$. The vote of m determines the outcome.

(c) Let M consists of an odd number ≥ 3 of board members, let $V = \{0, 1\}$ and for a vote $f : M \to V$ define $v(f) = \sum_{i \in M} f(i)$ modulo 2.

Then v is a fair voting, however it is not autocratic. $\mathcal{F}$ is not an ultrafilter, for example, because $\mathcal{F}$ contains all singletons $\{m\}$ for $m \in M$.

69. Take any ultrafilter $\mathcal{F}$ over M and for $f : M \to V$ define

$$v(f) = x \quad \Leftrightarrow \quad \{m \in M : f(m) = x\} \in \mathcal{F}$$

Write $I_x = \{m \in M : f(m) = x\}$. $M = \bigcup_{x \in V} I_x$ is a finite partition, thus exactly one $I_x \in \mathcal{F}$. Therefore v is well defined. A moment of thought shows that v is a fair voting scheme.

Conversely, let v be a fair voting scheme with $|V| \geq 3$ and finite. Then there exists an ultrafilter $\mathcal{F}$ according to Problem 68(a). Then for this $\mathcal{F}$ we have

$$v(f) = x \quad \Leftrightarrow \quad \{m \in M : f(m) = x\} \in \mathcal{F}$$

for all $f : M \to V$.

70. As K_n^I contains a complete n-graph (the diagonal), its chromatic number is at least n. On the other hand, let c_i be an n-coloring of the i-th factor for some $i \in I$. Then c_i induces an n-coloring of the product by giving each vertex the color of its i-th component. Therefore the chromatic number is exactly n.

Take an ultrafilter U over I and for every $i \in I$ let c_i be a correct n-coloring of the i-th factor of the product. Define $v : K_n^I \to n$ as

$$v(a) = k \quad \Leftrightarrow \quad \{i \in I : c_i(a(i)) = k\} \in U.$$

Then v is well defined as exactly one member of the finite partition $I = \bigcup_{k=1}^{n} \{i \in I : c_i(a(i)) = k\}$ belongs to the ultrafilter. We call this coloring induced by the c_i's and U. It is easy to check that v is a correct coloring of K_n^I.

We show that every correct coloring of K_n^I is given by an induced coloring by some c_i's and an ultrafilter U. Note first that a correct coloring of K_n^I can be identified with a homomorphism $v : K_n^I \to K_n$ which in turn can be considered as a voting scheme.

This voting scheme is fair up to a permutation π of K_n: if every vote goes to the same alternative, then we speak about the diagonal embedding of K_n which is isomorphic to K_n. This isomorphism gives rise to the required permutation: if $f(i) = x$ for all $i \in I$, then $\pi \circ v(f) = x$. Now, if every vote is changed, then the outcome should change as well: if $f, g \in K_n^I$ are everywhere different, then they are connected in the product graph, therefore a correct coloring gives them different colors.

According to Problem 69 there must exist an ultrafilter U such that

$$\pi \circ v(f) = x \quad \Leftrightarrow \quad \{i \in I : f(i) = x\} \in \mathcal{F}$$

for all $f : I \to K_n$. This is the same as saying there are permutations π_i such that

$$v(f) = x \quad \Leftrightarrow \quad \{i \in I : \pi_i \circ f(i) = x\} \in \mathcal{F}$$

12.3 FORMAL LANGUAGES AND AUTOMATA

71. (a) Let $L = \{w^i : i < N\}$ be a finite language and suppose $w^i = x_1^i x_2^i \cdots x_{n_i}^i$. We describe a DFA that accepts L. For each $i < N$ and $j < n_i$ let q_j^i be the states and let q_λ be the starting state. Halting states are the $q_{n_i}^i$, and transitions are $\delta(q_j^i, x_{j+1}^i) = q_{j+1}^i$ (for $j < n_i$) and $\delta(q_\lambda, x_1^i) = q_1^i$. (The states $q_j^{i_1}$ and $q_j^{i_2}$ are the same if the first j symbols of w_{i_1} and if w_{i_2} are the same.)

(b) Every finite set is the finite union of its singletons, all of which are regular by definition.

72. Let $L_1 = \{a\}$, $L_2 = \{aa\}$. Then $(L_1 \cap L_2)^* = \emptyset^* = \emptyset \neq L_1^* \cap L_2^*$.

> **Remark.** $(L_1 \cap L_2)^* \subseteq L_1^* \cap L_2^*$ holds for arbitrary languages, as $L_1 \cap L_2 \subseteq L_1, L_2$, therefore $(L_1 \cap L_2)^* \subseteq L_1^*, L_2^*$.

73. (a) $a^* b(a|b)^*$. (b) $a^*|a^* ba^*$. (c) $(a|b)^* bbb$.

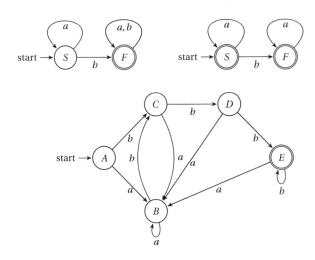

> **Remark.** The DFA in (c) figured above originates from a much simpler non-deterministic automaton presented below. We used the method in Problem 79 to convert this NFA into a DFA.

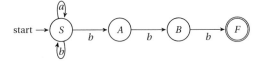

74. (a)

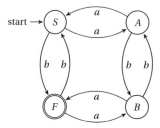

(b) Let L_1 be the set of words which contain an even number of a' and an even number of b's. Split $w \in L_1$ into segments of two letters; it has ab or ba at an even number of places, thus the following expression defines L_1:

$$L_1 = \big((aa|bb)^* (ab|ba)(aa|bb)^* (ab|ba)\big)^* (aa|bb)^*.$$

Let $w \in L$ and consider the last place where $b \in w$ is preceded by a word w' with an even number of a's and b's. If b follows immediately w', then w is in the language $L_1 b L_1$. Otherwise w starts as $w' ab$, followed by a block of b's followed by an a (making the number of a's even). If the block of b's is even, then this is a longer sequence of even numbers of a's and b's. Thus the block must contain an odd number of b's, and then w is in the language $L_1 ab(bb)^* a L_1$. The required expression can be

$$L_1 \big(b | (ab(bb)^* a)\big) L_1.$$

75. Let p be larger than $|Q|$, where Q is the set of states in $\mathcal{A}$. Suppose $|w| \geq p$ and $\mathcal{A}$ accepts uwv. While reading w, eventually $\mathcal{A}$ returns to a state that has already been visited. Let the first such state be q. Clearly, there are $r < |Q|$ steps while $\mathcal{A}$ returns to q. Let x be the part of w that was read before we first reached q, let y be the part of w which was read before $\mathcal{A}$ returned to q, and let z be the remaining part of w.

76. If L is accepted by a DFA, then by the Pumping lemma 3.6 there exists some p such that every $w \in L$ can be written as $w = xyz$ with $|y| > 0$, $|xy| < p$ such that $xy^n z \in L$ for all $n \geq 0$. Choose a word of length $|w| = p^2$ and let $w = xyz$ be its decomposition. Then $xy^2 z$ has length $p^2 < |xy^2 z| < (p+1)^2$, therefore $xy^2 z \notin L$, contradicting the pumping lemma.

77. Let $L = \{a^n : n \in A\}$ be accepted by a DFA. By the Pumping lemma 3.6 there is p such that if $n \in A$, then either $n < p$ or there is $0 < k < p$ such that $n + t \cdot k \in A$ for all $t \geq 0$. Note that there might be several different $0 < k_1 < \ldots < k_\ell < p$ with the property that $n + t \cdot k_i \in A$. Thus A must be of the form

$$A = F \cup \{n + t \cdot k_i : n \in A, n \geq p, \ t \geq 0, \ 1 \leq i \leq \ell\}$$

for some and finite set F that contains numbers $< p$. In short: A is eventually periodic with block size at most the least common multiple of $k_1 \ldots, k_\ell$.

That all such sets A can be accepted by a DFA is easy to see. As an illustration we design a DFA having two periods $k_1 = 2$ and $k_2 = 5$ for $n > 3$ and below 3 only $1 \in A$. We give the DFA as a regular expression:

$$a|a(aa)^*|a(aaaaa)^*.$$

Remark. The following are equivalent for a language L over a single letter:

- L is regular.
- L is context-free.
- A is eventually periodic.
- For $a_i = \chi_{i \in A}$ the number $0.a_0 a_1 \ldots$ is rational.
- The generating function $\sum_{i \in A} x^i$ is a rational function (i.e. the quotient of two polynomials).

78. (a) If L is accepted by a DFA, then by Theorem 3.6 there is p such that whenever $|a^n b^n| \geq p$ then $a^n b^n$ can be written as xyz with y not empty, $|xy| < p$ and such that for all $i \geq 0$, $xy^i z \in L$. But clearly there is i so that $xy^i z$ is not of the form $a^\ell b^\ell$ for some i.

(b) Suppose L_f is accepted by a DFA. The Pumping lemma 3.6 gives us the constant p. If there is $n \in \omega$ such that $f(n) \geq p$, then decompose $a^n b^{f(n)} \in L$ into $a^n b^i b^p$. Using the pumping lemma there is $0 < k < p$ such that for all $t \geq 0$ the word $a^n b^i b^p b^{t \cdot k}$ belongs to L. But then $f(n) = f(n) + t \cdot k$ which is a contradiction.

Consequently, f is bounded ($f(x) < p$ for the p given by the pumping lemma). We claim $f(n)$ is periodic for large enough n. According to the pumping lemma there is $0 < k < p$ such that for each $n \geq p$ every word of the form $a^{n+t \cdot k} b^{f(n)}$ belongs to L ($t \geq 0$). Hence $f(n) = f(n + t \cdot k)$ for all $t \geq 0$, provided $n \geq p$.

Conversely, if $f(n)$ is bounded and periodic for large enough n, then L_f is accepted by a DFA. We illustrate the design when the period is 3, and $f(0 + 3t) = 2$, $f(1 + 3t) = 4$, $f(2 + 3t) = 2$.

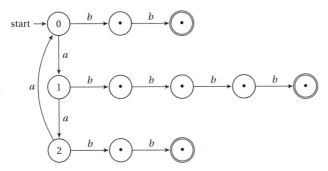

We note that for any bounded and almost periodic f it is easy to construct L_f as a regular expression. As an illustration take

$$abbbb|aab|aaa(aa)^* bb|aaaa(aa)^* b.$$

This corresponds to $f(1) = 4$, $f(2) = 1$, $f(3 + 2t) = 2$ and $f(4 + 2t) = 1$.

79. As DFA's are special NFA's, any languages accepted by some DFA are accepted by some NFA. Conversely, for a NFA $\mathcal{N} = \langle Q, S, F, \delta \rangle$ construct the DFA $\mathcal{A} = \langle Q', F', \delta' \rangle$, where Q' is the collection of non-empty subsets of Q,

$$F' = \{H \subseteq Q \mid Q \cap F \neq \varnothing\},$$

i.e. the sets of states that include at least one halting state of $\mathcal{N}$, and for $H \in Q'$,

$$\delta'(H, a) = \{\delta(p, a) : p \in H\}.$$

If $\mathcal{N}$ accepts the language L, then so does $\mathcal{A}$.

80. We assume that the alphabets are the same: simply add a sink state to $\mathcal{A}$ which absorbs all letters which are not in Σ_A. So let $\mathcal{A} = \langle Q_A, S_A, F_A, \delta_A \rangle$ and $\mathcal{B} = \langle Q_B, S_B, F_B, \delta_B \rangle$. In each case we construct the automaton $\mathcal{M} = \langle Q, S, F, \delta \rangle$ over the alphabet Σ, which might be non-deterministic.

(a) The basic idea is to construct an automaton that runs $\mathcal{A}$ and $\mathcal{B}$ in parallel on a word $w \in \Sigma^*$. At the end the final state (r_1, r_2) indicates whether $w \in L_A$ or $w \in L_B$. Set $\Sigma = \Sigma_A \cup \Sigma_B$, $Q = Q_A \times Q_B$, $S = (S_A, S_B)$, and

$$F = \{(q_A, q_B) \in Q : \text{ either } q_A \in F_A \text{ or } q_B \in F_B\},$$
$$\delta((q_A, q_B), x) = (\delta_A(q_A, x), \delta_B(q_B, x)).$$

(b) Same as in (a) except for the halting states which are $F = F_A \times F_B$.

(c) Idea: run $\mathcal{A}$ first and then $\mathcal{B}$. The main difficulty is to figure out where the L_A prefix ends and the L_B suffix starts in a word. This can be achieved by creating a non-deterministic automaton. Take $\mathcal{M} = \mathcal{A} \cup \mathcal{B}$ (disjoint union), the starting state of $\mathcal{M}$ is S_A, and its final states are those in F_B. Add all transitions from S_B as non-deterministic choices to every $\mathcal{A}$-final state $q \in F_A$. Also, if S_B is a final state in $\mathcal{B}$, then mark states in F_A as final states of $\mathcal{M}$.

(d) Q, Σ, δ, S are the same as in $\mathcal{A}$, but as for the halting states we let $F = Q_A \smallsetminus F_A$.

(e) Similar to (c). Copy all transitions from S_A as non-deterministic possibilities to every final state in F_A. Also add S_A as a final state so the new automaton accepts the empty word.

81. Regular languages are precisely the languages generated from finite languages by the operations concatenation, union, and $*$. It is straightforward that $\varnothing, \{\lambda\}, \{a\}$ for $a \in \Sigma$ can be generated by DFA. Apply Problem 80(a,c,e) to complete the proof.

82. Let $\mathcal{A} = \langle Q, S, F, \delta \rangle$ be a DFA over the alphabet Σ. The language accepted by $\mathcal{A}$ is $\bigcup \{L_{S,q} : q \in F\}$, thus it suffices to show that all languages L_{q_1, q_2} are regular. We need a slightly stronger induction hypothesis: the transition function δ can be undefined on some letters in Σ, in those cases the input is automatically rejected.

If $\mathcal{A}$ has a single state q, then $L_{q,q}$ is the language $(a|b|c)^*$ where a, b, c are the letters for which there is a transition from q to itself.

Now suppose the statement is true for $|Q|-1$ states. Let B be the automaton when the state $q \in Q$ is deleted (and all transitions to q are illegal in B). The language which is accepted by A starting from q, visiting states different from q, then returning to q is

$$L = \bigcup \{a L^B_{q_1,q_2} b : \delta(q,a) = q_1,\ \delta(q_2,b) = q,\ a,b \in \Sigma,\ q_1,q_2 \in Q_B\}.$$

This is a finite union, $L^B_{q_1,q_2}$ is equivalent to a regular expression by induction, thus L can also be specified as a regular expression. Consequently

$$L^A_{q,q} = \left(L \cup \{a \in \Sigma : \delta(q,a) = q\} \right)^*$$

is also a regular expression. Now if A starts at $q_1 \in B$ and stops at q, then there is a first visit to q from some $q_2 \in B$, thus

$$L^A_{q_1,q} = \left(\bigcup_{q_2,a} \{L^B_{q_1,q_2} a : a \in \Sigma,\ \delta(q_2,a) = q\} \right) L^A_{q,q}.$$

Similar expressions work for the remaining cases.

83. Define $L_1 = \{c^k a^i b^j : k \neq 1,\ i,j \geq 0\}$, $L_2 = \{c a^n b^n : n \geq 0\}$, and let $L = L_1 \cup L_2$. L can be pumped (it satisfies the conclusion of Theorem 3.6). The class of regular languages is closed under intersection (Problem 80(b)) and L_1 is regular but L_2 is not. Hence, L cannot be regular as $L \cap L_1 = L_2$.

84. Solution 1. Take a DFA $A = \langle Q, S, F, \delta \rangle$ which accepts L. We create a non-deterministic automaton $B = \langle Q \cup \{T\}, T, \{S\}, \delta^* \rangle$, that is, the states of B are $Q \cup \{T\}$, the starting state is T, and the only accepting state is $S \in Q$. For each $a \in \Sigma$ and $q \in Q$, if $\delta(q,a) \in F$ then add q as a possible next state to $\delta^*(T,a)$. For states $q \in A$, define δ^* as the inverse of δ. (If $\delta^*(q,a)$ were empty for some $a \in \Sigma$, create a new sink state for B.) It is clear that an accepting run of B is just a reverse of an accepting run of A.

Solution 2. Clearly $(L_1 | L_2)^R = L^R_1 | L^R_2$, $(L_1 L_2)^R = L^R_2 L^R_1$, and $(L^*)^R = (L^R)^*$. As every regular language L is a regular expression, L^R is also a regular expression.

85. Using the congruence property of $\sim$ each run of the machine $A/\sim$ where the word w is accepted can be lifted up to a run of A which accepts w. The converse clearly holds.

86. Take equivalent states $q_1 \sim q_2$ witnessed by the words w_1 and w_2, that is, $A(w_1) = q_1$, $A(w_2) = q_2$ while $B(w_1) = B(w_2)$.

(i) Consider the transitions $q_1 \xrightarrow{\alpha} q'_1$ and $q_2 \xrightarrow{\alpha} q'_2$. Then $A(w_1 a) = q'_1$ and $A(w_2 a) = q'_2$ while $B(w_1 a) = B(w_2 a)$. Therefore $q'_1 \sim q'_2$.

(ii) Suppose q_1 is a halting state. Then A accepts w_1, hence $B(w_1)$ should be a halting state as A and B accept the same languages. As $B(w_2) = B(w_1)$, B accepts the word w_2, hence so does A. Therefore $A(w_2) = q_2$ is a halting state.

87. Suppose $\mathcal{B}$ accepts the same language and has the same (minimal) number of states. Define the congruence relation $\sim$ from Problem 86 on the states of $\mathcal{A}$. If $\sim$ is a proper congruence, then $\mathcal{A}/\sim$ would be an automaton accepting the same language (Problem 86) having a smaller number of states. This is impossible by the choice of $\mathcal{A}$. Therefore $\sim$ must be a trivial congruence (each state is congruent to only itself), which implies that $\mathcal{A}$ and $\mathcal{B}$ are isomorphic.

88. Make $\mathcal{A}$ deterministic by the algorithm of Problem 79. Take a maximal congruence $\sim$ on $\mathcal{A}$ (the set of congruences of $\mathcal{A}$ is finite, so there is at least one maximal congruence) and let $\mathcal{B} = \mathcal{A}/\sim$. Problem 86 implies that $\mathcal{B}$ is minimal: for if not, then there would be a minimal $\mathcal{B}'$ defining a congruence on $\mathcal{B}$. But congruences of $\mathcal{A}/\sim$ can be obtained from congruences of $\mathcal{A}$ (by the second homomorphism theorem of universal algebra), thus $\sim$ could not be maximal.

89. Let $\mathcal{A}$ be a DFA which accepts L. For each state $q \in Q$ let $s_q(n)$ be the number of words of length n accepted by the automaton which starts from q rather than from S. Thus $s_q(0) = 1$ if q is a halting state, otherwise $s_q(0) = 0$. Clearly $s_q(n+1) = \sum\{s_r(n) :$ there is a transition $\delta(q,a) = r\}$. Let $\mathbf{s}_n$ be the (column) vector $\langle s_{q_1}(n), s_{q_2}(n),\ldots\rangle$, therefore there is a 0–1 matrix M such that $\mathbf{s}_{n+1} = M\mathbf{s}_n$, and then $\mathbf{s}_{n+d} = M^d\mathbf{s}_n$. Consider the minimal polynomial of M. It has rational coefficients and can be written as

$$M^d = c_1 M^{d-1} + c_2 M^{d_2} + \cdots + c_{d-1}M + c_d.$$

Multiplying this equation by $\mathbf{s}_n$ we get

$$\mathbf{s}_{n+d} = c_1\mathbf{s}_{n+d-1} + \cdots + c_{d-1}\mathbf{s}_{n+1} + c_d\mathbf{s}_n.$$

In particular this holds for the coordinate of $\mathbf{s}_n$ indexed by the starting state S, which gives the claim.

90. $L = (a|ab)^*$. If $w \in L$ ends with an a, then removing this letter the truncated word is also in L. If w ends with b, then the last two letters are ab, and removing them we get another word in L. Thus $s(n) = s(n-1) + s(n-2)$, as required.

91. Assume L is regular and let $\mathcal{A} = \langle Q, S, F, \delta\rangle$ be a DFA over the alphabet $\Sigma \cup \{a\}$ that accepts L. Consider a word $w \in L_a$. There are $w_1, w_2 \in (\Sigma \,|\, a)^*$ such that $w_1 awaw_2 \in L$. Take the path $q_1,\ldots q_n$ from the initial state to the halting state that is labelled by $w_1 awaw_2$. Let q_i be the last state visited until $w_1 a$ and let q_j be the first state after passing $w_1 aw$. Then the path in between q_i and q_j is labelled by w. q_i is reached by an edge labelled by a and we leave q_j using an edge labelled by a.

Take all the possible paths from q_i to q_j that do not use the label a. These paths induce a labelled subgraph of $\mathcal{A}$, let us denote it by $\mathcal{B}_w$. $\mathcal{B}_w$ is a DFA if we set q_i the starting state and q_j the halting state. Observe that $\mathcal{B}_w$

accepts the word w, moreover for every word u accepted by $\mathcal{B}_w$ we have $w_1 a u a w_2 \in L$. Therefore $L(\mathcal{B}_w) \subset L_a$.

Repeat the procedure for all possible $w \in L_a$. As $\mathcal{A}$ is finite, we get only finitely many different $\mathcal{B}_w$'s. For each such $\mathcal{B}_w$ we have $L(\mathcal{B}_w) \subset L_a$, and $L(\mathcal{B}_w)$ is regular. Therefore the union of these languages is regular, and clearly equals to L_a.

> **Notation.** Several production rules with the same nonterminal on the left-hand side are merged into a single rule separating the right-hand sides by |.

92. (a) $S \to SS$, $S \to (S)$, $S \to ()$.

(b) $S \to aSa$, $S \to bSb$, $S \to \lambda$.

(c) $S \to T \mid U$, $T \to VaT \mid VaV \mid TaV$, $U \to VbU \mid VbV \mid UbV$, $V \to aVbV \mid bVaV \mid \lambda$.

93. (a) Not regular (Problem 78(a)), but context-free given by the grammar $S \to aSb$, $S \to \lambda$.

(b) A similar argument as in Problem 78(a) shows that it is not regular, but context-free.

(c) Regular shown by the regular expression $abab(ab)^*$, hence context-free (Problem 94).

(d) Regular: $abab(ab)^*(bc)^*$, hence context-free.

(e) Context-free: $S \to bSbb \mid A$, $A \to aA \mid \lambda$.

94. Regular expressions can easily be converted into context-free production rules. We only hint some examples: $S \to Sa \mid \lambda$ generates a^*; $S \to Ab$, $A \to Aa \mid \lambda$ generates a^*b. $a(b \mid c)^*$ can be produced by $S \to aX$, $X \to bX \mid cX \mid \lambda$; etc.

95. Suppose the two context-free languages are represented by production rules with starting symbols S_1 and S_2, respectively. To get the union add the rule $S \to S_1 \mid S_2$. To get concatenation add the rule $S \to S_1 S_2$. Finally, to get the Kleene star, add the rule $S \to S_1 S \mid \lambda$.

96. Add a new starting symbol S' and add the rule $S' \to S$; this takes care of the first restriction. Delete all occurrences of λ in the right-hand side strings. If none of them becomes empty, then we are done. Otherwise we add new production rules which handle the cases when the rule $A \to$ empty is used. If this non-terminal A was processed before, then simply delete this rule (as it adds no new possibilities). Otherwise on the right-hand side of each production rule either keep or delete each occurrence of A, and then mark A as processed. (If a right-hand side contains k A's, then this means adding $2^k - 1$ new production rules.)

97. The proof is similar in spirit to that of Theorem 3.6, see Solution 75. A parse tree is a tree labeled by symbols of the context-free grammar. The root is labeled by the start symbol S. Leaves are labeled by terminal symbols or λ. Interior nodes are labeled by variables (nonterminal symbols) and children of a node are labeled by the right side of a production rule whose left side is the variable of the parent. The yield of a parse tree is the concatenation of the labels of the leaves in left-to-right order. Now $w \in L$ if and only if w is the yield of some parse tree.

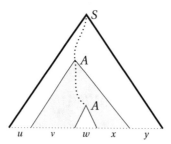

Figure 12.1: Parse tree

If the set of production rules satisfy the conditions in Problem 96 then no leaf is labeled by λ. Suppose the production rules use k variables and let $s \in L$ be long enough, say $|s| \geq 2^k$. Then the *minimal size* parse tree with yield s must have a path of length at least $k+1$. Consider a longest path. As there are only k variables, among the lowest $k+1$ we can find two nodes with the same label, say A. The parse tree then looks like on Figure 12.1. Here vx cannot be empty as otherwise there would be a smaller parser tree generating the word $uvwxy$. Removing or repeating the gray subtree gives the desired conclusion, see Figure 12.2.

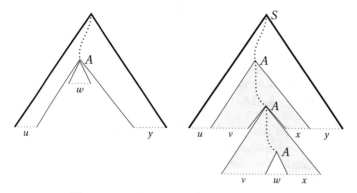

Figure 12.2: Parse tree after replacement

98. By way of contradiction, assume L is context-free and let p be the constant from the pumping lemma (Theorem 3.12). Each word $a^n b^n c^n$ can be written as $uvwxy$ with $|vx| \geq 1$, $|vwx| \leq p$ so that uv^2wx^2y must belong to L. Nevertheless, for large enough n it is not of the form $a^kb^kc^k$.

99. (a) Both $L_1 = \{a^mb^mc^n : m, n \geq 0\}$ and $L_2 = \{a^mb^nc^n : m, n \geq 0\}$ are context-free, however $L_1 \cap L_2 = \{a^nb^nc^n : n \geq 0\}$ is not.

(b) By Problem 95 the class of context-free languages is closed under taking union. As $A \cap B = (A' \cup B')'$, the same class cannot be closed under complementation.

(c) $L_1 = \{a^ib^jc^k : i, j, k \geq 0\}$ is regular (the regular expression $a^*b^*c^*$ generates L_1) and $L_2 = \{w \in \{a, b, c\}^* : w$ has the same number of a, b and $c\}$ is context-free, however $L_1 \cap L_2 = \{a^nb^nc^n : n \geq 0\}$ is not context-free.

100. (a) There is no regular language that is universal for regular languages. By way of contradiction, suppose U is universal for regular languages and U is regular. By the Pumping lemma (Theorem 3.6) there is a pumping constant p that corresponds to U. Take a regular language L such that $a^p \in L$ but for any $0 < k \leq p$ we have $a^{p+k} \notin L$. (It is straightforward to construct such an L.)

By assumption, L has a prefix w such that $x \in L$ iff $wx \in U$. Clearly then $wa^p \in U$, hence by the pumping lemma, $wa^{p+k} \in U$ for some $k \leq p$. But this contradicts the choice of L as $a^{p+k} \notin L$ for any $0 < k \leq p$.

(b) Very similar argument to (a) using the pumping lemma for context-free languages (Theorem 3.12).

(c) Assume $a, b \in \Sigma$ and take a bijection $f : \mathcal{F} \to a^*$. For each $L \in \mathcal{F}$ the prefix of L will be the word $f(L)b$. Define

$$U = \{f(L)bx : L \in \mathcal{F}, \text{ and } L \text{ accepts } x\}.$$

(d) Do the same as in (c) and let $\mathcal{F}^+ = \mathcal{F} \cup \{U\}$. Then U can be universal for $\mathcal{F}^+$ if we choose the prefix of U to be λ.

12.4 RECURSION THEORY

101. (a) $\mathbf{k}(n) = k$ is $S^{(k)}(\mathbf{0}(n))$, which is $\mathsf{Comp}(S, \mathsf{Comp}(S, \cdots \mathsf{Comp}(S, \mathbf{0})))$.

(b) Predecessor: $\mathsf{PrRec}(U_1^1, U_1^2)$ as $\delta(0) = 0$ and $\delta(m+1) = m = U_1^2(m, \delta(m))$.

Sign: $\mathsf{PrRec}(U_1^1, \mathsf{Comp}(\mathbf{1}, U_1^2))$ as $\mathrm{sgn}(0) = 0$ and $\mathrm{sgn}(n+1) = 1$.

Limited subtraction: $m \mathbin{\dot-} 0 = m$ and $m \mathbin{\dot-} (n+1) = \delta(m \mathbin{\dot-} n)$.

(c) Addition: $m + 0 = m$ and $m + (n+1) = (m+n) + 1 = S(m+n)$. More precisely, $f(m, n) = m + n$ is defined as

$$+ = \mathsf{PrRec}(U_1^2, \mathsf{Comp}(S, U_3^3)).$$

Multiplication: $m \cdot 0 = 0$ and $m \cdot (n+1) = (m \cdot n) + m$, or

$$\times = \mathsf{PrRec}(U_2^2, \mathsf{Comp}(+, U_3^3, U_1^3)).$$

Exponentiation is similar.

(d) $0! = 1$ and $(n+1)! = (n+1) \cdot n!$, or

$$\mathsf{PrRec}\left(S, \mathsf{Comp}(\times, \mathsf{Comp}(S, U_1^2), U_2^2)\right).$$

102. $K_<(x, y) = 1 \div (y \div x)$.
$E(x, y) = K_<(x, y) \cdot K_<(y, x)$.

103. (a) Use the primitive recursive operator for the functions f and $h(\vec{x}, n, u) = u + f(\vec{x}, y+1)$, as

$$\sigma_f(\vec{x}, 0) = f(\vec{x}, 0),$$
$$\sigma_f(\vec{x}, y+1) = \sigma_f(\vec{x}, y) + f(\vec{x}, y+1).$$

Similarly,

$$\pi_f(\vec{x}, 0) = f(\vec{x}, 0),$$
$$\pi_f(\vec{x}, y+1) = \pi_f(\vec{x}, y) \cdot f(\vec{x}, y+1).$$

(b) Let $h' = 1 \div (1 \div h)$, this takes zero where h is zero, and 1 otherwise. Thus $\pi_{h'}(\vec{x}, y)$ is zero if h takes zero between 0 and y (bounds included), and 1 otherwise. Thus adding up $\pi_{h'}(\vec{x}, y)$ as y runs from 0 to $g(\vec{x})$ gives the value of the bounded minimization:

$$f(\vec{x}) = \sigma_{\pi_{h'}}(\vec{x}, g(\vec{x})).$$

104. (a) $\chi_{A \wedge B} = \chi_A \cdot \chi_B$, $\chi_{\neg A} = 1 \div \chi_A$.

(b) See Problem 102 and use part (a) of this problem.

(c) Use Problem 103 (b).

105. $f(\vec{x}) = h_1(\vec{x}) \cdot \chi_{A_1}(\vec{x}) + \cdots + h_k(\vec{x}) \cdot \chi_{A_k}(\vec{x})$.

106. By the previous problems the following is a primitive recursive definition of $g(x, y) = [x/y]$:

$$g(0, y) = 0;$$
$$g(x+1, y) = \begin{cases} g(x) + 1 & \text{if } y > 0 \text{ and } (\exists i \leq x+1)(i \cdot y = x+1), \\ g(x) & \text{otherwise.} \end{cases}$$

107. (a) Divisibility is a primitive recursive relation: $m \mid n$ if and only if $n = m \cdot [n/m]$ (see Problem 106). Let $P(n)$ be the characteristic function of the set of primes. P is primitive recursive: n is prime iff

$$(1 < n) \wedge (\forall x < n)(x = 1 \vee x \nmid n).$$

Recall that if p is prime, then there is a prime between p and $p! + 1$. Also note that the factorial function is primitive recursive ($0! = 1$ and $(n+1)! = (n+1) \cdot n!$). Now let

$$h(z) = \min\{y \leq z! + 1 : z < y \text{ and } P(y) = 1\}.$$

Then h is primitive recursive by Problem 103. Now we can define the function $p(n)$ giving the n-th prime number as follows:

$$p(0) = 2, \quad \text{and} \quad p(n+1) = h(p(n)).$$

(b) Using Problem 104 (c) this function can be calculated as

$$\min\{y < x : p(n)^{y+1} \nmid x\}.$$

108. (a) Consider the primitive recursive definition

$$f(x,0) = x,$$
$$f(x,n+1) = g(f(x,n)).$$

As g is primitive recursive, so is f. Clearly, $f(x,n) = g^{(n)}(x)$, thus $g^{(x)}(x) = f(x,x)$.

(b) $g_0(x)$ and $h(x,y) = x^y$ are primitive recursive, and define $a(x,k)$ as

$$a(x,0) = g_0(x),$$
$$a(x,n+1) = h(a(x,n),x).$$

Then $g_x(x) = a(x,x)$ is primitive recursive.

109. (a) Write $g^{(1)}(x) = g(x)$ and $g^{(n+1)}(x) = g(g^{(n)}(x))$. Induction on k shows $A_k(x) = g^{(k!)}(x)$. Now $g^{(n)}(x) = f(x,n)$ for the primitive recursive f from Solution 108. Thus $A(x) = f(x,x!)$ is primitive recursive as $x!$ is primitive recursive.

(b) Here $g^{(n)}(x) = x + n$, thus $A(10) = A_{10}(10) = g^{(10!)}(10) = 10 + (10!)$.

(c) By (a), $A(5) = g^{(120)}(5)$. Now $g(5) = 6^6$, $g(g(5)) = (6^6 + 1)^{(6^6+1)} > 6^{6^6}$, $g^3(5) = g(g^2(5)) > 6^{g^2(5)}$, thus

$$A(5) = g^{(120)}(5) > 6^{6^{6^{\cdot^{\cdot^{\cdot^6}}}}} \left.\right\} 121$$

110. By induction on the complexity of f. For initial functions we have
- $0 < A(1,n)$,
- $S(n) = A(0,n) < A(1,n)$,
- $U_i^k(\vec{m}) < A(1,m_1 + \cdots + m_k)$.

For composition and primitive recursion one can take N to be $2 \cdot \max + 4$ of the corresponding N's of the subfunctions.

111. By Problem 110 $A(m,m)$ grows faster than any primitive recursive function. If $f(m) = A(m,m)$ were primitive recursive, then for some $N \in \omega$ we would have $A(m,m) < A(N,m)$ for all m, yielding a contradiction when $m = N$.

112. Suppose by contradiction that $U(i,\vec{x})$ is a universal primitive recursive function. Then

$$U(x_1, x_1, x_2, \ldots, x_n) + 1$$

is an n-variable primitive recursive function, therefore there is a $k \in \omega$ such that

$$U(k, x_1, x_2, \ldots, x_n) = U(x_1, x_1, x_2, \ldots, x_n) + 1.$$

Choosing $x_1 = k$ we obtain

$$U(k, k, x_2, \ldots, x_n) = U(k, k, x_2, \ldots, x_n) + 1$$

for all $x_2, \ldots, x_n$, which is impossible.

113. There are continuum many continuous functions (they are determined by their values at rationals, and $|\mathbb{R}|^\omega = 2^\omega$), enumerate them.

If $U(x,y)$ is continuous, then so is $U(x,x)+1$ which differs from $U(x,y)$ for all y.

114. (a) $K_<(x,x)$ is one for all $x \in \omega$, thus the constant one function $\mathbf{1}$ is $\mathsf{Comp}(K_<, U_1^1, U_1^1)$. $K_<(x,x+1)$ is constant zero, thus $\mathbf{0}$ is

$$\mathsf{Comp}(K_<, U_1^1, \mathsf{Comp}(+, U_1^1, \mathbf{1})).$$

(b) $\mathrm{sgn}(x) = K_<(x,1)$; and $x \mathbin{\dot-} y$ is the smallest u for which $x < u+y+1$, or, for which $K_<(x, u+y+1) = 0$. Using the μ operator, the function $x \mathbin{\dot-} y$ is

$$\mu\big(\mathsf{Comp}(K_<, U_1^3, \mathsf{Comp}(+, \mathsf{Comp}(+, U_3^3, U_2^3), \mathsf{Comp}(\mathbf{1}, U_1^3)))\big).$$

115. Apply the μ operator to the recursive function $1 \mathbin{\dot-} \chi_A$.

116. (a), (b) See Solution 104 (a), (b).

(c) Let $A \subseteq \omega^{n+1}$ be recursive. By (a) and (b) the relation

$$A^* = \{\langle \vec{x}, n, i\rangle : \langle \vec{x}, i\rangle \in A \text{ or } i \geq n\}$$

is recursive. For each $\langle \vec{x}, n\rangle$ there is an i with $\langle \vec{x}, n, i\rangle \in A^*$, thus Problem 115 gives the recursive function $g(\vec{x}, n)$ whose value is the minimal $i < n$ for which $\langle \vec{x}, i\rangle \in A$, or n if no such an $i < n$ exists. The recursive relation $g(\vec{x}, n) < n$ is the same as the relation

$$\{\langle \vec{x}, n\rangle : (\exists i < n)\langle \vec{x}, i\rangle \in A\}.$$

117. By Problem 115 as each function is the minimal value satisfying a recursive relation:

$$[x/y] = \min\{u : y = 0 \text{ or } (u+1)\cdot y > x\},$$
$$[\sqrt{x}] = \min\{u : (u+1)\cdot(u+1) > x\},$$
$$\mathrm{rem}(x,y) = \min\{z : y = 0 \text{ or } x = [x/y]\cdot y + z\}.$$

118. The function g which takes 0 everywhere except for finitely many places is recursive (linear combination of characteristic functions of single-element sets). Then $(f + g_1) \mathbin{\dot-} g_2$ is recursive as well.

119. Write

$$f_n(x) = \begin{cases} f(x) & \text{if } x \leq n \\ 0 & \text{otherwise.} \end{cases}$$

Then f_n is recursive (e.g. by Problem 118) and $\lim_{n\to\infty} f_n = f$.

120. There are countably many recursive functions and $\omega < 2^\omega$. This proves the existence of a non-recursive function. A specific example could be given using the halting problem, see Problem 230.

121. The relation $\{(x, y) : f(x) = y\}$ is recursive by Problem 116 (b), and $f^{-1}(y) = \mu\{x : f(x) = y\}$.

122. Take any non-recursive function $g : \omega \to \omega$ (see Problem 120), and put

$$f_i(k) = \begin{cases} g(i) & \text{if } k = i \\ 0 & \text{otherwise.} \end{cases}$$

Then $f_i(i) = g(i)$ and f_i can clearly be obtained by changing the constant 0 function at a single point, thus f_i is recursive (see Problem 118).

123. $f(\vec{x}) = h_1(\vec{x}) \cdot \chi_{A_1}(\vec{x}) + \cdots + h_k(\vec{x}) \cdot \chi_{A_k}(\vec{x})$.

124. (a) Choose $u = (x + y + 1)^2 + x$. (b) By Problem 115 as K and L are recursive functions.

125. Observe that $b(i + 1) + 1$ are pairwise coprime for $0 \le i < n$ if b is divisible by $n!$. Use the Chinese Remainder Theorem.

126. (a) There are countably many finite sequences, enumerate them.

(b) Use Gödel's β function with $m = K(u)$ and $b = L(u)$ from Problem 124 with $\text{Len}(u) = \beta(m, b, 0)$ and $\text{Elem}(u, i) = \beta(m, b, i + 1)$.

For the additional property observe that $\text{Elem}(i, u) \le m$. If $\text{Len}(u) > 0$ then $m \ge 1$ (as $\beta(0, b, i) = 0$), and $m = K(u) < u$ when $1 \le K(u)$.

(c) u is a sequence code if all $v < u$ codes a different sequence. This happens if either $\text{Len}(v) \ne \text{Len}(v)$, or $\text{Len}(v) = \text{Len}(u)$ but for some $i < \text{Len}(u)$, $\text{Elem}(v, i) \ne \text{Elem}(u, i)$. Use that recursive relations are closed for bounded quantifiers, see Problem 116 (c).

(d) The relation "v codes a sequence which has length one more than $\text{Len}(u)$, the last element of v is z, and for $i < \text{Len}(u)$ the i-th element of u and v are the same" is recursive by Problem 116. Thus the "append" function is recursive by Problem 115.

127. Using the recursive coding functions Elem and Len, let

$$\text{Elem}^*(u, i) = \begin{cases} \text{Elem}(u \dot{-} 1, i) & \text{if } u > 0, \\ 1 & \text{otherwise;} \end{cases}$$

$$\text{Len}^*(u) = \begin{cases} \text{Len}(u \dot{-} 1) & \text{if } u > 0, \\ 1 & \text{otherwise.} \end{cases}$$

They are recursive by Problem 123, and $\text{Len}^*(0) = 1$, while $\text{Elem}^*(0, 0) = 1 > 0$.

128. $[\sqrt{x}]$ is primitive recursive as it is the smallest $y \le x$ which satisfies the primitive recursive relation $(y + 1)^2 > x$, see Problem 103(b). Therefore functions $K(u)$ and $L(u)$ are also primitive recursive. Similarly, Gödel's β functions is primitive recursive (a bounded value satisfying a primitive recursive relation), which means that $\text{Elem}(u, i)$ and $\text{Len}(u)$ are primitive recursive. To show that $u ^\frown z$ is also primitive recursive, it is the smallest v which satisfies the primitive recursive relation using a bounded quantifier

$$\text{Len}(v) = \text{Len}(u) + 1 \; \wedge \; (\forall i < n)\,\text{Elem}(u, i) = \text{Elem}(v, i) \; \wedge \; \text{Elem}(v, \text{Len}(u)) = z.$$

This v is bounded by the primitive recursive function $(u + z + 1)^{2(u+z+1)}$.

129. Divisibility is a recursive relation (see Problem 116), and the set of primes $P \subset \omega$ is recursive by using bounded quantifiers:

$$P = \{n : n \geq 2 \text{ and } (\forall k < n)(k \mid n \to k \leq 1)\}.$$

(a) This is the characteristic function of the recursive relation

$$\{n : n \geq 1 \wedge (\forall k < n)((P(k) \wedge k \mid n) \to k = 3)\}.$$

(b) Let R be the relation in (a). Then

$$f_b(x) = \mu\{y : x < y \wedge R(y)\},$$

which is recursive by Problem 115.

(c) $z = k \cdot x + y$ for the smallest 3-power k which is bigger than y. Using the relation R in part (a) we can write

$$f_c(x, y) = \mu\{n : R(n) \wedge y < n\} \cdot x + y.$$

(d) x can be split as $x_1 x_2$ where x_1 is $22 \cdots 20$. Thus f_d is the characteristic function of the recursive relation

$$\{x : (\exists x_1 < x)(\exists x_2 < x)(x = f_c(x_1, x_2) \wedge x_1 > 1 \wedge x_1 + 3 = f_b(x_1))\}.$$

130. (a) x "contains" y if

- there is an $s < x$ which is at least 2, its base 3 representation is a sequence of 2's followed by a zero (a recursive relation by Problem 129 (d));
- x starts with s, that is, $x = f_c(s, x')$ for some $x' < x$;
- y does not contain s: $(\forall y_1 < y)(\forall y_2 < y)(\forall y_3 < y)$ if $y = f_c(y_1, f_c(y_2, y_3))$ then $y_2 \neq s$ (actually, we need more: y differs from s, and y neither starts nor ends with s);
- x contains the sequence sys: $(\exists x_1 < x)(\exists x_2 < x)(x = f_c(x_1, f_c(t, x_2))$, where $t = f_c(s, f_c(y, s))$.

This is clearly a Boolean combination and bounded quantification of recursive relations.

(b) Write each y_i in base 3 as v_i, choose a separator s which has one more digit than the longest y_i sequence. Let x be the value of $s v_1 s v_2 \ldots v_n s$ in base 3.

(c) For each x and z there is an x' which satisfies the following recursive relation:

- for all $u < x$, if x "contains" u then x' "contains" u; and
- x' contains z; and
- for all $v < x'$ if x' "contains" v then $v = z$ or x "contains" v.

The minimal such an x' is a recursive function of x and z, see Problem 115. From here use induction on n.

131. The function $p(n)$ is primitive recursive by Problem 107. The minimal $i < u$ satisfying the following primitive recursive relation is again primitive recursive by Problem 103:

$$(\forall j < u)\,(j \le i \ \lor \ p(j) \nmid u).$$

It returns the index of the largest prime dividing u (as it is always smaller than u when $u \ge 2$), and this is just the length of the sequence coded by u.

The function which returns the exponent of the i-th prime in u is primitive recursive by Problem 107, thus the set of code numbers is primitive recursive; define $\mathsf{Len}(u) = 0$ if u is not a code number.

The append function is clearly primitive recursive when $\mathsf{Len}(u) = 0$, otherwise it is $p(\mathsf{Len}(u)+1) \cdot [u / p(\mathsf{Len}(u))] \cdot p(\mathsf{Len}(u))^2$, which is primitive recursive as integer division and exponentiation are primitive recursive.

132. If $0 < \mathsf{Len}(0)$, then $\mathsf{Elem}(0,0) < 0$, which is impossible.

133. (a) $f_a(u)$ is the minimal $v \in \omega$ which satisfies the following recursive relation (definition by cases and bounded quantifiers), thus it is recursive:

> if $\mathsf{Len}(u) = 0$ then $\mathsf{Len}(v) = 0$,
> if $\mathsf{Len}(u) > 0$ then $\mathsf{Len}(v) = \mathsf{Len}(u) \doteq 1$ and $(\forall i < \mathsf{Len}(u))\big((u)_i = (v)_i\big)$.

(b) $\mathsf{Len}(u) \le \mathsf{Len}(v) \ \land \ (\forall i < \mathsf{Len}(u))\big((u)_i = (v)_i\big)$.

(c) $f_c(u)$ is the minimal element v satisfying the recursive relation

$$\mathsf{Len}(v) = \mathsf{Len}(u) \ \land$$
$$(\forall i < \mathsf{Len}(u))\big(\mathsf{Len}((v)_i) = i \ \land \ R_b((v)_i, u)\big).$$

134. Following the hint, $u \in \omega$ is a code of $\langle 1, 2, \ldots, 2^{n-1}\rangle$ if the following holds:

- either $\mathsf{Len}(u) = 0$ (empty sequence),
- or $\mathsf{Len}(u) > 0$, and $(u)_0 = 1 \ \land \ (\forall i < \mathsf{Len}(u) \doteq 1)\big((u)_{i+1} = 2 \cdot (u)_i\big)$.

This $R \subseteq \omega$ is a recursive relation. Let

$$g(n) = \mu\{u : u \in R \text{ and } \mathsf{Len}(u) > n\},$$

which is recursive by Problem 115 as for each n there exists such a u. Finally, the n-th element of $g(n)$ is $2^n = ((g(n))_n$.

135. Let $f = \mathsf{PrRec}(g, h)$, and $R_f \subseteq \omega$ be the set of integers $u \in \omega$ which satisfy $(u)_i = f(i)$ for all $i < \mathsf{Len}(u)$. Similarly to Solution 134, the relation R_f is recursive, consequently

$$g_f(n) = \mu\{u : u \in R_f \text{ and } \mathsf{Len}(u) > n\}$$

is a recursive function. Finally $f(n) = (g_f(n))_n$.

136. The two-place function $g(i,b)$ takes value n if n is the $(b+1)$-st place where $f(n) = i$. Then $g(i,i)$ is the desired function, thus it suffices to show that g is recursive. Let $h(i,u)$ be the smallest $v \geq u$ where $f(v) = i$; this is recursive by Problem 115. Then g can be defined by primitive recursion as

$$g(i,0) = h(i,0),$$
$$g(i,b+1) = h(i,g(i,b)+1).$$

137. Let $g(i) = \mu\{u : (\forall j \leq i)\,(f(j) \leq u)\}$. There are infinitely many i such that $f(j) \leq f(i)$ for all $j \leq i$.

138. Consider the recursive relation $(u,v) \in R$ defined as

$$\mathsf{Len}(v) = \mathsf{Len}(u) + 1 \ \wedge$$
$$(v)_0 = 1 \ \wedge$$
$$(\forall i < \mathsf{Len}(i))\big((v)_{i+1} = (v)_i \cdot f((u)_i)\big).$$

Then $h(u) = \mu\{v : (u,v) \in R\}$ is recursive, and $g(u) = (h(u))_{\mathsf{Len}(u)}$.

139. Let $F(\vec{x}, n)$ be the code of the sequence $\langle f(\vec{x},0), \ldots, f(\vec{x},n)\rangle$. F can be defined by primitive recursion as follows: $F(\vec{x},0) = 0^\frown G(\vec{x}, u_0)$ (as 0 is the code of the empty sequence), and $F(\vec{x}, n+1) = F(\vec{x},n)^\frown G(\vec{x}, F(\vec{x},n))$. Finally, $f(\vec{x},n) = (F(\vec{x},n))_n$.

140. Solution 139 works in this case.

141. There are countably many recursive functions, enumerate them as f_0, f_1, etc., and let $H(i) = i \cdot \max\{f_j(i) : j < i\}$.

142. Let $g(b,k)$ be the minimal v such that $u^\frown z \leq v$ for all $u \leq b$ and $z \leq k$. This is a recursive function. Let $f(0,k) = 0$ (the code of the empty sequence), and $f(n+1,k) = g(f(n,k),k)$.

143. The celebrated Ramsey theorem says that the number $f(k,r)$ exists. The point of this exercise is to show that this function is recursive.

First, we say that $u \in \omega$ encodes an r-coloring of an n-graph if the elements of u are triplets $\langle i,j,c\rangle$ with $i < j < n$ indicating that the edge ij has color $c < r$; and for every $i < j$ there is a unique triplet specifying the edge color. This is a recursive relation which can be written quite easily using logical operators and bounded quantifiers. For example, the condition

$$(\forall i < n)\,(\forall j < i)\,(\exists k < \mathsf{Len}(u))\big(((u)_k)_0 = i \wedge ((u)_k)_1 = j \wedge ((u)_k)_2 < r\big)$$

expresses that every edge has a (not necessarily unique) color.

A *subgraph of size* k is an increasing sequence v of length k such that the last element of the sequence is less than n. The relation that the subgraph defined by v is homogeneous in the coloring defined by u is similarly recursive. The problem asks for the minimal n such that for every r-coloring u of an n-graph

there is a size k subgraph v

which is not homogeneous.

To finish the demonstration that f is recursive we need to show that both quantifiers in the sentence above are bounded by recursive functions: for a given n every r-coloring u is smaller than some recursive function $f(n)$, and, similarly, each size k subgraph v is smaller than some other recursive function $g(n)$. Both of them follow from Problem 142.

144. The first problem is that not every number is of the form $y^\frown z$, thus the recursive definition does not define $C(x, y)$ for every y. A more serious problem is that $y^\frown z$ can be strictly smaller than y, thus the Rec operator is not applicable.

Use the idea of Solution 138: $C(x, y)$ is the minimal u, which satisfies the recursive relation

$$\mathsf{Len}(u) = \mathsf{Len}(x) + \mathsf{Len}(y), \quad \text{and}$$
$$(\forall i < \mathsf{Len}(x))\,(u)_i = (x)_i, \quad \text{and}$$
$$(\forall j < \mathsf{Len}(y))\,(u)_{\mathsf{Len}(x)+j} = (y)_j.$$

145. This is the minimal v which satisfies

$$\mathsf{Len}(v) = \mathsf{Len}(u) \quad \text{and}$$
$$(\forall i < \mathsf{Len}(u))\,(v)_i = (u)_{\mathsf{Len}(u) \dotminus (i+1)}.$$

146. A sequence of triplets $\langle x_i, y_i, z_i \rangle$ is an *evaluation of the Ackermann function*, if looking at the i-th triplet in the sequence, one of the following possibilities hold (see the Definition 4.4 on page 21):

- $x_i = 0$ and $z_i = y_i + 1$ (first line of the definition);
- $x_i > 0$ and $y_i = 0$ and there is a $j < i$ such that in the j-th triplet $x_j = x_i - 1$, $y_j = 1$, and $z_i = z_j$ (second line);
- both x_i, y_i are positive and there is a $j < i$ such that $x_j = x_i$, $y_j = y_i - 1$, and there is a $k < i$ such that $x_k = x_i - 1$, $y_k = z_j$, and $z_k = z_i$ (third line).

It is clear that the set of sequences which are "evaluations" is a recursive relation, moreover for each n, m there is an evaluation which contains a triplet where the first two elements are n and m, respectively (meaning that the value of the Ackermann function can be computed at that place). Consequently $B(n, m) = \mu\{u : u$ is an evaluation of the Ackermann function, and if v is the last element of u then $(v)_0 = n$ and $(v)_1 = m\}$ is a recursive function. Then $A(n, m)$ is the third element of the last element of $B(n, m)$.

147. (a) See Problem 115.

(b) The set of primitive recursive functions is not closed for the μ operator, thus there must be some primitive recursive f for which $\mu(f)$ is not primitive recursive. Let $f(\vec{x}, y)$ be such a function. Then the relation $R(\vec{x}, u)$ defined by

$$f(\vec{x}, u) = 0 \ \wedge \ (\forall v < u)\, f(\vec{x}, v) \neq 0$$

is primitive recursive, see Problem 104, and it is the graph of $\mu(f)$.

148. Let Len and Elem be primitive recursive coding functions (see Problem 131), and $f: \omega \to \omega$ be a not primitive recursive function whose graph is primitive recursive (see Problem 147 (b)). Functions $K(u)$ and $L(u)$ from Problem 124 are primitive recursive. For $u^* \in \omega$ we let $u = K(u^*)$ and $y = L(u^*)$. Define the functions Len^* and Elem^* as follows:

$$\text{Len}^*(u^*) = \begin{cases} \text{Len}(u) & \text{if } f(\text{Len}(u)) = y, \\ 0 & \text{otherwise;} \end{cases}$$

$$\text{Elem}^*(u^*, i) = \text{Elem}(u, i).$$

Len^* and Elem^* are primitive recursive as the relation $f(x) = y$ is primitive recursive; they are clearly coding functions. We claim that the corresponding append function is not primitive recursive. If it were so, then the function

$$F(0) = \text{code of the empty sequence,}$$
$$F(n+1) = F(n) \frown 0$$

returning the code of the all-zero sequence of length n would be primitive recursive. As $\text{Len}^*(F(n)) = n$, we have $f(n) = L(F(n))$ for $n \geq 1$, contradicting that f is not primitive recursive.

149. Recall that 0 is the code of the empty sequence. If A is enumerated by $f_1, \ldots, f_n$, then let

$$f(i) = 0 \frown f_1(i) \frown \cdots \frown f_n(i).$$

The other direction is clear.

150. The function $C(x, y)$ which returns the concatenation of the sequences x and y is recursive, see Problem 144. Suppose the unary f enumerates A and the unary g enumerates B. Then $g(u) = C(f((u)_0), g((u)_1))$ enumerates $A \times B$.

151. Suppose f is defined on ω^n, and τ from Problem 150 enumerates ω^n. f and the unary function $f((\tau(u))_0, \ldots (\tau(u))_{n-1})$ has the same range.

152. This is clear if A is finite. Otherwise fix the unary recursive function τ enumerating ω^n (see Problem 150) and let $g\tau(u) = g((\tau(u))_0, \ldots, (\tau(u))_{n-1})$. Define h by primitive recursion as $h(0) = \mu\{u : g\tau(u) = 0\}$ and $h(n+1) = \mu\{u : g\tau(u) = 0 \wedge u > h(n)\}$.

153. (a) The relation A is the zero-set of the function $1 \dot- \chi_A$. The claim follows from Problem 152.

(b) The enumerating function g must avoid all values it has taken earlier, thus the Rec operator comes handy. Using the function τ enumerating ω^n, the function

$$G(u) = \mu\{j : \tau(j) \in A \wedge (\forall i < \text{Len}(u)) \, j \neq (u)_i\}$$

returns the minimal element of A which is not listed by any member of the sequence u. As A is infinite, this is recursive, and $\mathrm{Rec}(G)$ is the required enumeration.

(c) Suppose the unary h enumerates $A \subseteq \omega^n$, where each $h(i)$ is a sequence code of length n. First define g using the Rec operator such that the values $h(g(j))$ are all different:

$$G(u) = \mu\{j : (\forall i < \mathrm{Len}(u))\, h(j) \neq h((u)_i)\}$$

(that is, $G(u)$ is the first place where h takes a value not encountered in u). The final function is $f(j) = h(g(j))$.

154. Suppose $A_i \subseteq \omega^n$ is enumerated by the unary f_i for $i = 1, 2$. The union is enumerated by

$$f(i) = \begin{cases} f_1([i/2]) & \text{if } i \text{ is even,} \\ f_2([i/2]) & \text{if } i \text{ is odd.} \end{cases}$$

Let $\vec{a} \in A_1 \cap A_2$ arbitrary (if the intersection is empty, there is nothing to prove), and $a \in \omega$ be the code of the sequence $\vec{a}$. For $i, j \in \omega$ let $g(i, j) = f_1(i)$ if $f_1(i) = f_2(j)$, and $g(i, j) = a$ otherwise. It is clear that g is recursive and the range of g is $A_1 \cap A_2$.

155. Assume B is not empty, let $\vec{b} \in B$ arbitrary, and suppose the unary g enumerates A. The recursive function

$$h(\vec{x}, u) = \begin{cases} \langle \vec{x} \rangle & \text{if } f(\vec{x}) = g(u), \\ \langle \vec{b} \rangle & \text{otherwise} \end{cases}$$

enumerates B.

156. Let the unary recursive g enumerate the graph of f, and let $H(\vec{x}) = \mu\{u : \text{the first } n \text{ elements of } g(u) \text{ encode } \vec{x}\}$ returning the smallest u which codes $\langle \vec{x}, f(\vec{x}) \rangle$. H is clearly recursive, and $f(\vec{x})$ is the last element of $g(H(\vec{x}))$.

157. Let f and g enumerate A and its complement, respectively. For each $\vec{x} \in \omega^n$ the code x of the tuple $\vec{x}$ (computed as $x = 0^\frown x_1 {}^\frown \cdots {}^\frown x_n$) is in the domain of either f or g (but not in both). Thus the function

$$H(\vec{x}) = \mu\{j : x = f(j) \ \lor \ x = g(j)\}$$

is recursive (for each $\vec{x}$ the indicated recursive relation holds for some $j \in \omega$). The characteristic function of A is the recursive relation

$$f(H(\vec{x})) = x.$$

158. If $A \cup B$ were recursive, then both A and the complement of A relative to $A \cup B$ would be recursively enumerable, and thus we could apply the same argument as in 157.

159. Let A be infinite and the range of the recursive function f. Define g by recursion as $g(0) = f(0)$ and $g(n+1) = f(\mu\{y : f(y) > g(n)\})$. The range of g is recursive and is an infinite subset of A.

160. Enumerate all infinite recursive sets as G_0, G_1, ... (there are count-ably many of them). Choose distinct elements a_i, b_i, our set will be $X = \{b_0, b_1, ...\}$. At the i-th step choose $a_i \in G_i$ (as G_i is infinite, there is an element different from all previously chosen elements), and any b_i different from all the other elements. At the end $a_i \notin X$, thus G_i is not a subset of X. On the other hand all $b_i \in X$, thus X is infinite.

161. The Comp and μ operators on Ω^* are extensions of the same operators on Ω.

162. Let $g(\vec{x}, u)$ be 0 if $\vec{x} \in A$ (independently of u), and 1 otherwise; it is recursive. The function $\mu(g)$ has domain A and it takes 0 there (the smallest u where $g(\vec{x}, u) = 0$). Thus $h = (1 + \mu(g)) \cdot f$ is partial recursive.

163. Let A be the set of points where the function is defined. As A is finite, it is recursive. The function f which takes the given values at A and zero everywhere else is recursive. Apply Problem 162.

164. $h(\vec{x}) = 0 \cdot g(\vec{x}) + f(\vec{x})$.

165. Pick a (total) recursive function that takes every value infinitely often. Such a function is, e.g., the excess function $K(x) = x \div [\sqrt{x}]^2$. Then $g(n) = f(K(n))$ is suitable.

166. Assume that the unary recursive $g(u)$ enumerates the graph of f (see Problem 149). For $\vec{x} \in \omega^n$ let $H(\vec{x}, u)$ be zero if the first n elements of the tuple $g(u)$ give $\vec{x}$, and 1 otherwise. As H is recursive, $h = \mu(H)$ is partial recursive. If f is not defined at $\vec{x}$, then $h(\vec{x}) = \uparrow$ (as $H(\vec{x}, u) = 1$ for all u); if f is defined at $\vec{x}$, then $h(\vec{x})$ is the tuple where the first n elements code $\vec{x}$, and the last element is the value of f. Thus $f(\vec{x}) = (h(\vec{x}))_n$ is partial recursive.

167. By induction on the definition of partial recursive functions in Definition 4.14. All initial functions are recursive, thus their graphs are enumerable by Problem 153 (a).

The composition operator. Suppose that the graphs of g and $h_1 ..., h_\ell$ are all enumerable. If any of these graphs is empty, or the composite function f has an empty domain, then there is nothing to prove. Thus pick $a \in \omega$ coding an arbitrary element of the graph of f. Let the unary recursive functions $\tilde{g}$, $\tilde{h}_i$ enumerate the corresponding graphs. The following $\ell + 1$-variable recursive function defined on $(y_1, ..., y_\ell, z)$ enumerates the graph of f:

> $\tilde{g}(z)$ encodes a point in the graph of g, and for $1 \le i \le \ell$ $\tilde{h}_i(y_i)$ encodes a point in the graph of h_i. If all sequences $\tilde{h}_i(y_i)$ agree on their first n elements (encoding the argument $\vec{x}$), and $\tilde{g}(z)$ encodes g at the place determined by $(n+1)$-st elements of the tuples $\tilde{h}_i(y_i)$, then the functions returns the encoding of $\vec{x}$ followed by the last element of $\tilde{g}(z)$. Otherwise the function returns a.

The μ operator. Assume that $\tilde{g}$ enumerates the graph of the function g, and the domain of $\mu(g)$ is not empty. At $u \in \omega$ check if $\text{Len}(u) > 0$, and for each

$i < \mathrm{Len}(u)$, $\tilde{g}((u)_i)$ encodes g at $(\vec{x}, i)$ for the same $\vec{x}$; furthermore all function values are positive except for the last one which is zero (the last coordinates of $\tilde{g}((u)_i)$). If this is the case, the result is the encoding of $\langle \vec{x}, \mathrm{Len}(u) \doteq 1 \rangle$, otherwise the result is the fixed element from the graph.

168. (a) Suppose $\mathrm{dom}(f_1) = A$ and $\mathrm{dom}(f_2) = \omega - A$. Then $\mathrm{dom}(g_1) = \emptyset$ while $\mathrm{dom}(g_2) = \omega$.

(b) g_1 is partial recursive as Ω^* is closed for addition and multiplication. By Problems 166 and 167 it is enough to check that the graph of g_2 is enumerable. In even / odd steps take the next element from the enumeration of f_1 and f_2, respectively, and check if the enumerated argument is in A (it is not in A).

169. (a) $\Rightarrow$ (b), (c): If A is enumerable, then so is $A \times A$, which is the graph of the partial recursive function with $\mathrm{dom}(f) = \mathrm{ran}(f) = A$.

(b), (c) $\Rightarrow$ (a): If $A \subseteq X \times Y$ is enumerable, then so is the projection $\pi_X(A) = \{x \in X : (x, y) \in A$ for some $y \in Y\}$.

(b) $\Rightarrow$ (d): If $\mathrm{dom}(f') = A$ then $f(x) = f'(x) \cdot 0$.

(d) $\Rightarrow$ (b): $\mathrm{dom}(f) = A$.

Remark. The equivalence (b) $\Leftrightarrow$ (c) also follows from Problem 203.

170. The graph of a partial recursive function is enumerable by Problem 167. If this is the graph of a total function, then the function is recursive by Problem 156.

171. (a) Any sequence code determines uniquely the length and the elements of the sequence. Thus $\alpha(\sigma)$ determines uniquely σ.

(b) Let $C(u)$ be the characteristic function of expression codes. Assuming $C(u)$ has been defined for all $v < u$, $C(u) = 1$ if the following recursive relation (referring to values C takes for smaller arguments) holds, otherwise $C(u) = 0$

$$u = \langle 0 \rangle, \text{ or}$$
$$\mathrm{Len}(u) = 2 \wedge (u)_0 = 1 \wedge (u)_1 < |\Sigma|, \text{ or}$$
$$\mathrm{Len}(u) = 3 \wedge (u)_0 = 2 \wedge C((u)_1) = 1 \wedge C((u)_2) = 1, \text{ or}$$
$$\mathrm{Len}(u) = 3 \wedge (u)_0 = 3 \wedge C((u)_1) = 1 \wedge C((u)_2) = 1, \text{ or}$$
$$\mathrm{Len}(u) = 2 \wedge (u)_0 = 4 \wedge C((u)_1) = 1.$$

172. (a) $f(n, u, v)$ is the smallest w for which the following four recursive relations hold. The first two say that w contains every word of length at most n which is in either u or v; the third one says that every word in w has length at most n; and the last one says that every element of w comes from either v or w:

$$(\forall i < \mathrm{Len}(u)) \left(\mathrm{Len}((u)_i) \leq n \rightarrow (\exists k < \mathrm{Len}(w)) (w)_k = (u)_i \right),$$
$$(\forall j < \mathrm{Len}(v)) \left(\mathrm{Len}((v)_i) \leq n \rightarrow (\exists k < \mathrm{Len}(w)) (w)_k = (v)_j \right),$$
$$(\forall k < \mathrm{Len}(w)) (\mathrm{Len}((w)_k) \leq n),$$
$$(\forall k < \mathrm{Len}(w)) \left((\exists i < \mathrm{Len}(u)) (w)_k = (u)_i \vee (\exists j < \mathrm{Len}(v)) (w)_k = (v)_j \right).$$

(b) Similar to (a); concatenating two sequences is recursive by Problem 144.

173. If $w \in L^*$ has length at most n, then $w \in \{\lambda\} \cup L \cup \cdots \cup L^n$. Thus it suffices to show that there is a recursive function $h(n, i, u)$ which encodes the length $\leq n$ words in the language $\{\lambda\} \cup L \cup \cdots \cup L^i$; then $f(n, i) = h(n, n, u)$ works. $h(n, i, u)$ can be defined by primitive recursion on i using the functions in Problem 172.

174. Observe that

$$(\sigma\tau) \cap \Sigma^{<n} \subseteq (\sigma \cap \Sigma^{<n})(\tau \cap \Sigma^{<n}),$$
$$(\sigma^*) \cap \Sigma^{<n} \subseteq (\sigma \cap \Sigma^{<n})^*, \text{ and}$$
$$(\sigma|\tau) \cap \Sigma^{<n} = (\sigma \cap \Sigma^{<n})|(\tau \cap \Sigma^{<n}).$$

Thus we can keep n fixed and generate $\sigma \cap \Sigma^{<n}$ by "structural recursion" from the constituent restricted languages $\tau \cap \Sigma^{<n}$ using the recursive functions in Problems 172 and 173.

175. Trivial from Problem 174.

176. Similar to Solution 171 (b).

177. Lengthy but straightforward case by case checking. For example, if $(u)_i = \langle c_i, x_i, y_i \rangle$ and c_i codes $\mathsf{PrRec}(g, h)$ as $c_i = \langle \ell + 1, 2, \alpha(g), \alpha(h) \rangle$, then we must have $\mathsf{Len}(x_i) = \ell + 1$ (saying that the arguments for $\mathsf{PrRec}(g, h)$ are $\langle \vec{x}, n \rangle$ as coded by x_i). If the last element of x_i is zero, then we must have $y_i = g(\vec{x}, 0)$, and then some $(u)_j$ must code this computation:

$$(x_i)_\ell = 0 \rightarrow (\exists j < i)\big(((u)_j)_0 = \alpha(g) \wedge ((u)_j)_1 = x_i \wedge ((u)_j)_2 = y_i\big).$$

If the last element of x_i is positive, then $y_i = h(\vec{x}, n-1, f(\vec{x}, n-1))$, thus we must have an earlier triplet which computes $z = f(\vec{x}, n-1)$:

$$(\exists j < i)(u)_j = \langle c_i, \langle \vec{x}, n-1 \rangle, z \rangle,$$

and another earlier triplet which computes $y_i = h(\vec{x}, n-1, z)$:

$$(\exists k < i)(u)_k = \langle \alpha(h), \langle \vec{x}, n-1, z \rangle, y_i \rangle.$$

178. If i is a code of a primitive recursive function with arity n, then let $H_n(i, \vec{x})$ be the minimal justified computation u (Problem 177) which ends with the triplet $\langle i, \langle \vec{x} \rangle, y \rangle$. Such a u exists as the function can be computed at $\vec{x}$. Let $W_n(i, \vec{x})$ be the last element of the last element of $H_n(i, \vec{x})$.

179. Otherwise there would be an $i \in \omega$ such that $W_1(i, x) = W_1(x, x) + 1$ for all $x \in \omega$, in particular for $x = i$.

180. Let $g(x)$ be the unary primitive recursive function which takes 2 everywhere. Then $f_{i+1} = \mathsf{PrRec}(g, \mathsf{Comp}(f_i, U_2^2))$. Thus $\alpha(f_{i+1}) = \langle 1, 2, \alpha(g), z \rangle$, where $z = \langle 2, 1, \alpha(f_i), \alpha(U_2^2) \rangle$. As both $\alpha(g)$ and $\alpha(U_2^2)$ are fixed natural numbers (they do not depend on i), $\alpha(f_{i+1})$ is a recursive function of $\alpha(f_i)$. Use primitive recursion to define $i \mapsto \alpha(f_i)$.

181. By course-of-values recursion: one can tell whether $u \in \omega$ is a function code if the same is known for all $v < u$. See also Problems 176 and 171.

182. This is the function $g(x) \cdot 0$. Let $\alpha(\mathbf{0}) \in \omega$ be the code of the unary all-zero function (see Problem 114 (a)). To get the required function apply the Comp operator to the multiplication, g, and $\mathbf{0}$ functions. Thus $f(i)$ can be the code of the sequence $\langle 1, 1, \alpha(\cdot), i, \alpha(\mathbf{0}) \rangle$, which can be computed with the append function as

$$0 \frown 1 \frown 1 \frown \alpha(\cdot) \frown i \frown \alpha(\mathbf{0}).$$

183. Define f by primitive recursion. The identically zero and one functions are $\mathbf{0}$ and $\mathbf{1}$, their codes are concrete natural numbers. Let $f(0) = \alpha(\mathbf{0})$, and $f(i+1)$ be the code of $\mathsf{Comp}(+, f(i), \mathbf{1})$, that is,

$$f(i+1) = 0 \frown 1 \frown 1 \frown \alpha(+) \frown f(i) \frown \alpha(\mathbf{1}).$$

184. Let $C_i(x)$ be the function which takes i everywhere, then $g_i(x)$ is just the composition $g(x, C_i(x))$. It has code $\langle 1, 1, \alpha(g), \alpha(U_1^1), \alpha(C_i) \rangle$, which is a recursive function of $\alpha(g)$ and i as the function $i \mapsto \alpha(C_i)$ is recursive by Problem 183.

185. (a) If the function in the triplet $\langle c_i, x_i, y_i \rangle$ encodes the $\mu(g)$ operation, then for each $j \le y_i$ there must be an earlier triplet computing g at $x_i \frown j$, giving a non-zero result for $j < y_i$, and zero for $j = y_i$.

(b) As in Solution 177, only handling the μ operation is indicated. Thus let $(u)_i = \langle c_i, x_i, y_i \rangle$ where $c_i = \langle \ell, 2, \alpha(g) \rangle$ codes $\mu(g)$. Then for each $j \le y_i$ g must be computed earlier at $x_i \frown j$ yielding correct results:

$$(\forall j \le y_i)(\exists k < i) \big($$
$$(u)_k = \langle \alpha(g), x_i \frown j, z_j \rangle \wedge (j < y_i \to z_j > 0) \wedge (j = y_i \to z_j = 0) \big).$$

186. Let $a = \langle \vec{x}, g(\vec{x}) \rangle$ be a point in the graph of g. If u is a justified computation ending with a triplet $\langle \alpha(g), x_i, y_i \rangle$, then output $x_i \frown y_i$, otherwise output a.

187. $g(\vec{x}) = y$ if and only if there is a justified computation u ending with the triplet $\langle \alpha(g), \langle \vec{x} \rangle, y \rangle$, see Definition 4.15. Thus let

$$H_n(e, \vec{x}, u) = \begin{cases} 0 & \text{if } u \text{ is a justified computation ending with } \langle e, \langle \vec{x} \rangle, y \rangle, \\ 1 & \text{otherwise}, \end{cases}$$

and let $U(t)$ be the last element of the last element of t. The number e in the theorem is the code of g.

188. Let e be the code of g, then the graph of g is

$$\{ \langle \vec{x}, G(u) \rangle : H_n(e, \vec{x}, u) = 0 \wedge (\forall v < u) H_n(e, \vec{x}, v) \neq 0 \}.$$

As the condition is clearly a recursive one, this set can be enumerated by checking whether for a given tuple $(\vec{x}, u)$ it holds or not. If yes, the next point of the enumeration is $\langle \vec{x}, G(u) \rangle$, if not, the next point is some fixed point of the graph.

189. Using the Normal Form Theorem 4.16 for $n = 2$, let $h(e, x, y) = G(\mu\{u : H_2(e, x, y, u) = 0\})$. If the code of g is e, then $h(e, x, y) = g(x, y)$. Using the recursive function $\tilde{g}$ in the hint, $e = (\tilde{g}(x))_0$, and $x = (\tilde{g}(x))_1$, thus the function $H(z, y) = h((z)_0, (z)_1, y)$ works.

190. Observe that if the partial recursive function f does not depend on x, then $f(y) = H(e, y)$ for the constant $e = \tilde{f}(0)$.

Let $g(x, y) = H(H(x, x), y) = H(\tilde{g}(x), y)$. By the above, there is a constant e such that $H(e, x) = h(\tilde{g}(x))$ with the function h given in the problem. Then $H(H(e, e), y) = H(\tilde{g}(e), y)$ on one hand, and $H(e, e) = h(\tilde{g}(e))$ on the other. Thus $m = \tilde{g}(e)$ works.

191. The proof is analogous to Solution 112. Suppose on the contrary that $U : \omega^{n+1} \to \omega$ is total recursive and universal for total recursive functions. Then $U(x_1, x_1, \dots, x_n) + 1$ is total recursive, hence for some index i we have

$$U(i, x_1, \dots, x_n) = U(x_1, x_1, \dots, x_n) + 1.$$

Letting $x_1 = i$ yields to a contradiction.

192. The existence of universal functions is clear from Theorem 4.16: $U_n(e, \vec{x}) = G(\mu\{u : H_n(e, \vec{x}, u) = 0\})$ is universal. The proof that they form a coherent family is indicated in Problem 184. Fix the values of $\vec{x}$ and consider g as an m-variable function $g_{\vec{x}}(\vec{y})$. Then $g_{\vec{x}}$ can be obtained by using the composition

$$g_{\vec{x}} = g(f_1(\vec{y}), \dots, f_n(\vec{y}), \vec{y}),$$

where $f_i(\vec{y})$ takes the value x_i for all $\vec{y}$. The code for $f_i(\vec{y})$ is a recursive function of x_i by Problem 183 (actually, one has to compose that unary function with the projection function U_1^m to make it an m-variable function). Thus the code for $g_{\vec{x}}$ can be created by a recursive function from the code of g and from all x_i.

193. If g has index i, then $g(y, x)$ is the function $f(i, x, y) = U_2(i, y, x)$. This f is a partial recursive function of three variables, thus it has an index, say e: $f(i, x, y) = U_3(e, i, x, y)$. By the s-m-n theorem $U_3(e, i, x, y) = U_2(S_2^1(e, i), x, y)$, thus $g(y, x)$ has index $S_2^1(e, i)$. As e is a fixed number (does not depend on i), this is a recursive function of i.

194. Clearly $g' = g \cdot 0$, that is, we need $\varphi_{r(i)}(x) = \varphi_i(x) \cdot 0$. Let $f(i, x) = \varphi_i(x) \cdot 0$, and let e be its index: $f(i, x) = U_2(e, i, x)$. By the s-m-n theorem this is $U_1(S_1^1(e, i), x)$. Finally, set $r(i) = S_1^1(e, i)$.

195. $\varphi_i(x) + \varphi_j(x)$ as a three-variable partial recursion function has some index e. Then, by applying the s-m-n theorem we get a recursive function S so that

$$\varphi_i(x) + \varphi_j(x) = \varphi_{S(e, i, j)}(x).$$

Set $r(i, j) = S(e, i, j)$.

196. By the s-m-n theorem we have

$$\varphi_i(\varphi_j(x)) = U^3(e, i, j, k) = \varphi_{S(e,i,j)}(x),$$

thus $c(i, j) = S(e, i, j)$ works.

197. Let i_1, i_2 be the index of f_1, f_2, respectively. The recursive function

$$r(i) = \begin{cases} i_1 & \text{if } i \in A, \\ i_2 & \text{otherwise} \end{cases}$$

works.

198. This is the s-m-n theorem in disguise. Each partial recursive $g(x, y)$ is of the form $U_2(e, x, y)$ for a fixed e. By the s-m-n theorem $U_3(e, x, y) = U_1(S_1^1(e, x), y)$, and the recursive function $\tilde{g}(x) = S_1^1(e, x)$ works.

199. $h(x, y) = \varphi_{f(x)}(y)$ is partial recursive, thus it has an index e. Then using the s-m-n theorem

$$h(x, y) = \varphi_{f(x)}(y) = U_2(e, x, y) = \varphi_{S(e,x)}(y),$$

and $f(x) = S(e, x)$ is total recursive.

200. $h(j, x, y) = \varphi_{\varphi_j(x)}(y)$ is partial recursive, let e be its index. Using the s-m-n theorem

$$h(j, x, y) = U_3(e, j, x, y) = \varphi_{S(e,j,x)}(y),$$

and, using the s-m-n theorem again, let $F(j)$ be the recursive function with $\varphi_{F(j)}(x) = S(e, j, x)$.

201. The function $g(x, y) = x$ is partial recursive, thus by Problem 198 $g(x, y) = \varphi_{\tilde{g}(x)}(y)$ for some recursive function $\tilde{g}$. Then $\varphi_{\tilde{g}(x)}(y) = x$ for all x.

202. The graph of the partial recursive function $U^1(i, x)$ is $\{\langle i, x, y \rangle \in \omega^3 : \varphi_i(x) = y\}$. By Kleene's Normal Form Theorem 4.16, there is an $e \in \omega$ such that

$$U_1(i, x) = G(\mu\{u : H_2(e, i, x, u) = 0\}),$$

where G and H_2 are recursive. The function H below works:

$$H(i, x, y, u) = \begin{cases} 0 & \text{if } y = G(u) \text{ and } H_2(e, i, x, u) = 0 \text{ and} \\ & (\forall v < u)\, H_2(e, i, x, v) \neq 0, \\ 1 & \text{otherwise.} \end{cases}$$

203. By Problem 202 there is a four-variable recursive relation H such that $\varphi_i(x) = y$ iff $H(i, x, y, u)$ holds for some $u \in \omega$. Let

$$d(i, y) = \min\{u : H(i, (u)_0, y, (u)_1) = 0\}.$$

It is clear that $y \in \mathrm{dom}(d(i,\cdot))$ iff $y \in \mathrm{ran}(\varphi_i)$. Also, let

$$R(i,u) = \begin{cases} u & \text{if } H(i,(u)_0,(u)_1,(u)_2) = 0, \\ \min\{v : H(i,(v)_0,(v)_1,(v)_2) = 0\} & \text{otherwise,} \end{cases}$$

and $r(i,u) = (R(i,u))_0$. It is clear that $r(i,\cdot)$ is total if $\mathrm{dom}(\varphi_i)$ is not empty, and $\mathrm{ran}(r(i,\cdot)) = \mathrm{dom}(\varphi_i)$.

The existence of the functions h follows from Problem 198.

204. (a) The function $g(x) = \varphi_x(x) + 1$ does not take the same value as φ_i if $i \in \mathrm{dom}(\varphi_i)$.

(b) The function $g(x) = 1 \dotminus \varphi_x(x)$ works as well.

(c) For all $k \in \mathrm{dom}(f)$ we have $f(k) = f(g(\min\{l : g(l) \geq k\}))$, and the function on the right hand side is partial recursive and total.

205. Let A be the domain of $\varphi_x(x) + 1$. It is infinite, as it contains the index of every (total) recursive function, and recursively enumerable by Problem 169. The function f enumerates A in increasing order. This f cannot be recursive by Problem 204.

206. The domain of $\varphi_x(x) + 1$ is such a set. If it were recursive, then the function f defined in Problem 205 would be recursive.

207. (a) Let $g(n) = \varphi_n(n) + 1$. This g is partial recursive. If f is total recursive, then $f = \varphi_i$ for some index i, hence $g(i)$ is defined and we have $g(i) = \varphi_i(i) + 1 > \varphi_i(i) = f(i)$.

(b) $g(n) = \varphi_{(n)_0}(n) + 1$ works as for each i there are infinitely many n such that $(n)_0 = i$.

208. If f is partial recursive, then $\mathrm{dom}(f)$ is recursively enumerable (Problem 169). The domain is infinite, thus the recursive enumeration function h can be chosen to be injective (Problem 153(c)). Define

$$H(i) = \mu\{u : (\forall j \leq i)(u \geq h(j))\}.$$

This is a recursive function, and clearly $H(i) = \max\{h(0),\ldots,h(i)\}$. Let $g(i) = f(H(i)) + 1$. As $H(i) \in \mathrm{dom}(f)$, this is total, consequently recursive. To show that $g(i) > f(i)$ for infinitely many values, it suffices to check that $H(i) = h(i)$ infinitely often.

As $\mathrm{dom}(f) = \mathrm{ran}(h)$ is infinite, the maximum of $\{h(0),\ldots,h(i)\}$ must jump infinitely often. When it jumps, the newly added value must be the new maximum, as required.

209. Take an enumeration $g_0, g_1, \ldots$ of the recursive functions (there are countably many of them). Then

$$f(n) = \max\{g_0(n), g_1(n), \ldots, g_n(n)\} + 1$$

is as desired.

Such an f cannot be recursive, as in this case $f + 1$ would be recursive and clearly f cannot dominate $f + 1$.

210. Let f be such a function. Applying Problem 208 one gets a recursive function g which takes a larger value than f infinitely often. Clearly f could not dominate g.

211. By Problem 204(a) there is a partial recursive f which cannot be extended to a total recursive function. By Theorem 4.16 there are recursive functions $H(x, u)$ and $G(u)$ such that

$$f(x) = G(\mu\{u : H(x, u) = 0\}).$$

Let $g(x) = \mu\{u : H(x, u) = 0\}$. This is partial recursive, and $f(x) = G(g(x))$. We claim that no recursive function dominates $g(x)$. Suppose by contradiction that h dominates g. By changing the value of h at finitely many places it remains total recursive thus we may assume that $g(x) \le h(x)$ for all $x \in \mathrm{dom}(g) = \mathrm{dom}(f)$. Now

$$\tilde{g}(x) = \mu\{u : u \ge h(x) \text{ or } H(x, u) = 0\}$$

is recursive (for each x there is a u satisfying the recursive condition), and $\tilde{g}(x) = g(x)$ when $x \in \mathrm{dom}(g)$. Thus $G(\tilde{g}(x))$ is recursive and extends $f(x) = G(f(x))$, the required contradiction.

212. Let f be total recursive with $\mathrm{ran}(f) = A$ and write

$$h(i, x) = 1 + \mu\{u : (\forall k < i)\, (u \ge \varphi_{f(k)}(x)\}.$$

Then $g(x) = h(x, x)$ is total and dominates all $\varphi_i(x)$ for $i \in A$.

213. Similarly to Solution 211 fix the partial recursive f and use Theorem 4.16 to write it as

$$f(x) = G(\mu\{u : H(x, u) = 0\})$$

with recursive G and H. The relation R will be defined as

$$(x, u) \in R \text{ iff } u = 0 \text{ or } \left(H(x, u) = 0 \ \wedge \ (\forall v < u)\, H(x, v) \ne 0 \right).$$

It is clear that R is recursive (as $H(x, u)$ is), and for each $x \in \omega$ either one or two u satisfies $(x, u) \in R$. If $g(x) = \max\{u : (x, u) \in R\}$ is recursive, then so is $G(g(x))$ which takes the same value as f for all $x \in \mathrm{dom}(f)$. Thus g cannot be recursive when f is not the restriction of a total recursive function (Problem 204(a)).

214. Problem 175 states that the set of code pairs $\langle \alpha(\sigma), \alpha(w) \rangle$ where the word w is in the language generated by the regular expression σ is recursive.

215. Assume, by way of contradiction, that there is a recursive f such that

$$f(x) = \begin{cases} 1 & \text{if } \varphi_x(x) = \downarrow \\ 0 & \text{otherwise.} \end{cases}$$

Let

$$\rho(x) = \begin{cases} \uparrow & \text{if } \varphi_x(x) =\downarrow \\ 0 & \text{otherwise.} \end{cases}$$

Then ρ is partial recursive as $\rho(x) = \mu\{u : u + f(x) = 0\}$. Thus $\rho = \varphi_z$ for some z, but then

$$\rho(z) = \begin{cases} \uparrow & \text{if } \varphi_z(z) =\downarrow \\ 0 & \text{otherwise,} \end{cases}$$

which is a contradiction.

216. As K is the domain of the partial recursive function $\varphi_x(x)$, it is enumerable by Problem 169.

217. By Problem 169 A is the domain of a partial recursive function $h(x)$. Let $g(x, y) = h(x)$, by Problem 198 there is a recursive $\tilde{g}(x)$ such that $h(x) = g(x, y) = \varphi_{\tilde{g}(x)}(y)$ independently of the value of y. Thus $x \in \text{dom}(h)$ iff $\varphi_{\tilde{g}(x)}(\tilde{g}(x)) =\downarrow$.

218. Suppose there exists a recursive function f with $\text{ran}(f) = \{i : \varphi_i \text{ is total}\}$. Then $g(x) = \varphi_{f(x)}(x) + 1$ is total recursive, hence for some $k \in \omega$ we have $g = \varphi_{f(k)}$. But then $\varphi_{f(k)}(k) = g(k) = \varphi_{f(k)}(k) + 1$, a contradiction.

219. In this problem the set A contains indices of two-variable functions. Questions about one-variable functions can be transformed using the s-m-n theorem. For an index i of a one-variable function let $g(i, x, u) = 0 \cdot \varphi_i(x)$. This function has an index e

$$g(i, x, u) = U_3(e, i, x, u) = U_2(S(e, i), x, u) = \varphi_{S(e,i)}(x, u)$$

by the s-m-n theorem for some recursive function S. The unary $\varphi_i(x)$ is total if and only if $S(e, i) \in A$. If A were recursive, then so would be the set of unary total recursive functions, contradicting Problem 218.

220. (a) Let h be a recursive function with $\varphi_{g(\varphi_x(x))}(z) = \varphi_{h(x)}(z)$, see Problem 198. Let i be an index of h, and $m = h(i)$. Then $h(i) = \varphi_i(i)$, thus $\varphi_{g(m)} = \varphi_m$, as required.

   (b) If g is partial recursive, then it may happen that $g(m) =\uparrow$, and then $\varphi_{g(m)}$ is not defined. In this case, however, φ_m has empty domain.

221. Let f be in as Problem 201, that is $\varphi_{f(i)}(x) \equiv i$ is the constant i function. By the fixed point theorem there is m such that $\varphi_{f(m)}(x) = \varphi_m(x) = \varphi_{f(m)}(x) \equiv m$.

222. For $g(x, y) = x + y$ there is a recursive h so that $g(x, y) = \varphi_{h(x)}(y)$. Apply the fixed point theorem to h to get $i \in \omega$ with $\varphi_i(y) = \varphi_{h(i)}(y) = i + y$.

223. Recall that $\varphi_i(x)$ and $\varphi_i(x, y)$ abbreviate the universal functions $U_1(i, x)$ and $U_2(i, x, y)$. Using the s-m-n theorem one can find a recursive function $h(i, x)$ such that for all x

$$U_1(\varphi_i(\varphi_x(i, x)), z) = U_1(h(i, x), z).$$

Let j be the index of h: $h(i,x) = \varphi_j(i,x)$, and let $m(i) = h(i,j) = \varphi_j(i,j)$. The above equation with $x = j$ gives

$$\varphi_{\varphi_i(m(i))} \equiv \varphi_{m(i)}.$$

224. (a) Using the idea of Solution 223 let h be such that $\varphi_{h(x)} = \varphi_{f(h(x),x)}$ returning a fixed point of $\varphi_{f(i,x)}$. By the Fixed point theorem 4.19 there is j such that $\varphi_j = \varphi_{g(h(j),j)}$. Finally, put $i = h(j)$.

(b) Take constant functions f and g giving indices to different functions.

(c) If the recursive function $f(x)$ takes all values, then for every recursive h there is a recursive h' such that $h(x) = f(h'(x))$. Similarly to Solution 220, let $h(x)$ be such that $\varphi_{h(x)} = \varphi_{g(\varphi_x(x))}$, and $h'(x)$ be such that $h(x) = f(h'(x))$. Let m be an index of h', then $h(x) = f(\varphi_m(x))$ and $\varphi_{f(\varphi_m(x))} = \varphi_{g(\varphi_x(x))}$. Choosing $i = \varphi_m(m)$ gives $\varphi_{f(i)} = \varphi_{g(i)}$.

225. If f has a finite set of fixed points, then there exists a partial recursive g which is different from φ_i for any $i \in \mathrm{Fix}(f)$. Let h be the recursive function such that

$$\varphi_{h(x)} = \begin{cases} g & \text{if } x \in \mathrm{Fix}(f), \\ \varphi_{f(x)} & \text{otherwise.} \end{cases}$$

This h does not have any fixed point, contradicting Theorem 4.19.

226. Suppose by contradiction that A and $\omega \smallsetminus A$ are the fixed point sets of the f and g, respectively. In this case the recursive function h with the property

$$\varphi_{h(x)} = \begin{cases} \varphi_{g(x)} & \text{if } x \in A \\ \varphi_{f(x)} & \text{otherwise.} \end{cases}$$

has no fixed point, a contradiction.

227. $\mathrm{Fix}(f) \supseteq A$ is true, but equality does not necessarily hold. If $\mathrm{Fix}(f) = A$, then, by Problem 226, $\omega \smallsetminus A$ cannot be a set of fixed points of any recursive function. Thus, the statement must fail either for A or for $\omega \smallsetminus A$.

228. Following the hint $g(n,x,y)$ is partial recursive, and by the s-m-n theorem there is a recursive h such that $\varphi_{h(n)}(x,y) = g(n,x,y)$. This $\varphi_{h(n)}(x,y)$ satisfies the following equations:

$$\begin{aligned}
\varphi_{h(n)}(0,y) &= y+1 \\
\varphi_{h(n)}(x+1,0) &= \varphi_n(x,1) \\
\varphi_{h(n)}(x+1,y+1) &= \varphi_n(x,\varphi_n(x+1,y))
\end{aligned}$$

By the fixed point theorem 4.19 there is an m such that $\varphi_m = \varphi_{h(m)}$. Therefore φ_m is the Ackermann function. Thus, the Ackermann function is partial recursive, and since it is total, it is total recursive as well (see Problem 170).

229. For $z = 1$ and $z = 2$ the value of $B(x,y,z)$ can be obtained by induction of y.

To show that B is recursive, apply the trick of Solution 228. Write

$$\alpha(x,z) = \begin{cases} 0 & \text{if } z = 0, \\ 1 & \text{if } z = 1, \\ x & \text{otherwise,} \end{cases}$$

and consider the partial recursive function $g(n,x,y,z)$ defined as

$$g(n,x,y,z) = \begin{cases} x+y & \text{if } z = 0, \\ \alpha(x, z \dotdiv 1) & \text{if } y = 0 \text{ and } z > 0, \\ \varphi_n(x, \varphi_n(x, y \dotdiv 1, z), z \dotdiv 1) & \text{if } y > 0 \text{ and } z > 0. \end{cases}$$

By the s-m-n theorem there is a recursive h such that $\varphi_{h(n)}(x,y,z) = g(n,x,y,z)$, and by the fixed point theorem 4.19 there is an m such that $\varphi_m = \varphi_{h(m)}$. Now φ_m satisfies the defining equations of $B(x,y,z)$, thus B is partial recursive. It is easy to see that B is a total function, hence it is recursive by Problem 170.

230. Let $A \subset \omega$ be proper non-empty and pick $i \in A$ and $j \notin A$. If A is recursive then so is the function

$$g(x) = \begin{cases} i & \text{if } x \in A, \\ j & \text{if } x \notin A. \end{cases}$$

By the fixed point theorem Problem (220) there is m with $\varphi_m = \varphi_{g(m)}$. Thus, if A is an index property then $m \in A \Leftrightarrow g(m) \in A$. On the other hand

$$m \in A \quad \Rightarrow \quad g(m) = j \notin A,$$
$$m \notin A \quad \Rightarrow \quad g(m) = i \in A,$$

contradiction.

231. As the sets are proper index properties, they are undecidable (not recursive) by Rice's theorem 4.21.

(a) A is the domain of the partial recursive function $g(n) = \varphi_n(0)$, hence it is recursively enumerable (see Problem 169).

(b) The complement of B is recursively enumerable by Problem 203(b). Therefore B cannot be recursively enumerable, because that would imply the decidability of B.

(d) The complement of D is recursively enumerable as D is the domain of the partial recursive function $f(n) = \varphi_n(0)$. Therefore D cannot be recursively enumerable, because that would imply that D is decidable.

(c), (e), (f) The complement of the set is recursively enumerable, but the set is not recursive, thus is cannot be recursively enumerable.

232. Such a set is a non-trivial index property, thus undecidable by Rice's theorem. Finite sets are decidable.

233. The idea is that the identically zero function will have a single W-index. We achieve this by explicitly letting $W(0,x)$ to be zero for all x, and for all other indices we increase the value of the function somewhere. As all non-zero functions take a non-zero value somewhere (or not defined somewhere), all partial recursive functions will have an index. Thus let $W(0, x) = 0$, and for $i = \langle j, y \rangle \geq 1$ let

$$W(i, x) = \begin{cases} \varphi_j(x) & \text{if } x \neq y, \\ \varphi_j(x) + 1 & \text{if } x = y. \end{cases}$$

234. Let A be a recursively enumerable but not recursive set A. Let h be a total recursive function such that $\mathrm{ran}(h) = A$. Observe that the sequence

$$\frac{1}{2^{h(0)}}, \quad \frac{1}{2^{h(0)}} + \frac{1}{2^{h(1)}}, \quad \frac{1}{2^{h(0)}} + \frac{1}{2^{h(1)}} + \frac{1}{2^{h(2)}}, \cdots$$

converges to the real number represented by χ_A, i.e. the n-th binary digit of the limit point is $\chi_A(n)$. By choice χ_A is not recursive.

It only remains to show that the sequence above can be represented in the form $f(n)/g(n)$. But this is easy as

$$\frac{1}{2^{h(0)}} + \cdots + \frac{1}{2^{h(n)}} = \frac{\sum_{i \leq n} \prod_{j \neq i} 2^{h(i)}}{\prod_{j \leq n} 2^{h(j)}}$$

and both the enumerator and the denominator are recursive.

235. No. For total recursive f and g write $a(n) = \frac{f(n)}{g(n)}$ and suppose $|a(n) - \alpha| < \frac{1}{n}$. If α is rational, then $d(n)$ is recursive, so suppose α is irrational. For a real number x let $\{x\} = x - [x]$ denote its fractional part. The relation $\frac{m}{n} < \{m \cdot a(n)\} < 1 - \frac{m}{n}$ is total recursive and for a fixed m it holds for every large enough n. Let

$$k(m) = \mu\big(n : m/n < \{m \cdot a(n)\} < 1 - m/n\big).$$

Then $[m \cdot \alpha] = [m \cdot a(k(m))]$, and the n-th decimal digit of α is

$$d(n) = \mathrm{rem}([10^n \cdot \alpha], 10) = \mathrm{rem}([10^n \cdot a(k(10^n))], 10),$$

which is total recursive.

236. As the characteristic function of ω (i.e. the constant 1) is recursive, the only if part is clear. Suppose A and $\prec$ are recursive such that $\langle A, \prec \rangle$ is isomorphic to $\langle \alpha, < \rangle$. We need to find a recursive $\lhd$ on ω such that $\langle A, \prec \rangle$ and $\langle \omega, \lhd \rangle$ are isomorphic. Define $g : \omega \to A$ by

$$g(0) = \mu\{x : \chi_A(x) = 1\}$$
$$g(n + 1) = \mu\{x : \chi_A(x) = 1 \text{ and } g(n) < x\}$$

As A is infinite, g is a well defined, recursive bijection. For $x, y \in \omega$ write

$$x \lhd y \quad \text{if and only if} \quad g(x) \prec g(y).$$

Then $\lhd$ is recursive, and $\langle \omega, \lhd \rangle$ and $\langle A, \prec \rangle$ are isomorphic.

237. (a) Define $\prec$ on ω such that $n \prec 0$ for all $n > 0$ and $n \prec m$ for all $0 < n < m \in \omega$. Then $\langle \omega, \prec \rangle$ is isomorphic to $\langle \omega + 1, < \rangle$. As for $\omega + \omega$ write

$$x \lhd y \Leftrightarrow \begin{cases} x \text{ is odd and } y \text{ is even, or} \\ x \text{ and } y \text{ are both odd and } x < y, \text{ or} \\ x \text{ and } y \text{ are both even and } x < y. \end{cases}$$

Then $\langle \omega, \lhd \rangle$ is isomorphic to $\langle \omega + \omega, < \rangle$. The relations $\prec$ and $\lhd$ are clearly recursive.

(b) Let $A = \{p^k : k \in \omega, p \text{ is a prime}\}$. Define $\prec$ on A by $p^k \prec q^\ell$ iff $p < q$ or $p = q$ and $k < \ell$. Then the mapping $p_n^k \mapsto \omega \cdot n + k$ is an isomorphism between $\langle A, \prec \rangle$ and $\langle \omega^2, < \rangle$. (Here p_n is the n-th prime). By Problems 107 and 104 both A and $\prec$ are recursive (in fact, primitive recursive).

(c) Each ordinal $\alpha < \omega^\omega$ can be expressed as $\omega^n \cdot a_n + \cdots + \omega \cdot a_1 + a_0$ for suitable $a_0, \ldots, a_n \in \omega$ with $a_n > 0$. Recall that $\langle a_n, \ldots, a_0 \rangle$ denotes both the sequence itself and the code of this sequence (as an element of ω). Let

$$A = \{\langle a_n, \ldots, a_0 \rangle : n \in \omega, \ a_n > 0\} \subseteq \omega$$

Define $\prec$ on A by $\langle a_n, \ldots, a_0 \rangle \prec \langle b_m, \ldots, b_0 \rangle$ if and only if

- $n < m$ or
- $n = m$ and $\exists k \leq n$ so that $a_i = b_i$ for $i > k$ and $a_k < b_k$.

The mapping

$$\langle a_n, \ldots, a_0 \rangle \mapsto \omega^n \cdot a_n + \cdots + \omega \cdot a_1 + a_0$$

is an isomorphism between $\langle A, \prec \rangle$ and $\langle \omega^\omega, < \rangle$.

By Problem 126 the set of sequence codes is recursive. Comparison relations are also recursive (Problem 104), therefore the set A of sequence codes such that the first element of the corresponding sequence is non-zero is recursive as well. The relation $\prec$ can easily be defined using the recursive functions Len and Elem.

238. (a) Let $A \subset \omega$ be a recursive set and let $\prec$ be a recursive linear ordering on A such that $\langle A, \prec \rangle$ is isomorphic to $\langle \alpha, < \rangle$. If $\beta < \alpha$, then β is isomorphic to an initial segment of α. Thus there is an initial segment of $\langle A, \prec \rangle$ that is isomorphic to $\langle \beta, < \rangle$. Let this initial segment be $\langle B, \prec_B \rangle$. We need that B and $\prec_B$ are recursive. As $B = \{a \in A : a \prec d\}$ for some fixed $d \in A$, and thus $\chi_B(x) = \chi_A(x) \cdot \chi_\prec(x, d)$, B is recursive. Similar argument shows $\prec_B$ is recursive as well.

(b) Let $A, B \subset \omega$ be recursive sets with recursive well-orderings $\prec_A$, $\prec_B$, such that $\langle A, \prec_A \rangle \cong \langle \alpha, < \rangle$ and $\langle B, \prec_B \rangle \cong \langle \beta, < \rangle$. The ordinal $\alpha \cdot \beta$ is the order type of $\beta \times \alpha$ with the lexicographic ordering.

By Problem 126 the set of sequence codes is recursive. Write

$$C = \{\langle a, b \rangle : a \in A, \ b \in B\} \subseteq \omega$$

Then C is recursive as $x \in C$ if and only if x is a sequence code, $\text{Len}(x) = 2$, and $\text{Elem}(x, 0) \in A$ and $\text{Elem}(x, 1) \in B$. Define the ordering $\prec$ on C by

$x < y$ iff $x, y \in C$ and $\text{Elem}(x, 0) < \text{Elem}(y, 0)$, or $\text{Elem}(x, 0) = \text{Elem}(y, 0)$ and $\text{Elem}(x, 1) < \text{Elem}(y, 1)$.

(c) Follows from (a), (b) and $\alpha + 1 < \alpha \cdot 2$.

(d) By (a) the set of all recursive ordinals forms an initial segment of the class of ordinals. This initial segment must be countable as there are countably many recursive relations only. Thus, there must be a least non-recursive ordinal which is countable. By (c), ω_1^{CK} is a limit ordinal. An ordinal α is recursive if and only if $\alpha < \omega_1^{CK}$.

239. (a) The statement is false. As ω_1^{CK} is countable, there is a countable sequence of ordinals $\alpha_i < \omega_1^{CK}$ such that $\sup_{i \in \omega} \alpha_i = \omega_1^{CK}$. Each α_i is a recursive ordinal (as $\alpha_i < \omega_1^{CK}$), while ω_1^{CK} is not recursive.

(b) The statement is true, because the sequence is *uniformly recursive*. We show first that $\sum_{i \in \omega}(A_i, <_i)$ is a recursive well-order. By Problem 126 the set of sequence codes is recursive. Write

$$A = \{\langle n, m \rangle : \varphi_{f(n)}(m) = 1\} \subseteq \omega$$

and put $\langle n, m \rangle < \langle n', m' \rangle$ if and only if $n < n'$ or $n = n'$ and $m <_n m'$. Then $\langle A, < \rangle$ is a recursive well-ordering. If α is the order type of $\langle A, < \rangle$, then $\alpha_i \leq \alpha$ for all $i \in \omega$. As $\sup \alpha_i \leq \alpha$, by Problem 238(b), $\sup \alpha_i$ is a recursive ordinal.

240. To apply an inductive argument it is enough to prove that if α is a recursive ordinal, then so is α^{ω}. Assume A and $\prec$ are recursive and $\langle A, \prec \rangle \cong \langle \alpha, < \rangle$.

By Problem 126 the set of sequence codes is recursive: let us write $\text{Code}(x)$ if x is a sequence code. For each $n \in \omega$ define the set $A_n \subset \omega$ by

$$A_n = \{x \in \omega : \text{Code}(x), \text{Len}(x) = n, (\forall i < n) \text{Elem}(x, i) \in A\}$$

The ordering $\prec_n$ on A_n is given by $x \prec_n y$ if and only if $x, y \in A_n$ and $\exists k < n$ so that $\text{Elem}(x, i) = \text{Elem}(y, i)$ for $k < i < n$ and $\text{Elem}(x, k) < \text{Elem}(y, k)$. It is clear that both A_n and $\prec_n$ are recursive. What is more, the construction is uniform in n, thus there are recursive f and g such that $\varphi_{f(n)}$ is the characteristic function of A_n, and $\varphi_{g(n)}$ is the characteristic function of $\prec_n$.

Each $\langle A_n, \prec_n \rangle$ is isomorphic to $\langle \alpha^n, < \rangle$, hence Problem 239 concludes that α^{ω} is a recursive ordinal.

12.5 PROPOSITIONAL CALCULUS

241. (a) $\bigwedge_{i \neq j} \neg(A_{1,i} \wedge A_{1,j})$.

(b) $\bigvee_{i \neq j \neq k} (A_{2,i} \wedge A_{2,j} \wedge A_{2,k})$.

(c) $\bigwedge_{j=1}^{5} \bigvee_{i=1}^{3} A_{j,i}$.

(d) $\neg\left(\bigwedge_{j=1}^{11} \bigvee_{i=1}^{10} (A_{j,i}) \wedge \bigwedge_{k=1}^{10} \bigwedge_{i \neq j} \neg(A_{i,k} \wedge A_{j,k}) \right)$.

242. (a) $\bigwedge\limits_{v\in V}\left(\bigvee\limits_{c=1}^{4}(A_{v,c})\wedge\bigwedge\limits_{i\neq j}\neg(A_{v,i}\wedge A_{v,j})\right).$

(b) $\bigwedge\limits_{u\neq v\in V}\left(E(u,v)\rightarrow(\bigwedge\limits_{c=1}^{4}\neg(A_{u,c}\wedge A_{v,c}))\right).$

243. $\bigvee\limits_{v\in V}\bigwedge\limits_{w\in V-\{v\}}\left(A_{v,w}\vee\bigvee\limits_{z\in V-\{v,w\}}(A_{v,z}\wedge A_{z,w})\right).$

244. (a) $\left[\bigwedge\limits_{g\in G}\left(\bigvee\limits_{b\in B}(A_{g,b})\wedge\bigwedge\limits_{i\neq j\in B}\neg(A_{g,i}\wedge A_{g,j})\right)\right]\wedge$

$\qquad\qquad\left[\bigwedge\limits_{b\in B}\left(\bigvee\limits_{g\in G}(A_{g,b})\wedge\bigwedge\limits_{i\neq j\in G}\neg(A_{b,i}\wedge A_{b,j})\right)\right].$

(b) $\bigwedge\limits_{H\subseteq G}\bigwedge\limits_{F\in[B]^{<|H|}}\neg A_{H,F}$, where $A_{H,F}=\bigwedge\limits_{h\in H}\bigwedge\limits_{f\notin F}\neg A_{h,f}.$

245. An n-ary Boolean function is defined by giving all 2^n values the operation assumes on the possible combination of the arguments. For a 0-1 number ϵ write

$$x^{\epsilon}=\begin{cases}x & \text{if }\epsilon=1,\\ \neg x & \text{if }\epsilon=0.\end{cases}$$

Every function $f:\{0,1\}^n\rightarrow\{0,1\}$ (provided it is not the constant 0 function) can be represented by a *full disjunctive normal form*

$$f(x_1,\ldots,x_n)=\bigvee\limits_{f(\epsilon_1,\ldots,\epsilon_n)=1}x_1^{\epsilon_1}\wedge\ldots\wedge x_n^{\epsilon_n}.$$

One way to express the constant 0 function using $\wedge$ and $\neg$ is $x\wedge\neg x$. Similarly, every Boolean function f (provided it is not the constant 1 function) can be represented by a *full conjunctive normal form*

$$f(x_1,\ldots,x_n)=\bigwedge\limits_{f(\epsilon_1,\ldots,\epsilon_n)=0}x_1^{1-\epsilon_1}\vee\ldots\vee x_n^{1-\epsilon_n}.$$

Remark. A set F of Boolean functions is *functionally complete* if all Boolean functions can be obtained from elements of F by taking compositions. Therefore $\{\wedge,\neg\}$ and $\{\vee,\neg\}$ are functionally complete.

Remark. Problem 245 can be generalized to a broader class of structures: A finite algebra $\mathfrak{A}$ is called *primal* if every n-ary function $f:A^n\rightarrow A$ for every $n\geq 1$ is representable by a term of $\mathfrak{A}$, i.e. there is a term $p(x_1,\ldots,x_n)$ such that $f(a_1,\ldots,a_n)=p^{\mathfrak{A}}(a_1,\ldots,a_n)$ for all $a_1,\ldots,a_n\in A$. We proved in this exercise that the 2-element Boolean algebra is primal. (In fact, the 2-element Boolean algebra is the only primal Boolean algebra). Other primal algebras exist, for instance $\langle\mathbb{Z}_p,+,\cdot,-,0,1\rangle$ for a prime number p.

246. The number of Boolean functions on n variables is 2^{2^n}. Let $F_n(k)$ be the number of Boolean functions on n variables that can be expressed using at most k logical operators ($\wedge$, $\vee$ or $\neg$).

$$F_n(k)\leq(k^2)^k\cdot 3^k\cdot k^n\cdot k\cdot\frac{1}{k!}$$

Treating $\neg$ as a binary operation, every logical operation has two inputs (the output of two other operations) hence the first term $(k^2)^k$. 3^k is the number of possible operations. The number of choices for the input is k^n, the number of outputs is k. The final $\frac{1}{k!}$ is to avoid over-counting isomorphic expressions. Basic algebra shows

$$F_n(k) \leq 2^{k \log k + O(k + n \log k)}$$

We are seeking for a k such that $F_n(k) \geq 2^{2^n}$, i.e. $2^{k \log k + O(k + n \log k)} \geq 2^{2^n}$. Such k satisfies $k = O(2^n)$.

A specific example which requires exponential length expression using only $\wedge$, $\vee$ and $\neg$ is the n-ary Boolean function which takes value 1 if and only if an odd number of its arguments are equal to 1 (Johan Håstad). Interestingly, if one can use the logical operator $\leftrightarrow$ then the same function has a succinct expression

$$(\cdots((x_1 \leftrightarrow x_2) \leftrightarrow x_3) \leftrightarrow \cdots x_{n-1}) \leftrightarrow x_n.$$

247. Mimic the disjunctive normal form construction in Problem 245. First check that for each $a, b \in \{0, 1, 2\}$ the unary functions

$$H_{a,b}(x) = \begin{cases} b & \text{if } x = a, \\ 0 & \text{otherwise} \end{cases}$$

can be expressed as compositions. Given an n-tuple $\vec{a} \in \{0, 1, 2\}^n$ and $b \in \{0, 1, 2\}$ the function

$$G_{\vec{a}, b}(\vec{x}) = \min\{H_{a_i, b}(x_i) : 1 \leq i \leq n\}$$

takes zero everywhere except at $\vec{x} = \vec{a}$, where it takes b. (This function corresponds to the conjunction $x_1^{\epsilon_1} \wedge \ldots \wedge x_n^{\epsilon_n}$.) Given any ternary function f, the composition

$$\max\{G_{\vec{a}, f(\vec{a})}(\vec{x}) : \vec{a} \in \{0, 1, 2\}^n\}$$

takes the same value as f.

248. (a) As $f_1(x, y) = x \wedge y$ and $f_2(x) = \neg x$ the result follows from Problem 245.

(b) It is enough to express $x \wedge y$ using f_1 and f_2. But $x \wedge y = x(1 - y) + x$ (mod 2) $= f_1(x, f_2(y))$.

(c) Observe that the following class K of Boolean functions is closed under composition, and both f_1 and f_2 are members of K.

$$K = \{f : f(0, \ldots, 0) = 0\}.$$

As $1 - x$ does not belong to K, it cannot be expressed using f_1 and f_2.

249. (a) From the function $F(x, y) = \neg(x \wedge y)$ all other Boolean functions can be obtained by composition. To see this, by Problem 245, it is enough to express $x \wedge y$ and $\neg x$ using F. This can be done as follows.

$$x \wedge y \;\; = \;\; F\big(F(x, y), F(x, y)\big),$$

$$\neg x \;=\; F(x,x).$$

Out of the sixteen binary Boolean functions exactly two of them are functionally complete: $F(x, y) = \neg(x \wedge y)$ and $G(x, y) = \neg(x \vee y)$.

| **Remark.** $F(x, y)$ is usually denoted by $x|y$ and called the *Sheffer stroke*.

(b) There are only countably many functions which can be obtained by composition from a single (or even from countably many) functions, but the cardinality of the set of real functions is strictly greater than countable.

250. For each $a < k$ let $\chi_a(x)$ be the characteristic function of the single-element set $\{a\}$:

$$\chi_a(x) = \begin{cases} 1 & \text{if } x = a, \\ 0 & \text{otherwise,} \end{cases}$$

and write $c_a(x)$ for the constant function $c_a(x) = a$. Any function f can be represented as

$$f(x_1, \ldots, x_n) = \sum_{a_1, \ldots, a_n < k} c_{f(a_1, \ldots, a_n)}(x_1) \cdot \chi_{a_1}(x_1) \cdot \ldots \cdot \chi_{a_n}(x_n).$$

Note that the only properties of $+$ and $\cdot$ used were $x + 0 = 0 + x = x$, $x \cdot 0 = 0$ and $x \cdot 1 = x$.

251. The binary (and unary) functions $f_{i,j}$ $(i, j < k; i \neq j)$ can be merged into a single function with four variables as follows. Let $G(x_1, x_2, x_3, x_4)$ be the k-valued function

$$G(x, x, x, x) = x + 1 \quad (\text{mod } k),$$

and

$$\begin{aligned} G(x, x, x+1, x+i) &= c_i(x) \quad \text{for } i < k, \\ G(x, x, x+2, x+i) &= \chi_i(x) \quad \text{for } i < k. \end{aligned}$$

Finally, for each $i \neq j$ put

$$G(i, j, x, y) = f_{i,j}(x, y).$$

In order to prove that functions $f_{i,j}$ can be recovered from G it is enough to show that constant functions i, j can be recovered. But using the notation $F(x) = G(x, x, x, x)$ we have $i = G(x, x, F(x), F^i(x))$.

To complete the proof observe that all of the constant functions and characteristic functions are encoded in G in the form $G(x, x, F(x), F^i(x))$ and $G(x, x, FF(x), F^i(x))$, and one can merge $k(k-1) \geq 2$ additional functions into G.

252. Denote $\max\{x, y\} + 1 \pmod{k}$ by $F(x, y)$. Define the following sequence of functions (all additions are mod k):

$$x + 1 \;=\; F(x, x),$$

$$
\begin{aligned}
\max\{x, y\} &= F(x, y) + (k-1), \ (\text{add} +1 \ (k-1) \ \text{times}) \\
k-1 &= \max\{x, x+1, x+2, \ldots, x+(k-1)\}, \\
i &= (k-1) + (i+1), \\
(k-1)\chi_i(x) &= \max\{x+j : j \neq k-1-i\} + 1 \\
\chi_i(x) &= \max\{(k-1)\chi_j(x) : j \neq i\} + 1, \\
(k-1) - x &= \max\{\max\{(k-1)\chi_j(x), j\} + (k-j) : j < k\}, \\
\min\{x, y\} &= (k-1) - \max\{(k-1) - x, (k-1) - y\}.
\end{aligned}
$$

Applying Problem 250 with min and max playing the role of $\cdot$ and $+$, respectively, completes the proof.

253. There are homomorphisms $f : \mathbf{A}_1 \rightarrow \mathbf{A}_2$ and $g : \mathbf{A}_2 \rightarrow \mathbf{A}_1$ which are identical on G. Therefore the compositions $f \circ g$ and $g \circ f$ are also identical on G and since each homomorphism is uniquely determined by its values on the set of generators we get $f \circ g = \text{id}$ and $g \circ f = \text{id}$.

254. We treat partitions of $\mathbf{A}$ as equivalence relations $\vartheta \subseteq \mathbf{A}^2$ (i.e a and b are in the same partition if and only if $\langle a, b \rangle \in \vartheta$). ϑ is a congruence if whenever $\langle a_i, b_i \rangle \in \vartheta$ $(i < n)$ and f is a fundamental n-ary operation, we have

$$
\langle f^{\mathbf{A}}(a_0, \ldots, a_{n_1}), f^{\mathbf{A}}(b_0, \ldots, b_{n_1}) \rangle \in \vartheta.
$$

The smallest and largest congruences are respectively $\Delta = \{\langle a, a \rangle : a \in \mathbf{A}\}$ (each element forms a single block of the partition), and $\nabla = \mathbf{A}^2$ (there is only one block containing all elements). Greatest lower bound of ϑ_1 and ϑ_2 is $\vartheta_1 \wedge \vartheta_2 = \vartheta_1 \cap \vartheta_2$, while their smallest upper bound is the transitive closure of the relation $\vartheta_1 \cup \vartheta_2$

$$
\vartheta_1 \vee \vartheta_2 = \vartheta_1 \cup (\vartheta_1 \circ \vartheta_2) \cup (\vartheta_1 \circ \vartheta_2 \circ \vartheta_1) \cup (\vartheta_1 \circ \vartheta_2 \circ \vartheta_1 \circ \vartheta_2) \cup \ldots
$$

255. In the figure below points of the same bubble belong to the same partition.

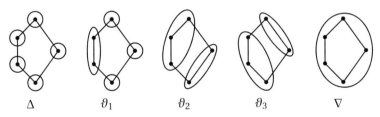

$$\Delta \qquad \vartheta_1 \qquad \vartheta_2 \qquad \vartheta_3 \qquad \nabla$$

The lattice of congruences is figured below.

256. Let $t(x, y, u, v)$ be a 4-place function symbol interpreted in the algebra $\mathfrak{A} = \langle X, t^{\mathfrak{A}} \rangle$ as

$$t^{\mathfrak{A}}(a, b, c, d) = \begin{cases} c & \text{if } (a, b) \in \vartheta \\ d & \text{otherwise.} \end{cases}$$

Then ϑ is a congruence of $\mathfrak{A}$. On the other hand, if $\sim$ is a congruence and a, b are elements such that $a \sim b$ but $(a, b) \notin \vartheta$, then

$$c = t^{\mathfrak{A}}(a, a, c, d) \sim t^{\mathfrak{A}}(a, b, c, d) = d$$

for any c and d, thus $\sim$ is ∇.

257. Let A be the collection of finite subsets of Σ, this is the set of voters. Take an ultrafilter U on A representing "majority", and consider the evaluation $f : V \to \{\top, \bot\}$ defined by what the majority voted for: for a variable $v \in V$ let $f(v) = \top$ if $\{a \in A : f_a(v) = \top\} \in U$, otherwise let $f_a(v) = \bot$. By induction on the complexity of φ this extends to every formula:

$$f(\varphi) = \top \iff \{a \in A : f_a(\varphi) = \top\} \in U.$$

The formula $\varphi \in \Sigma$ will evaluate to $\top$ iff the set $X_\varphi = \{a \in A : f_a(\varphi) = \top\}$ is in U. By the assumption the evaluation f_a makes true every formula in a, so $A_\varphi = \{a \in A : \varphi \in a\} \subseteq X_\varphi$. Consequently we are done if U can be chosen so that $A_\varphi \in U$ for every $\varphi \in \Sigma$. To finish the proof observe that the family $\{A_\varphi : \varphi \in \Sigma\}$ has the finite intersection property, thus there is an ultrafilter U over A that contains each A_φ (see Problems 7, 29).

258. Solution 1. The ultrafilter construction in Solution 257 works here as well. For a finite subset a of Σ, let f_a be the evaluation for which $\varphi(f_a) = 0$ for all $\varphi \in a$. Fix the ultrafilter U, and let $f(v) = i$ if $U_i = \{a \in A : f_a(v) = i\} \in U$. As the sets U_i form a k-partition of A, exactly one of them is an element of U (Problem 31), thus this definition is sound. From here proceed as before.

Solution 2. Use the Compactness Theorem 5.8 for propositional formulas. For each $v \in V$ and $i < k$ let $A_{v,i}$ be a propositional variable which is true if and only if the value of v is i. The propositional formulas $\neg(A_{v,i} \wedge A_{v,j})$ and $(A_{v,0} \vee \cdots \vee A_{v,k-1})$ ensure that for each v exactly one of $A_{v,i}$ is true. For each k-valued logical function $\varphi(v_{\alpha_1}, \ldots, v_{\alpha_n})$ there is a corresponding propositional formula which is true if and only if φ evaluates to zero using the values coded in the variables $A_{v,i}$. This set of propositional formulas is satisfiable if and only if Σ is satisfiable. By assumption, every finite subset is satisfiable, thus the compactness theorem gives the claim of this problem.

259. Consider the elements of the field as logical values, and the polynomials as k-valued logical formulas. Apply the compactness theorem for k-valued logic (Problem 258).

260. (a) Consider the following system of polynomial equations:

$$(x_0 - 1) - x_1^2 = 0,$$

$$(x_0 - 2) - x_2^2 = 0,$$

$$\vdots$$

Each finite sub-system is solvable: A solution for the first k equations is $x_0 = k$ and $x_i = \sqrt{k-i}$ for $1 \le i \le k$. However, the whole system does not have any solutions: For if $x_0 = s$ then for $n > s$ the equations $(x_0 - n) - x_n^2 = 0$ could not be solved in $\mathbb{R}$.

(b) Same as (a) but replace x_i^2 by the sum of four squares, and remark that every natural number can be written as the sum of four squares.

(c) Consider the system which contains the equation $(z - c)z_c - 1 = 0$ for each $c \in \mathbb{C}$. Each of its finite sub-systems can be solved: if $(z - d)z_d - 1 = 0$ does not belong to the finite sub-system, then $z = d$ and $z_c = (d - c)^{-1}$ are the solutions. Nevertheless, the whole system cannot be solved, because for each $z = c$ the equation $(z - c)z_c - 1 = 0$ cannot be satisfied.

> **Remark.** The last example has continuum many equations and variables. There is no such an example with countably many polynomial equations only, see Problems 656 and 680.

261. We know that $\Sigma \vDash \varphi$ iff $\Sigma \cup \{\neg\varphi\}$ is not satisfiable. So if $\Sigma' \nvDash \varphi$ then each finite subset of $\Sigma \cup \{\neg\varphi\}$ is satisfiable, contradicting Theorem 5.8.

262. (a) For a set Γ of formulas write U_Γ for the open set

$$U_\Gamma = \{e \in X : \text{for some } \varphi \in \Gamma \text{ we have } e(\varphi) = \top\}, \qquad U_\varphi = U_{\{\varphi\}}.$$

It is straightforward that $U_\Gamma = \bigcup_{\varphi \in \Gamma} U_\varphi$, therefore it is enough to check that $\{U_\varphi\}$ forms a basis of a topology. It is so as $U_{\varphi \wedge \neg\varphi} = \emptyset$ and $U_{\varphi \vee \neg\varphi} = X$, finally $U_\varphi \cap U_\psi = U_{\varphi \wedge \psi}$. As $X - U_\varphi = U_{\neg\varphi}$, this basis contains clopen sets only.

(b) Since X has a clopen basis, we may use the following description of compactness of a topological space: If $\{F_\alpha : \alpha < \beta\}$ is a family of basic closed sets having the FIP, then $\bigcap_{\alpha < \beta} F_\alpha \ne \emptyset$. A set Γ of formulas is satisfiable if and only if $\bigcap_{\gamma \in \Gamma} U_\gamma$ is not empty.

Let Γ be a set of formulas such that all finite $\Gamma' \subseteq \Gamma$ are satisfiable. Then the system $\{U_\varphi : \varphi \in \Gamma\}$ has the FIP, because for a finite Γ' we have

$$\bigcap_{\gamma \in \Gamma'} U_\gamma = U_{\wedge\Gamma'} \ne \emptyset.$$

If X is compact, then it follows that $\bigcap_{\gamma \in \Gamma} U_\gamma \ne \emptyset$, hence Γ is satisfiable.

For the other direction let $\{F_\alpha : \alpha < \beta\}$ be an FIP family of basic closed sets. For each F_α there is a corresponding formula γ_α such that $F_\alpha = U_{\gamma_\alpha}$. Note that every finite subset of $\Gamma = \{\gamma_\alpha : \alpha < \beta\}$ is satisfiable because

$$U_{\gamma_{\alpha_1}} \cap \ldots \cap U_{\gamma_{\alpha_n}} \ne \emptyset \quad \text{by FIP}.$$

By the compactness theorem Γ is satisfiable, consequently

$$\bigcap F_\alpha = \bigcap_{\gamma \in \Gamma} U_\gamma \ne \emptyset.$$

(c) X is Hausdorff: let x and y be distinct evaluations and $v \in V$ such that $x(v) = \top$ and $y(v) = \bot$. Then $x \in U_v$ and $y \in U_{\neg v}$ are disjoint neighborhoods. X is regular: Pick a closed set A and a point $x \in X - A$. For some set Φ of formulas we have $A = \bigcap_{\varphi \in \Phi} U_\varphi$, and there must exist $\varphi \in \Phi$ such that $a \notin U_\varphi$. Then $a \in U_{\neg\varphi}$ and $A \subseteq U_\varphi$.

(d) If $|V| \leq \aleph_0$, then each point $x \in X$ has a countable neighborhood-base: $\{U_\varphi : x(\varphi) = \top\}$. Since $|F| = |V| \cdot \aleph_0$ we have that $\{U_\varphi : \varphi \in F\}$ is a countable base if and only if $|V| \leq \aleph_0$. If $|V| > \aleph_0$, then X does not have a countable base: each such base avoids infinitely many $v \in V$, hence the open sets U_v are not generated. By Urysohn's metrization theorem (M_2+regular is metrizable) it follows that X is metrizable if and only if $|V| \leq \aleph_0$. If V is countable, then X is separable.

(e) As the underlying set of X and $^V 2 = {}^V\{\top, \bot\}$ is the same, it is enough to prove that the identity map $\mathrm{id} : {}^V 2 \to X$ and its inverse is continuous. To see that id is continuous, we shall prove that (the inverse image of) each basic open set U_φ of X is open in $^V 2$. Write φ in disjunctive normal form

$$\varphi \equiv \bigvee_i \bigwedge_j v_{ij}^{\epsilon_{ij}}.$$

For $v \in V$ denote by $\pi_v : {}^V 2 \to 2$ the v-th projection, and define $W_{ij} = \pi_{v_{ij}}^{-1}(\epsilon_{ij})$. Then

$$U_\varphi = \bigcup_i \bigcap_j W_{ij},$$

which is open in $^V 2$.

To see that id^{-1} is continuous, we shall prove that each basic open set $W = \prod_{v \in V} W_v$ of $^V 2$ is open in X. Write

$$\varphi = \bigwedge_{v : W_v = \{1\}} v \wedge \bigwedge_{v : W_v = \{0\}} \neg v.$$

Then $U_\varphi = W$.

263. Suppose the propositional variables in φ are $A_1, \ldots A_m$ and $B_1, \ldots, B_n$ where A_i does not occur in ψ while all B_j occur in ψ. If $n = 0$ then either ψ is a tautology, or $\neg\varphi$ must be a tautology, thus either $\top$ or $\bot$ works as the interpolant formula ϑ. Assume therefore that $n > 0$ and use induction on m. If $m = 0$ then $\vartheta = \varphi$ works. Otherwise denote φ as $\varphi(A_1, \ldots, A_m, B_1, \ldots B_n)$, and let

$$\varphi_0 = \varphi(B_1, A_2, \ldots, A_m, B_1, \ldots, B_n),$$
$$\varphi_1 = \varphi(\neg B_1, A_2, \ldots, A_m, B_1, \ldots, B_n).$$

Now $\varphi_0 \to \psi$ and $\varphi_1 \to \psi$ are tautologies, hence so are the formulas $(\varphi_1 \to \psi) \wedge (\varphi_0 \to \psi)$ and $(\varphi_0 \vee \varphi_1) \to \psi$. By induction there exists some interpolant ϑ between $\varphi_0 \vee \varphi_1$ and ψ for which $(\varphi_0 \vee \varphi_1) \to \vartheta$ and $\vartheta \to \psi$ are tautologies. But $\varphi \to \varphi_0 \vee \varphi_1$ is also a tautology, so we are done.

> **Remark.** If $\top$ and $\bot$ were not allowed as interpolants, then the tautology $(A \wedge \neg A) \rightarrow (B \vee \neg B)$ could not have interpolant.

264. $(F/\equiv, \leq)$ is a partial ordering: $[\varphi] \leq [\varphi]$ is trivial. If $[\varphi] \leq [\psi]$ and $[\psi] \leq [\varphi]$, then $f(\varphi) = f(\psi)$ for all f, hence $[\varphi] = [\psi]$. Finally, suppose $[\varphi] \leq [\psi]$ and $[\psi] \leq [\mu]$. Then we have $f(\varphi) \leq f(\psi) \leq f(\mu)$ for all f, in particular $f(\varphi) \leq f(\mu)$, therefore $[\varphi] \leq [\mu]$.

$(F/\equiv, \leq)$ is a distributive lattice: the least upper bound of $[\varphi]$ and $[\psi]$ is $[\varphi \vee \psi]$, while their greatest lower bound is $[\varphi \wedge \psi]$.

265. Solution 1 (Along the lines of Problem 242). Let $A_{v,e}$ be propositional variables for each vertex v and possible color e. Let Γ be the set of propositional formulas which expresses the fact that every vertex has exactly one color, and connected vertices have different colors. Γ is satisfiable if and only if G can be colored by n colors.

By the compactness theorem, Γ is satisfiable if and only if all finite subsets of Γ are satisfiable. If G' is a finite subgraph of G then the corresponding set Γ' is a finite subset of Γ. Moreover all finite subsets of Γ are contained in some Γ' (simply take the subgraph spanned by vertices mentioned in that subset). Thus all finite subgraphs of G are n-colorable if and only if all finite subsets of Γ are satisfiable, and we are done.

Solution 2 (Along the lines of the proof of compactness theorem). For all $j \in [V(G)]^{<\omega}$ fix a coloring c_j of the subgraph spanned by j. We wish to choose an ultrafilter U on $[V(G)]^{<\omega}$ such that the function $c: V(G) \rightarrow n$ defined below becomes a coloring of G.

$$c(v) = k \text{ if and only if } \left\{ j \in [V(G)]^{<\omega} : c_j(v) = k \right\} \in U.$$

Note that $\bigcup^*_{k<n} \{ j \in [V(G)]^{<\omega} : c_j(v) = k \}$ is a partition of $[V(G)]^{<\omega}$ and exactly one member of this partition belongs to U (Problem 31), therefore c is well defined. To ensure that c is defined on all vertices, we need for all $v \in V(G)$ that $\{ j \in [V(G)]^{<\omega} : v \in j \} \in U$. For $j \in [V(G)]^{<\omega}$ write $\hat{j} = \{ k \in [V(G)]^{<\omega} : j \subseteq k \}$. Observe that the set system $E = \{ \hat{j} : j \in [V(G)]^{<\omega} \}$ has the FIP: $j_1 \cup \dots \cup j_n \in \hat{j_1} \cap \dots \cap \hat{j_n}$. Therefore there is an ultrafilter U extending E (Problems 29 and 28) and clearly $\hat{v} = \{ j \in [V(G)]^{<\omega} : v \in j \} \in U$.

> **Remark.** Problem 626 points toward a more general approach (and in fact the original proof of Erdős and deBruijn).

266. Let $G = \langle V, E \rangle$ be a graph, all of its finite subgraphs are colorable by k colors. Put

$$\mathcal{H} = \{ \langle V, F \rangle : E \subseteq F \text{ and each finite subgraph of } \langle V, F \rangle \\ \text{can be colored by } k \text{ colors} \}.$$

For two elements $H_1 = \langle V, F_1 \rangle$ and $H_2 = \langle V, F_2 \rangle$ of $\mathcal{H}$ we say $H_1 \leq H_2$ if $F_1 \subseteq F_2$. We claim that the partial order $\langle \mathcal{H}, \leq \rangle$ satisfies the assumption of Zorn's lemma: If $\{ \langle V, F_i \rangle : i \in I \}$ is a linearly ordered subset of $\mathcal{H}$, then $H = \langle V, \bigcup_{i \in I} F_i \rangle$ belongs to $\mathcal{H}$. For if not, then there would exists a finite subgraph of H which cannot be colored by k colors. This finite subgraph must be a subgraph of some $\langle V, F_i \rangle$ in the chain; a contradiction. Therefore, by Zorn's lemma, there is a maximal element $H = \langle V, F \rangle$ of $\mathcal{H}$.

To see that non-adjacency of H is an equivalence relation, suppose by way of contradiction that $\langle a, b \rangle, \langle b, c \rangle \notin F$ while $\langle a, c \rangle \in F$. By maximality of H there are finite subgraphs H_1 and H_2 such that $H_1 + \langle a, b \rangle$ (a and b are vertices of H_1 and we add the edge $\langle a, b \rangle$) and $H_2 + \langle b, c \rangle$ are not k-colorable. We claim then $H_1 \cup H_2 + \langle a, c \rangle$ is not k-colorable: If γ were a k-coloring, then $\gamma(a) = \gamma(b)$ and $\gamma(b) = \gamma(c)$ would follow, since otherwise γ would be a k-coloring of $H_1 + \langle a, b \rangle$ and $H_2 + \langle b, c \rangle$. But then $\gamma(a) = \gamma(c)$ would hold which is impossible. Consequently some finite subgraph of H cannot be colored by k colors, which is a contradiction.

Equivalence classes of non-adjacency provide a coloring of H. To complete the proof it is enough to show that there are at most k equivalence classes. If a subgraph of H contains a vertex from each equivalence class, then it also contains a complete graph K_ℓ, where ℓ is the number of equivalence classes. As K_{k+1} is not k-colorable, we get that $\ell \leq k$.

267. Straightforward consequence of the Erdős–deBruijn theorem (Problems 265 or 266).

268. Partition the plane into blocks of 3×3 unit-diagonal squares, the interior of each square is colored by one of the $3 \cdot 3$ colors, accordingly. As for the second part, by Problem 265 we have to find finitely many points on the plane so that their distance graph cannot be colored with three colors. This can be done as pictured below (each edge has unit length).

> **Remark.** The Hadwiger–Nelson problem asks for the minimum number of colors required to color the plane so that any two points at distance 1 are colored differently. This problem narrows down the answer to one of the numbers $4 \leq n \leq 9$. The correct value is known to be one of 5, 6 or 7. In higher dimensions it is only known that this chromatic number is exponential: it is between 1.2^n and 3^n.

269. Let $A_{e,c}$ be propositional variables for each edge e and possible color $c \in \{0, 1\}$. Let Γ be the set of propositional formulas which expresses the fact that every edge has exactly one color, and there are no homogeneous induced subgraphs of size k:

$$\Gamma = \big\{ (A_{e,0} \wedge \neg A_{e,1}) \vee (A_{e,1} \wedge \neg A_{e,0}) : e \in E,$$
$$\neg \bigwedge_{i < \ell} A_{e_i,c} : e_i \in E, c \in \{0, 1\}, e_1, \ldots, e_\ell \text{ are edges}$$
$$\text{in an induced subgraph of size } k \big\}.$$

Γ is satisfiable if and only if G can be colored by two colors having no homogeneous subgraph of size $k < \omega$

By the compactness theorem, Γ is satisfiable if and only if all finite subsets of Γ are satisfiable. If G' is a finite induced subgraph of G then the corresponding set Γ' is a finite subset of Γ. Moreover all finite subsets of Γ are contained in some Γ' (take the subgraph spanned by vertices mentioned in that subset). Thus all finite subgraphs of G can be colored by two colors having no homogeneous subgraph of size $k < \omega$ if and only if all finite subsets of Γ is satisfiable; and we are done.

270. By Hall's theorem G and B cannot be finite. Let $G = \{g_1, g_2, \ldots\}$ and let $B = \{b_0, b_1, \ldots\}$ two infinite sets (observe, there is no g_0). For every $i > 0$ connect b_i to g_i, and connect b_0 to every g_i.

271. Let $N(X)$ denote the set of neighbors of X. For each $g \in G$ and $b \in B$ choose a propositional variable $A_{g,b}$. There is a match between G and B iff the following formula set is satisfiable:

$$\text{for each } g \in G \text{ exactly one of } \{A_{g,b} : b \in N(g)\} \text{ is true;}$$
$$\text{for each } b \in B \text{ exactly one of } \{A_{g,b} : g \in N(b)\} \text{ is true.}$$

As $N(g)$ and $N(b)$ are finite sets, these statements can be expressed as propositional formulas. By compactness it suffices to show that every finite subset is satisfiable. Thus let G' and B' be the vertex sets mentioned in this finite subset, $G'' = G' \cup N(B')$ and $B'' = B' \cup N(G')$. The combinatorial lemma which ensures that the formula set is satisfiable can be worded as

Lemma. Let (G'', B'') be a (finite) bipartite graph, $G' \subseteq G''$, $B' \subseteq B''$. Suppose that for all $A \subseteq G'$ and for all $A \subseteq B'$ we have $|A| \leq |N(A)|$. Then there is a matching (independent set of edges) which covers $G' \cup B'$.

A proof similar to that of Hall's theorem works. If $|A| < |N(A)|$ for all relevant subsets, then leave out any edge (g, b) and apply induction. If $|A| = |N(A)|$ for some A, then the $(A, N(A))$ pair satisfies the conditions of Hall's theorem, thus it has a perfect matching. Deleting these vertices the remaining graph also satisfies the same conditions.

272. (a) Each vertex v with maximal out-degree is a king. For, let u_i $(i < n)$ be those vertices that are reachable from v in one step and suppose w cannot be reached from v by a directed path of length at most two. Then there must be edges directed from w to all the u_i's and to v, meaning that the out-degree of w is greater than that of v; a contradiction.

(b) The set of vertices is ω and an edge is directed from i to j if and only if $j < i$.

273. Call a tournament an *all-kings* tournament if every vertex is a king. For a vertex x we denote by $d^+(x)$ and $d^-(x)$ the out-degree and in-degree of x. For a subset X of vertices denote respectively by $\Gamma^+(X)$ and $\Gamma^-(X)$

the set out-neighbors and the set of in-neighbors of elements of X. Thus $d^+(x) = |\Gamma^+(\{x\})|$.

(a) We prove by induction on n that if there exists an all-kings tournament on n vertices ($n \geq 4$), then there is such a tournament on $n+1$ vertices, as well. Let G be an all-kings tournament on $n \geq 4$ vertices. Pick $x \in G$ with $d^+(x) \geq d^-(x)$. As $n \geq 4$ we have $d^+(x) \geq 2$. Add a new vertex y to the graph G. Draw an edge from y to x, from y to vertices in $\Gamma^-(x)$, and from vertices in $\Gamma^+(x)$ to y.

If there is no edge from $\Gamma^+(x)$ to $\Gamma^-(x)$, then this graph is an all-kings tournament. Otherwise, let $Z \subset \Gamma^-(x)$ be the non-empty set of vertices that can be reached from some vertex in $\Gamma^+(x)$. As $d^-(x) - |Z| < d^+(x)$, there must be $w \in \Gamma^+(x)$ so that for every $z \to v$ with $z \in Z$ and $v \in \Gamma^-(x) \smallsetminus Z$ there exists a 2-path from v to z in $G \smallsetminus \{w\}$. Then reverse all edges between Z and w. The resulting graph is an all-kings graph.

To complete the proof we only need to show one all-kings tournament. To construct one all-kings tournament having 5 vertices is straightforward.

(b) The set of vertices is $\mathbb{Z}$ and for $i < j$ we direct an edge from i towards j if and only if $|i - j|$ is odd. That is, edges to the right correspond to odd distances and edges to the left to even distances.

274. For a subset X of vertices denote the set of in-neighbors of the elements of X by $N(X)$.
Solution 1. Let A_v be a propositional variable for each vertex v. The tournament has a king if and only if the following formula set is satisfiable:

$$\neg(A_u \wedge A_v) \qquad\qquad \text{for all } u \neq v,$$
$$A_u \vee \bigvee\{A_v : v \in N(u)\} \vee \bigvee\{A_w : w \in N(N(u))\} \quad \text{for all vertex } u.$$

(We have used that $N(X)$ is finite when X is finite.) The first line says that at most one A_u is true (the king); the second line says that u can be reached in one or two steps from the king. By problem 272 every finite subset is satisfiable, thus it is also satisfiable.

Solution 2. Let u be a vertex with the smallest $|N(u)|$. We claim that u is a king. Indeed, if $v \notin N(u)$ then uv is a directed path of length 1. If $v \in N(u)$, then there must be a $w \in N(v)$ which is not in $N(u)$; and then uwv is a directed path of length 2.

275. Consider the tournament on three vertices without sinks or sources and adjoin a new vertex from which all the other three vertices can be reached by a directed edge. The resulting tournament has a source (the new vertex) but does not have any sink.

As for the second part, define the tournament T on ω so that an edge is directed from i to j if $j < i$. Then 0 is a sink in T, but there are no sources.

276. Perform Step 1 until applicable, then Step 2 until applicable, finally Step 3, until it can be applied. Step 1 is not applicable if each negation symbol is just ahead of the propositional variable. Step 2 disregards repeated applications of negation; finally Step 3 ends only if the formula is conjunctions of disjunctions of propositional variables, or negation of propositional variables. That the process halts for each formula is straightforward for Steps 1 and 2. For Step 3 define the "norm" $\|\varphi\|$ of propositional formulas as follows. The norm of variables and their negations is 1, and in general

$$\|\varphi \vee \psi\| = \|\varphi\|^2 \cdot \|\psi\|^2,$$
$$\|\varphi \wedge \psi\| = \|\varphi\| + \|\psi\|.$$

After applying Step 3 the norm decreases proving that Step 3 halts as well:

$$\|\psi \vee (\vartheta_0 \wedge \vartheta_1)\| = \|\psi\|^2 \cdot (\|\vartheta_0\| + \|\vartheta_1\|)^2 >$$
$$> \|\psi\|^2 \|\vartheta_0\|^2 + \|\psi\|^2 \|\vartheta_1\|^2 = \|(\psi \vee \vartheta_0) \wedge (\psi \vee \vartheta_1)\|.$$

As for the second statement: each step preserves this property, thus the claim follows by induction.

277. If $\mathcal{C} \cup \{c\}$ is satisfiable, then so is $\mathcal{C}$. For the other direction let $c = \mathbb{R}(c_0, c_1, \ell)$ and suppose $\mathcal{C}$ is satisfiable, i.e. the evaluation $f : V \to \{\top, \bot\}$ gives at least one true literal in each member of $\mathcal{C}$. Then some literal from c_0 is true, and some from c_1 are true. Since at most one of ℓ and $\neg\ell$ can be true, there still remains a true literal in c.

278. The set of clauses is $\{\{\neg A, \neg B, C\}, \{\neg A, B, \neg C\}, \{A, \neg B, \neg C\}, \{A, B, C\}\}$.

279. The conjunctive normal form of the negation of the formula is

$$(\neg A \vee B) \wedge (\neg C \vee A) \wedge C \wedge \neg B.$$

A possible derivation tree is

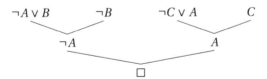

280. The conjunctive normal form of the negation of the formula is

$$(\neg A \vee \neg B \vee C) \wedge (\neg A \vee \neg B \vee D) \wedge (\neg A \vee B) \wedge A \wedge (\neg C \vee \neg D).$$

Using resolution we can derive the empty clause $\Box$ as shown:

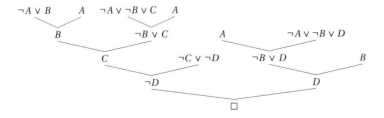

Consequently the original formula is a tautology.

281. Suppose $c_i \subseteq c_i' \subseteq c_i \cup \{\neg\ell\}$ for $i = 0, 1$, and let $c_2 = \mathbb{R}(c_0, c_1, k)$, $c_2' = \mathbb{R}(c_0', c_1', k)$. Then

$$c_2 \subseteq c_2' \subseteq c_2 \cup \{\neg\ell\}.$$

If c_0 is the one-element clause $\{\ell\}$, then k must be the same as ℓ, and $c_2 = c_1 \smallsetminus \{\neg\ell\}$. Choosing $c_2' = c_1'$ we have

$$c_2 \subseteq c_2' \subseteq c_2 \cup \{\neg\ell\}.$$

Similarly, if c_1 is $\{\ell\}$, then k is $\neg\ell$, $c_2 = c_0 \smallsetminus \{\neg\ell\}$, and $c_2' = c_0'$ gives

$$c_2 \subseteq c_2' \subseteq c_2 \cup \{\neg\ell\}.$$

It follows by induction that for every clause c derivable from $\mathcal{C} \cup \{\ell\}$ there exists a clause c' derivable from $\mathcal{C}$ such that $c \subseteq c' \subseteq c \cup \{\ell\}$.

282. Clearly, cannot both $\{v\}$ and $\{\neg v\}$ belong to $\mathcal{C}$. If $\{v\} \notin \mathcal{C}$ then by maximality the empty clause $\square$ can be derived from $\mathcal{C} \cup \{v\}$. Then by Problem 281, either $\mathcal{C} \vdash^R \square$ or $\mathcal{C} \vdash^R \{\neg v\}$.

283. If $\mathcal{C}$ is satisfiable, then by Problem 277, $\mathcal{C} \nvdash^R \square$. In the other direction suppose $\mathcal{C} \nvdash^R \square$. By Zorn's lemma, $\mathcal{C}$ can be assumed to be maximal. By Problem 282 either $\{v\} \in \mathcal{C}$ or $\{\neg v\} \in \mathcal{C}$ for each propositional variable v. Define the evaluation f by stipulating $f(v) = \top$ if $\{v\} \in \mathcal{C}$. We claim that each clause $c = \{\ell_1, \ldots, \ell_k\} \in \mathcal{C}$ has at least one true literal. If not, then $f(\ell_i) = \bot$ for each i, which means $\{\neg\ell_i\} \in \mathcal{C}$. Using k resolvents all literals from c can be eliminated, meaning $\mathcal{C} \vdash^R \square$, a contradiction.

284. (i)$\Rightarrow$(ii) follows from Problem 277. For the reverse implication let $c = \{\ell_1, \ldots, \ell_k\}$, and assume $\mathcal{C} \nvdash^R c'$ for subsets of c. Adding the k one-element clauses $\{\neg\ell_i\}$ to $\mathcal{C}$ we get $\mathcal{C}^*$. If $\mathcal{C}^* \vdash^R \square$, then by repeated application of the Deduction lemma (Problem 281) one gets $\mathcal{C} \vdash^R c'$ for some $c' \subseteq c$, contradicting the assumption. Thus $\mathcal{C}^* \nvdash^R \square$, and by Solution 283 there is an evaluation which satisfies all clauses in $\mathcal{C}^*$. This shows that (ii) does not hold either.

285. Let us denote the propositions in the premises as A =balloonists, B =carrying umbrellas, C =dancing on tight ropes, D =eating penny-buns, E =fat, F =liable to giddiness, G =looking ridiculous, H =may lunch in public, J =old, K =pigs, L =treated with respect, M =wise. One has to take "young" to be the negation of "old". Each premise translates to s single clause:

$$
\begin{array}{rll}
(1) & (\neg C \wedge \neg D) \rightarrow J & \Rightarrow\ C \vee D \vee J, \\
(2) & (K \wedge F) \rightarrow L & \Rightarrow\ \neg K \vee \neg F \vee L, \\
(3) & (M \wedge A) \rightarrow B & \Rightarrow\ \neg M \vee \neg A \vee B, \\
(4) & (G \wedge D) \rightarrow \neg H & \Rightarrow\ \neg G \vee \neg D \vee \neg H, \\
(5) & (\neg J \wedge A) \rightarrow F & \Rightarrow\ J \vee \neg A \vee F, \\
(6) & (E \wedge G \wedge C) \rightarrow H & \Rightarrow\ \neg E \vee \neg G \vee \neg C \vee H, \\
(7) & (M \wedge F) \rightarrow \neg C & \Rightarrow\ \neg M \vee \neg F \vee \neg C, \\
(8) & (K \wedge B) \rightarrow G & \Rightarrow\ \neg K \vee \neg B \vee G, \\
(9) & (\neg C \wedge L) \rightarrow E & \Rightarrow\ C \vee \neg L \vee E.
\end{array}
$$

Using resolution the required $\neg M \vee J \vee \neg K\ \neg A$ can be derived as follows:

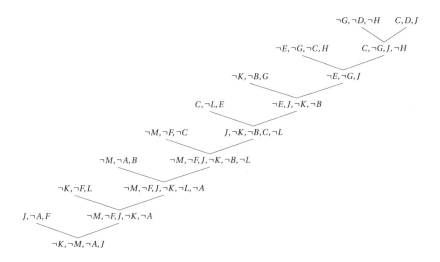

Consequently, $(K \wedge M \wedge A) \rightarrow J$, that is, "no wise young pigs go up in balloons" follows from the set of premises.

286. (a) The two clauses are $\{\ell, \ell\}$ and $\{\neg \ell, \neg \ell\}$. Each resolution step eliminates at most two literals from the union, thus every clause of the derivation will contain at least two literals.

(b) It used that if $c' \subseteq c \cup \{\neg \ell\}$, then $c' \subseteq (c \smallsetminus \{\neg \ell\}) \cup \{\neg \ell\}$, which is not true for multisets.

287. (a)

$$
\cfrac{
\cfrac{
\cfrac{A \vdash A, B}{A \vdash A \vee B}\ \vdash \vee
}{\emptyset \vdash \neg A,\ A \vee B}\ \vdash \neg
}{\emptyset \vdash \neg A \vee (A \vee B)}\ \vdash \vee
$$

(b)

$$
\frac{\dfrac{B, A \vdash A, B \qquad B, A \vdash A, B}{B, A \vdash A \wedge B} \vdash\wedge}{\dfrac{\dfrac{A \vdash \neg B, A \wedge B}{A \vdash \neg B \vee (A \wedge B)} \vdash\neg}{\dfrac{\varnothing \vdash \neg A, \neg B \vee (A \wedge B)}{\varnothing \vdash \neg A \vee (\neg B \vee (A \wedge B))} \vdash\neg} \vdash\vee}
$$

(c)

$$
\cfrac{
\cfrac{
\cfrac{
\neg A \vee (\neg B \vee C), \neg A \vdash \neg A, C
}{
\neg A \vee (\neg B \vee C), \neg A \vdash \neg A \vee C
}\ \vdash\vee
\qquad
\cfrac{
\cfrac{
\cfrac{B, \neg A \vdash \neg A, C}{\neg A \vdash \neg B, \neg A \vee C}\vdash\neg
\quad
\cfrac{\neg B \vdash \neg B, \neg A \vee C \qquad \cfrac{\cfrac{A, B, C \vdash C}{B, C \vdash \neg A, C}\vdash\neg}{\cfrac{B, C \vdash \neg A \vee C}{C \vdash \neg B, \neg A \vee C}\vdash\neg}\vdash\vee}{\neg B \vee C \vdash \neg B, \neg A \vee C}\ \vee\vdash
}{\neg A \vee (\neg B \vee C) \vdash \neg B, \neg A \vee C}\ \vee\vdash
}{
\cfrac{\neg A \vee (\neg B \vee C), B \vdash \neg A \vee C}{}
}\ \neg\vdash
}{
\neg A \vee (\neg B \vee C), \neg A \vee B \vdash \neg A \vee C
}\ \vee\vdash
}{
\cfrac{\neg A \vee B \vdash \neg(\neg A \vee (\neg B \vee C)), (\neg A \vee C))}{\cfrac{\neg A \vee B \vdash \neg(\neg A \vee (\neg B \vee C)) \vee (\neg A \vee C))}{\cfrac{\varnothing \vdash \neg(\neg A \vee B), \neg(\neg A \vee (\neg B \vee C)) \vee (\neg A \vee C))}{\varnothing \vdash \neg(\neg A \vee B) \vee \neg(\neg A \vee (\neg B \vee C)) \vee (\neg A \vee C))}\vdash\vee}\vdash\neg}\ \vdash\neg
}
$$

288. Consider a derivation, which is a tree of sequents, where every node of the tree is either an axiom or is formed using one of the inference rules. $\varnothing \vdash \varphi$ is derived if this sequent is the root of the tree. Let $\Gamma = \{\vartheta_1, \ldots, \vartheta_k\}$ and $\Delta = \{\psi_1, \ldots, \psi_n\}$. Interpret the sequent $\Gamma \vdash \Delta$ as the formula $\bigwedge \Gamma \to \bigvee \Delta$, that is,

$$\top \wedge \vartheta_1 \wedge \cdots \wedge \vartheta_k \;\to\; \bot \vee \psi_1 \vee \cdots \vee \psi_n.$$

Then every axiom $\Gamma, \varphi \vdash \varphi, \Delta$ is a tautology, and the rules of inferences transform tautologies into tautologies. Therefore, if $\varnothing \vdash \varphi$ is derived, then φ must be a tautology.

On the other hand, consider a tautology φ. We decompose φ step by step and build the "reverse" derivation tree (known as the *reduction tree*) as follows. Start with the sequent $\varnothing \vdash \varphi$ and in every step apply the inverse of one of the inference rules. For example, from $\varphi \wedge \psi \vdash \vartheta$ we make $\varphi, \psi \vdash \vartheta$ (we applied the inverse of $\wedge\vdash$), or from $\varphi \vee \psi \vdash \vartheta$ we get to sequents $\varphi \vdash \vartheta$ and $\psi \vdash \vartheta$, etc. So, by a series of steps, the right side of $\vdash$ can be processed until it includes only propositional letters and then the same is done for the left side.

The leaves of the tree we obtained contain only propositional letters. As each derivation rule preserves tautologies, the leaves should be axioms, that is, the propositional letters on the right side of $\vdash$ should appear on the left sides as well. Therefore the tree we constructed is a Gentzen-style deduction of $\varnothing \vdash \varphi$.

289. Suppose the evaluation f satisfies Σ. By induction on the length of the derivation $\Sigma \vdash \varphi$ check that $f(\varphi) = \top$.

If φ appears in the derivation as an element of Σ, or as an axiom, then $f(\varphi) = \top$ by assumption. If φ is a conclusion of MP, then $\Sigma \vdash \psi$ and $\Sigma \vdash \psi \to \varphi$ have shorter proofs, thus $f(\psi) = \top$ and $f(\psi \to \varphi) = \top$ by induction. It gives $f(\varphi) = \top$, as required.

290. Suppose $\Sigma \vDash \varphi$. By the Compactness Theorem 5.10 there are finitely many elements $\{\varphi_1, \ldots, \varphi_k\}$ of Σ such that $\{\varphi_1, \ldots, \varphi_k\} \vDash \varphi$. By definition of $\vDash$, the formula

$$\varphi_1 \to \big(\varphi_2 \to (\cdots \to (\varphi_k \to \varphi)\cdots)\big)$$

is a tautology (for every evaluation f, if all φ_i evaluates to $\top$, then $f(\varphi) = \top$). Using this tautology as an axiom, φ_i as an element of Γ, k instances of MP derives φ. If Σ is empty (or $k = 0$), then φ is a tautology, thus it is an axiom, and then derivable from the empty set.

291. ($\Leftarrow$) If $\Sigma \vdash \psi \to \varphi$ then by monotonicity of $\vdash$ we have $\Sigma, \psi \vdash \psi \to \varphi$, and then by MP we get $\Sigma, \psi \vdash \varphi$.

($\Rightarrow$) Since φ is a member of a derivation it is either (i) an axiom, or (ii) a member of Σ, or (iii) same as ψ, or (iv) the conclusion of an MP.

(i) and (ii) In this case $\Sigma \vdash \varphi$. Using Ax$_1$ we can derive $\psi \to \varphi$ from Σ as follows.

1. φ (given)
2. $\varphi \to (\psi \to \varphi)$ (axiom)
3. $\psi \to \varphi$ (MP:1,2)

(iii) $\varphi \to \varphi$ is just an instance of Ax$_2$.

(iv) Since φ is a conclusion of MP, there is ϑ so that both ϑ and $\vartheta \to \varphi$ occur previously in the proof, and so the inductive hypothesis gives $\Sigma \vdash \psi \to \vartheta$ and $\Sigma \vdash \psi \to (\vartheta \to \varphi)$. From here the derivation can be finished as

1. $\psi \to \vartheta$ (given)
2. $\psi \to (\vartheta \to \varphi)$ (given)
3. $(\psi \to (\vartheta \to \varphi)) \to ((\psi \to \vartheta) \to (\psi \to \varphi))$ (axiom)
4. $(\psi \to \vartheta) \to (\psi \to \varphi)$ (MP:2,3)
5. $\psi \to \varphi$ (MP:1,4)

292. ($\Rightarrow$) By the deduction lemma (Problem 291) we have $\Sigma \vdash \neg\varphi \to \bot$. Then apply MP to Ax$_4$ to derive φ from Σ.

($\Leftarrow$) By assumption we have $\Sigma \vdash \varphi$. Use Ax$_5$ and MP to get $\Sigma \vdash \neg\varphi \to \bot$. From here the deduction lemma yields $\Sigma, \neg\varphi \vdash \bot$.

293. If both φ and $\neg\varphi$ is in Σ, then Ax$_5$ and two applications of MP derives $\bot$, contradicting $\Sigma \nvdash \bot$.

Suppose $\neg\varphi \notin \Sigma$. By maximality, $\Sigma, \neg\varphi \vdash \bot$. From here Problem 292 gives $\Sigma \vdash \varphi$, which implies $\varphi \in \Sigma$.

294. The claim is true for propositional variables by definition. Suppose it is true for φ and check for $\neg\varphi$. By Problem 293 exactly one of φ and $\neg\varphi$ is in Σ, and exactly one of $\neg\varphi$ and $\neg\neg\varphi$ is in Σ. Thus $\varphi \in \Sigma$ if and only if $\neg\neg\varphi \in \Sigma$, which implies the claim for $\neg\varphi$.

Next, suppose the claim for φ and ψ. If either φ or ψ is in Σ, then using one of the axioms Ax_6 or Ax_7 we get $\varphi \vee \psi \in \Sigma$. If neither φ nor ψ is in Σ, then 293 gives $\neg\varphi \in \Sigma$ and $\neg\psi \in \Sigma$, thus axiom Ax_8 and two applications of MP yields $\neg(\varphi \vee \psi) \in \Sigma$. this, together with $f_\Sigma(\varphi \vee \psi) = \bot$ finishes the induction.

Axioms for other connectives can be generated similarly. For example, the axioms for $\wedge$ could be

Ax_9 $\varphi \to (\psi \to (\varphi \wedge \psi))$

Ax_{10} $\neg\varphi \to \neg(\varphi \wedge \psi)$

Ax_{11} $\neg\psi \to \neg(\varphi \wedge \psi)$

295. If Σ is satisfiable, then $\Sigma \nvdash \bot$ by Problem 289. If $\Sigma \nvdash \bot$, then extend Σ to a maximal consistent set. The evaluation f_Σ satisfies Σ by Problem 294.

296. Immediate from weak completeness and Problem 292.

297. (a)

1. $\psi \to \vartheta$	(given)
2. $(\psi \to \vartheta) \to (\varphi \to (\psi \to \vartheta))$	(Ax_1)
3. $\varphi \to (\psi \to \vartheta)$	(MP: 1,2)
4. $(\varphi \to (\psi \to \vartheta)) \to ((\varphi \to \psi) \to (\varphi \to \vartheta))$	(Ax_3)
5. $(\varphi \to \psi) \to (\varphi \to \vartheta)$	(MP: 3,4)
6. $\varphi \to \psi$	(given)
7. $\varphi \to \vartheta$	(MP: 5,6)

(b)

1. $\neg\varphi$	(given)
2. $\neg\varphi \to (\neg\varphi \vee \psi)$	(Ax_6)
3. $\neg\varphi \vee \psi$	(MP: 1,2)
4. $\varphi \to \psi$	(same as 3, shorthand)
5. φ	(given)
6. ψ	(MP: 4,5)

(c) If φ occurs in the proof sequence, then Ax_1 and MP gives $\psi \to \varphi$. These steps are put on a single line below. X is the formula $\varphi \to (\psi \to \vartheta)$, and Y is $(\varphi \to \psi) \to (\varphi \to \vartheta)$.

1. $\varphi \to (\psi \to \vartheta) = X$	(given)
2. $X \to ((\varphi \to \psi) \to (\varphi \to \vartheta))$	(Ax_3)
3. $\psi \to X$	(Ax_1 and 1)
4. $\psi \to (X \to Y)$	(Ax_1 and 2)
5. $(\psi \to (X \to Y)) \to ((\psi \to X) \to (\psi \to Y))$	(Ax_3)

6. $(\psi \to X) \to (\psi \to Y)$ (MP: 4,5)

7. $\psi \to Y$ (MP: 3,6)

8. $(\psi \to (\varphi \to \psi)) \to (\psi \to (\varphi \to \vartheta))$ (Ax$_3$ and 7)

9. $\psi \to (\varphi \to \psi)$ (Ax$_1$)

10. $\psi \to (\varphi \to \vartheta)$ (MP: 8,9)

(d) Derivation similar to part (a) is not repeated here.

1. $\neg\neg\varphi \to (\neg\neg\varphi \vee \bot)$ (Ax$_6$)

2. $\neg\neg\varphi \to (\neg\varphi \to \bot)$ (same as 1, shorthand)

3. $(\neg\varphi \to \bot) \to \varphi$ (Ax$_4$)

4. $\neg\neg\varphi \to \varphi$ (part (a): 2,3)

298. Go through the steps which led to the completeness. There is no change in the syntactic deduction lemma (Problem 291) which gives axioms Ax$_1$, Ax$_2$, and Ax$_3$.

The next step is the equivalence of weak and strong completeness, Problem 292. As $\bot$ is not available, consistency should be defined differently. One possibility is to fix an arbitrary formula (or propositional variable) Φ, and use $\neg(\Phi \to \Phi)$ in place of $\bot$. The needed axiom schemes are

Ax$_4^*$ $(\neg\varphi \to \neg(\Phi \to \Phi)) \to \varphi$,

Ax$_5^*$ $\varphi \to (\neg\varphi \to \neg(\Phi \to \Phi))$.

Define Σ to be consistent if $\Sigma \nvdash \neg(\Phi \to \Phi)$. Problem 293 goes through without changes. Finally the evaluation f_Σ works for $\neg$ as before, and for $\to$ the following axioms suffice:

Ax$_6^*$ $\neg\varphi \to (\varphi \to \psi)$,

Ax$_7^*$ $\psi \to (\varphi \to \psi)$,

Ax$_8^*$ $\varphi \to (\neg\psi \to \neg(\varphi \to \psi))$.

299. Solution 1. Using completeness, $\vdash \varphi \leftrightarrow \psi$ if and only if φ and ψ takes the same value for every evaluation. Then Φ and Ψ also takes the same value for every evaluation, thus $\vdash \Phi \leftrightarrow \Psi$ by completeness again.

Solution 2. A direct "synthetic" proof can go by induction on the complexity of Φ (and by the deduction lemma, Problem 291). Using the connectives $\neg$ and $\vee$ only, the following derivations should be provided à là Problem 297:

(a) $\varphi \leftrightarrow \psi \vdash (\neg\varphi) \leftrightarrow (\neg\psi)$,

(b) $\varphi_1 \leftrightarrow \psi_1, \varphi_2 \leftrightarrow \psi_2 \vdash (\varphi_1 \vee \psi_1) \leftrightarrow (\varphi_2 \vee \psi_2)$.

300. (a) Suppose φ and $\varphi \to \psi$ are all–1 formulas ($\varphi \equiv 1$ and $\neg\varphi \vee \psi \equiv 1$). Then $\neg\varphi$ is all–0, thus we have to guarantee that $0 \vee \psi \equiv 1$ implies $\psi \equiv 1$. Therefore the $\vee$ table should satisfy $0 \vee 1 = 1, 0 \vee 0 \neq 1$ and $0 \vee * \neq 1$. Nothing changes if $\neg * = 1$. Note that independent of what $\neg$ is we need $\vee$ to satisfy $\neg 1 \vee \psi \equiv 1$ if and only if $\psi \equiv 1$.

(b) The truth table of the first three axioms with the first (second) definition of $\neg$:

φ	ψ	$\varphi \to \psi$	$\varphi \vee \varphi \to \varphi$	$\varphi \to \varphi \vee \psi$	$\varphi \vee \psi \to \psi \vee \varphi$
0	0	1 (1)	1 (1)	1 (1)	1 (1)
0	$*$	1 (1)	1 (1)	1 (1)	$*$ (1)
0	1	1 (1)	1 (1)	1 (1)	1 (1)
$*$	0	$*$ (1)	$*$ (1)	$*$ (1)	$*$ (1)
$*$	$*$	$*$ (1)	$*$ (1)	$*$ (1)	$*$ (1)
$*$	1	1 (1)	$*$ (1)	1 (1)	1 (1)
1	0	0 (0)	1 (1)	1 (1)	1 (1)
1	$*$	$*$ ($*$)	1 (1)	1 (1)	1 (1)
1	1	1 (1)	1 (1)	1 (1)	1 (1)

With none of the two definitions of $\neg$ can the value of the fourth axiom be 0:

$$(\varphi \to \psi) \to (\vartheta \vee \varphi \to \vartheta \vee \psi) = 0 \Leftrightarrow \begin{cases} \varphi \to \psi = 1, \\ (\vartheta \vee \varphi \to \vartheta \vee \psi) = 0 \Leftrightarrow \begin{cases} \vartheta \vee \varphi = 1, \\ \vartheta \vee \psi = 0 \Leftrightarrow \begin{cases} \vartheta = 0, \\ \psi = 0. \end{cases} \end{cases} \end{cases}$$

By $\vartheta = 0$ and $\vartheta \vee \varphi = 1$ we get $\varphi = 1$; impossible as $\varphi \to \psi$ should be 1.

With the first definition of $\neg$, the fourth axiom can be either 1 or $*$, shown by $\varphi = \psi = \vartheta = 1$ and $\varphi = \psi = \vartheta = *$, but it cannot take the value $*$ using the second definition of $\neg$:

$$(\varphi \to \psi) \to (\vartheta \vee \varphi \to \vartheta \vee \psi) = * \Leftrightarrow \begin{cases} \varphi \to \psi = 1, \\ (\vartheta \vee \varphi \to \vartheta \vee \psi) = * \Leftrightarrow \begin{cases} \vartheta \vee \varphi = 1, \\ \vartheta \vee \psi = *. \end{cases} \end{cases}$$

Here none of ϑ and ψ can be 1, thus φ should be 1, but then $\varphi \to \psi$ cannot be equal to 1.

(c) Definition of $\neg$ is the first one, and the $\vee$ table is as follows:

x	0	0	0	$*$	$*$	$*$	1	1	1
y	0	$*$	1	0	$*$	1	0	$*$	1
$x \vee y$	0	0	1	0	0	1	1	1	1

301. The effect of $\neg$ is given by $\neg 0 = 1$, $\neg * = *$ and $\neg 1 = 0$. Consider the following $\vee$-tables.

	x	0	0	0	$*$	$*$	$*$	1	1	1
	y	0	$*$	1	0	$*$	1	0	$*$	1
A	$x \vee y$	0	0	1	0	0	1	1	1	1
B	$x \vee y$	0	$*$	1	$*$	1	1	1	1	1
C	$x \vee y$	0	0	1	1	0	1	1	1	1

In each case (i.e. when $\vee$ is defined according to row A, B or C) MP derives from all–1 formulas only all–1 formulas (see Problem 300(a)). In case A only the second axiom is not all–1 ($* \rightarrow * \vee 0 = 0$), in case B only the first axiom is not all–1 ($* \vee * \rightarrow * = *$) and in case C only the third axiom is not all–1 ($* \vee 0 \rightarrow 0 \vee * = 0$).

302. (a) As a subformula has a smaller code than the formula itself, the characteristic function of the propositional formula codes can be defined by a straightforward course-of-value recursion.

(b) For example, instances of the first axiom scheme $\varphi \rightarrow (\psi \rightarrow \varphi)$ are the formulas of the form $\neg\varphi \vee (\neg\psi \vee \varphi)$. Now $i \in \omega$ is the code of such formula if and only if

- i is a formula code, and
- there are formula codes $j < i$ and $k < i$ such that $i = \langle 3, \langle 2, j \rangle, \langle 3, \langle 2, k \rangle, j \rangle \rangle$.

(c) This is just the set of triplets $\langle i, \langle 3, \langle 2, i \rangle j \rangle, j \rangle$ where $i, j \in \omega$ are formula codes.

303. Define G by course-of-value recursion.

304. If the variable v_i occurs in φ, then $i < \alpha(\varphi)$. Thus φ is a tautology if $f_a(\varphi) = \top$ for $a < \alpha(\varphi)$ where f_a is defined as $f_a(v_i) = \top$ if the i-th binary digit of a is 1. As the exponential function is recursive and checking whether φ evaluates to $\top$ is recursive by Problem 303, we are done.

305. x codes a derivation of y if all elements of x are formula codes, y is the last element of x, and every element of x is

- an element of Σ, or
- an instance of an axiom scheme, or
- the conclusion of a MP with premises appearing earlier.

By the assumption and by Problem 302 all of the conditions are recursive.

306. (a) $\Sigma \vdash \varphi$ iff there is a derivation from Σ ending with φ. By Problem 305 checking whether x is a derivation of y is recursive. Thus the function

$$g(u) = \begin{cases} (u)_1 & \text{if } (u)_0 \text{ is a derivation from } \Sigma \text{ of } (u)_1, \\ \alpha(\neg\bot) & \text{otherwise} \end{cases}$$

enumerates formulas derivable from Σ.

(b) Suppose F enumerates Σ. The ternary relation "x is a derivation of y from the first n elements of Σ" is recursive, which can be used to enumerate all consequences of Σ.

307. Let $A \subseteq \omega$ be a recursively enumerable but not recursive set (see Problem 216), and assume that the recursive function g enumerates it. The i-th formula in Σ is $v_k \vee v_k \vee \cdots \vee v_k$ (($i+1$) times) where $k = g(i)$. This Σ is recursive as g is a recursive function, and $\Sigma \vdash v_k$ if and only if $k \in A$. Consequently the consequences of Σ is not a recursive set.

12.6   FIRST-ORDER LOGIC

308. Each variable symbol is a term, from here the lower bound follows. Let E_0 be the set of variable and constant symbols and for $k \in \omega$ put

$$E_{k+1} = E_k \cup \{f(t_1,\ldots,t_n) : f \in \tau \text{ and } t_1,\ldots,t_n \in E_k\}.$$

By induction $|E_k| \leq |\tau| \cdot \omega$ and $E(\tau) = \bigcup \{E_k : k \in \omega\}$.

309. Exactly the same way as in Problem 308.

310. By induction on the complexity of the formula φ. The only case that might need clarification is when φ is of the form $\exists x \psi$. Suppose e_0 and e_1 agree on the free variables of $\exists x \psi$, and pick $a \in A$ arbitrarily. As $V(\exists x \psi) = V(\exists x \psi) \smallsetminus \{x\}$, for each $y \in V(\psi)$ we have $e_0(x/a)(y) = e_1(x/a)(y)$. Applying the inductive hypothesis to ψ and the evaluations $e_0(x/a)$, $e_1(x/a)$ we get

$$\mathfrak{A} \vDash \psi[e_0(x/a)] \text{ if and only if } \mathfrak{A} \vDash \psi[e_1(x/a)].$$

Thus $\mathfrak{A} \vDash \exists x \psi[e_0]$ iff $\mathfrak{A} \vDash \exists x \psi[e_1]$.

311. $\mathfrak{A} \vDash \forall x \varphi[e]$ exactly if for all $a \in A$, $\mathfrak{A} \vDash \varphi[e(x/a)]$. Thus if $\mathfrak{A} \vDash \varphi$, then $\mathfrak{A} \vDash \forall x \varphi$. The converse direction follows from the fact that e occurs among the evaluations $e(x/a)$ with the choice $a = e(x)$.

312. Take $\varphi \equiv \exists y(x = y)$ and $t \equiv y + 1$ Then $\vDash \varphi$, on the other hand $\nvDash \varphi[x/t]$.

313. (a) If the substitution $\varphi[x/t]$ is admissible, then, by the Substitution Lemma 6.7, $\mathfrak{A} \vDash (\varphi[x/t])[e]$ if and only if $\mathfrak{A} \vDash \varphi[e(x/t^{\mathfrak{A}}[e])]$, therefore the task is to give examples for $\mathfrak{A} \vDash (\varphi[x/t])[e]$ and $\mathfrak{A} \nvDash (\varphi[x/t])[e]$.

For the first one let $\varphi \equiv x = x$, $t \equiv y$, then $\varphi[x/t] \equiv y = y$, and both are true in any structure for any evaluation. Similarly, the second formula can be $\varphi \equiv x \neq x$ with the same term. Then $\varphi[x/t] \equiv y \neq y$, and both formulas are false.

(b) In the examples below the type is $\tau = \langle 0, 1, + \rangle$ where 0 and 1 and constant symbols, $+$ is a binary function symbol. The universe of the structure $\mathfrak{A}$ is the set ω of natural numbers, the interpretation of 0 and 1 is zero and one respectively, and the interpretation of $+$ is the usual addition.

Take $\varphi \equiv \exists y(x = y)$ and $t \equiv y + 1$. The substitution $\varphi[x/t]$ is not admissible. Note that $\varphi[x/t] \equiv \exists y(y + 1 = y)$.

- With $e(y) = 0$ we have $\mathfrak{A} \nvDash (\varphi[x/t])[e]$ (as $\mathfrak{A} \nvDash \exists y(y + 1 = y)[e]$) yet $\mathfrak{A} \vDash \varphi[e(x/t^{\mathfrak{A}}[e])]$ (as $\mathfrak{A} \vDash \exists y(1 = y)$).

- Let $\mathfrak{B}$ be the same as $\mathfrak{A}$ except for the interpretation of $+$ as $a +^{\mathfrak{B}} b = a$, and let $e(y) = 1$. Then $\mathfrak{B} \vDash (\varphi[x/t])[e]$ (as $\mathfrak{B} \vDash \exists y(y + 1 = y)[e]$) and $\mathfrak{B} \vDash \varphi[e(x/t^{\mathfrak{A}}[e])]$ (as $\mathfrak{B} \vDash \exists y(1 = y)$).

Let now $\varphi \equiv \forall y(x \neq y)$ and $t \equiv y + 1$. The substitution $\varphi[x/k]$ is not admissible, and $\varphi[x/t] \equiv \forall y(y + 1 \neq y)$.

- Let $e(y) = 0$. Then we have $\mathfrak{A} \vDash (\varphi[x/t])[e]$ (as $\mathfrak{A} \vDash \forall y(y+1 \neq y)[e]$) yet $\mathfrak{A} \not\vDash \varphi[e(x/t^{\mathfrak{A}}[e])]$ (as $\mathfrak{A} \not\vDash \forall y(1 \neq y)$).

As for the last possibility, write $\psi \equiv \varphi \wedge (z \neq z)$. Then whatever $\mathfrak{B}$ and e are we have $\mathfrak{B} \not\vDash (\psi[x/t])[e]$ and $\mathfrak{B} \not\vDash \psi[e(x/t^{\mathfrak{A}}[e])]$.

314. $\Gamma = \emptyset$, $\mathfrak{A}$ is any model with at least two elements and φ is the formula $x = y$.

315. 1. $\Gamma \vDash \varphi \rightarrow \psi$ and $\Gamma \cup \{\varphi\} \vDash \psi$. Take $\varphi \equiv \psi \equiv \forall x(x = x)$.
2. $\Gamma \not\vDash \varphi \rightarrow \psi$ and $\Gamma \cup \{\varphi\} \vDash \psi$. Put $\Gamma = \{\forall x R(x) \rightarrow \forall x P(x)\}$, $\varphi \equiv R(x)$ and $\psi \equiv \forall x P(x)$.
3. $\Gamma \vDash \varphi \rightarrow \psi$ and $\Gamma \cup \{\varphi\} \not\vDash \psi$ is impossible by monotonicity of $\vDash$ and modus ponens. For if $\Gamma \vDash \varphi \rightarrow \psi$, then $\Gamma, \varphi \vDash \varphi \rightarrow \psi$, and also $\Gamma, \varphi \vDash \varphi$, thus $\Gamma, \varphi \vDash \psi$.
4. $\Gamma \not\vDash \varphi \rightarrow \psi$ and $\Gamma \cup \{\varphi\} \not\vDash \psi$. Put $\varphi \equiv \forall x \forall y(x = y)$ and $\psi \equiv \exists x \exists y(x = y)$.

316. In general no, but $\vDash$ and $\overset{*}{\vDash}$ are equivalent for closed formulas. Clearly $\Gamma \overset{*}{\vDash} \varphi$ implies $\Gamma \vDash \varphi$. To see that the converse direction does not hold, let $\Gamma = \{\forall x R(x) \rightarrow \forall x P(x), R(x)\}$ and $\varphi \equiv \forall x P(x)$.

317. (a) If $\Gamma \vDash \varphi$ and $\varphi \notin \Gamma$, then $\Gamma \cup \{\varphi\}$ would be a proper consistent extension of Γ, contradicting maximality.

(b) If $\Gamma \vDash \varphi \vee \psi$ but neither $\Gamma \vDash \varphi$ nor $\Gamma \vDash \psi$ holds, then one of $\Gamma \cup \{\varphi\}$ or $\Gamma \cup \{\psi\}$ would be a proper consistent extension of Γ, contradicting maximality.

(c) As φ is closed, $\varphi \notin \Gamma$ iff $\Gamma \not\vDash \varphi$. Then there is a model $\mathfrak{A} \vDash \Gamma \cup \{\neg\varphi\}$, thus $\Gamma \cup \{\neg\varphi\}$ is consistent. As Γ is maximal we must have $\neg\varphi \in \Gamma$.

318. If $\mathfrak{A}$ is a model of Γ, then $\Gamma \subseteq \mathrm{Th}(\mathfrak{A})$, thus it suffices to show that $\mathrm{Th}(\mathfrak{A})$ is maximal consistent. If $\varphi \notin \mathrm{Th}(\mathfrak{A})$, then $\mathfrak{A} \not\vDash \varphi$, and then by Problem 311, $\mathfrak{A} \not\vDash \bar{\varphi}$, thus $\mathfrak{A} \vDash \neg\bar{\varphi}$ as $\bar{\varphi}$ is closed. If $\mathrm{Th}(\mathfrak{A}) \cup \{\varphi\}$ were consistent, say $\mathfrak{B} \vDash \mathrm{Th}(A) \cup \{\varphi\}$, then $\mathfrak{B} \vDash \bar{\varphi}$ and $\mathfrak{B} \vDash \neg\bar{\varphi}$, a contradiction.

319. No such a closed formula exists. Suppose $\mathfrak{A}$ has at least two elements, $\Gamma = \mathrm{Th}(\mathfrak{A})$ (which is maximal consistent by Problem 318), and let φ be $x = y$. Neither $\mathfrak{A} \vDash x = y$ (as $\mathfrak{A}$ has at least two elements), nor $\mathfrak{A} \vDash \neg(x = y)$ (as each element of A is equal to itself). See also Problem 314.

320. (a) For each n-variable function symbol $f \in \tau$ add a new relation symbol $r_f(x_1, \ldots, x_n, z)$ with the intended meaning that $z = f(x_1, \ldots, x_n)$. Add the following formulas to Γ:

$$\forall x_1 \ldots \forall x_n \exists z\, r_f(x_1, \ldots, x_n, z),$$
$$\forall x_1 \ldots \forall x_n \forall z_1 \forall z_2 \left(r_f(x_1, \ldots, x_n, z_1) \wedge r_f(x_1, \ldots, x_n, z_2) \rightarrow z_1 = z_2 \right)$$

For each term $t(\vec{x}) \in E(\tau)$ create a formula $\varphi_t(\vec{x}, z)$ expressing that its value is z by induction on the complexity of t:

- if t is the variable symbol x, then $\varphi_t(x, z) \equiv x = z$,

- if t the constant symbol $c \in \tau$, then $\varphi_t(z) \equiv c = z$,
- if t is $f(t_1, \ldots, t_n)$, then $\varphi_t(\vec{x}, z)$ is

$$\forall z_1 \ldots \forall z_n \big(\varphi_{t_1}(\vec{x}, z_1) \wedge \cdots \wedge \varphi_{t_n}(\vec{x}, z_n) \rightarrow r_f(z_1, \ldots, z_n, z)\big).$$

Rewrite the formulas in Γ by replacing every atomic formula $t_1 = t_2$ by

$$\forall z_1 \forall z_2 \big(\varphi_{t_1}(\vec{x}, z_1) \wedge \varphi_{t_2}(\vec{x}, z_2) \rightarrow z_1 = z_2\big),$$

and every atomic formula $r(t_1, \ldots, t_n)$ with $r \in \tau$ by

$$\forall z_1 \ldots \forall z_n \big(\varphi_{t_1}(\vec{x}, z_1) \wedge \cdots \wedge \varphi_{t_n}(\vec{x}, z_n) \rightarrow r(z_1, \ldots, z_n)\big).$$

It is clear that every model of Γ can be turned to a model of the new formula set and the other way around.

(b) Suppose no function symbol occurs in Γ. Add a new binary relation $\approx$ to τ, and add the following formulas to Γ:

$$x \approx x, \quad x \approx y \rightarrow y \approx x, \quad x \approx y \wedge y \approx z \rightarrow x \approx z,$$
$$\bigwedge_{1 \leq i \leq n} x_i \approx x_i' \rightarrow (r(\vec{x}) \leftrightarrow r(\vec{x}')) \quad \text{for all } r \in \tau.$$

The first line says that $\approx$ is an equivalence relation; the second line that each relation is compatible with it. Finally replace each equality symbol in Γ by $\approx$. Factoring a model of the new formula set preserves validity and makes $\equiv$ to be the real equality relation.

321. Let φ_n express the fact that it is not true that the structure has exactly n elements:

$$\varphi_n \equiv \neg \exists x_1 \ldots \exists x_n \forall y \big(\bigwedge_{i \neq j} x_i \neq x_j \wedge \bigvee_i y = x_i\big)$$

Clearly $\nvDash \varphi_n$ as there are structures having exactly n elements. On the other hand there must be some n for which both $\mathfrak{A} \vDash \varphi_n$ and $\mathfrak{B} \vDash \varphi_n$.

322. (a) For each $r \in \mathbb{R}$ let f_r be a unary function symbol and set

$$f_r^{\mathfrak{A}}(q) = \begin{cases} r & \text{if } q \in \mathbb{R} \setminus \mathbb{Z}, \\ q & \text{if } q \in \mathbb{Z}. \end{cases}$$

Additionally, let s and p be unary function symbols such that $s^{\mathfrak{A}}(r) = r + 1$ and $p^{\mathfrak{A}}(r) = r - 1$.

(b) Let f_r ($r \in \mathbb{R}$) be as above and set $d^{\mathfrak{A}}(r) = r + 2$, $h^{\mathfrak{A}}(r) = r - 2$, and

$$g^{\mathfrak{A}}(x, y) = \begin{cases} \pi & \text{if } x \text{ and } y \text{ have different parity,} \\ 0 & \text{if } x \text{ and } y \text{ are even,} \\ 1 & \text{if } x \text{ and } y \text{ are odd.} \end{cases}$$

323. The intersection of all substructures containing X is a substructure (as it is not empty) which contains X; this is the generated substructure.

Let $X_0 = X \cup \{c^{\mathfrak{A}} : c$ is a constant symbol in $\tau\}$, and for $n \in \omega$ put

$$X_{n+1} = X_n \cup \{f^{\mathfrak{A}}[X_n] : f \in \tau \text{ is a function symbol}\},$$

where $f^{\mathfrak{A}}[X_n]$ is the image of X_n under the function $f^{\mathfrak{A}}$. Then $B = \bigcup_{n \in \omega} X_n$ is the ground set of a substructure which contains X. (Actually this is the generated substructure.) Each X_i has cardinality $\leq |X| \cdot |\tau|$, thus $|B| \leq \omega \cdot |X| \cdot |\tau|$.

324. (a) Let $\mathfrak{A}$ be any structure with empty type. Every non-empty subset of A is the universe of a substructure.

(b) The two substructures in Problem 322 (b).

325. Straightforward induction on the complexity of the formulas using that $t^{\mathfrak{B}}[e] = t^{\mathfrak{A}}[e]$ for all terms t (again by induction on the complexity of t).

326. Yes. For a non-empty subset $X \subseteq A$ let $\mathcal{F}(X) = \bigcap\{B \in \mathcal{F} : X \subseteq B\}$. By assumption (a), $X \subseteq \mathcal{F}(X) \in \mathcal{F}$, and it is clear that $X \subseteq Y$ implies $\mathcal{F}(X) \subseteq \mathcal{F}(Y)$.

We claim that $\mathcal{F}(X) = \bigcup\{\mathcal{F}(N) : N \in [X]^{<\omega}\}$ for non-empty subsets of A. This is clear when X is finite; otherwise prove if by induction on the cardinality of X. If $|X| = \kappa$, then write X as $\{x_\alpha : \alpha < \kappa\}$, and let $X \restriction \alpha = \{x_\beta : \beta < \alpha\}$. By condition (b), $\bigcup\{\mathcal{F}(X \restriction \alpha) : \alpha < \kappa\}$ is in $\mathcal{F}$, thus it is equal to $\mathcal{F}(X)$. Using the induction hypothesis, $\mathcal{F}(X \restriction \alpha)$ is the union of $\mathcal{F}(N)$ as N runs over the finite subsets of $X \restriction \alpha$. As any finite subset of X is a subset of $X \restriction \alpha$ for some $\alpha < \kappa$, we get the claim for X.

For each $N \in [A]^{<\omega}$ and $b \in \mathcal{F}(N)$ define the $|N|$-variable function $f_{N,b}$ as

$$f_{N,b}(x_1, \ldots, x_n) = \begin{cases} b & \text{if } N = \{x_1, \ldots, x_n\}, \\ x_1 & \text{otherwise.} \end{cases}$$

With these functions the universe of the substructure generated by $X \subseteq A$ is $X \cup \bigcup\{\mathcal{F}(N) : N \in [X]^{<\omega}\} = \mathcal{F}(X) \in \mathcal{F}$, and for $X \in \mathcal{F}$, $\mathcal{F}(X) = X$.

327. Solution 1. Note that every automorphism preserves the interpretation of all the constant symbols. Thus if all elements of the ground set are named by some constant (with interpretation of that element), then this structure has only one automorphism, the identity. This solution has been excluded by requiring the type to be finite.

(a) The structure has ground set the natural numbers, a single constant symbol denoting zero, and a unary function symbol f for the $x \mapsto x + 1$ function. Any automorphism preserves zero, and then $f(0)$, $f(f(0))$, etc.

(b) Let the ground set be the real numbers $\mathbb{R}$. In the type have constant symbols denoting 0 and 1, and function symbols for addition, multiplication and division. (The latter one is not a function, but extend division to be a function by saying that $x/0 = x$.) Any automorphism preserves the rational numbers. To extend it to the reals, add the ordering (a binary relation). Solution 2. Any well-ordered structure (the signature consists of a single binary relation symbol) has trivial automorphism only.

328. $\mathfrak{A}$ is a graph if the edges are undirected, and there are no loops. These properties can be expressed as

$$E(x, y) \rightarrow E(y, x) \quad \text{it is undirected, and}$$
$$\neg E(x, x) \qquad\qquad \text{there are no loops.}$$

329. (a) Every vertex has at least three different neighbors, but not more:

$$\forall x \exists y_1 \exists y_2 \exists y_3 \left(y_1 \neq y_2 \wedge y_2 \neq y_3 \wedge y_3 \neq y_1 \wedge \right.$$
$$E(x, y_1) \wedge E(x, y_2) \wedge E(x, y_3) \wedge$$
$$\left. \forall z (E(x, z) \rightarrow (z = y_1 \vee z = y_2 \vee z = y_3)) \right).$$

(b) Some vertex is connected to all others: $\exists x \forall y (x=y \vee E(x, y))$, and neighbors of x are not connected: $\forall x \forall y_1 \forall y_2 (E(x, y_1) \wedge E(x, y_2) \rightarrow \neg E(y_1, y_2))$.

(c) $\forall x_1 \forall x_2 \forall x_3 \neg (E(x_1, x_2) \wedge E(x_2, x_3) \wedge E(x_3, x_1))$.

(d) The graph has no edges, as x_1 and x_3 can be the same vertex.

330. (a) $\Gamma = \{\varphi_k : k \in \omega\}$ where for each $k \in \omega$, φ_k is the formula

$$\neg \exists x_1 \cdots \exists x_k \left(\bigwedge_{i \neq j} x_i \neq x_j \wedge \bigwedge_{i < k} E(x_i, x_{i+1}) \wedge E(x_k, x_1)\right)$$

(b) $\neg \exists x_1 \cdots \exists x_{16} \left(\bigwedge_{i \neq j} x_i \neq x_j \wedge \bigwedge_{i=1}^{15} E(x_i, x_{i+1})\right)$

(c) A graph is bipartite iff it does not contain cycles of odd length, thus $\Gamma = \{\varphi_k : k \text{ is odd}\}$ suffices, where φ_k is from part (a).

331. For every x there is a y which is connected to every vertex x is connected to, and a new one:

$$\forall x \exists y \left(\forall u (E(x, u) \rightarrow E(y, u)) \wedge \exists v (E(y, v) \wedge \neg E(x, v))\right).$$

Such a graph must be infinite as there can be no maximal degree vertex. To construct such a graph take vertices v_i, w_i for $i \in \omega$, and edges $\langle v_i, w_j \rangle$ where $i \leq j$.

332. $(x \cdot y) \cdot z = x \cdot (y \cdot z)$, $\quad x \cdot e = x$, $\quad e \cdot x = x$, $\quad x \cdot x^{-1} = e$, $\quad x^{-1} \cdot x = e$.

333. (a) The formula $\exists x_1 \ldots \exists x_n \bigvee_{i \neq j} x_i \neq x_j$ is true in a structure if it has at least n elements. Let Γ be an infinite set of such formulas.

(b) For $n \geq 1$ define the term x^n by induction as $x^1 \equiv x$, and $x^{n+1} \equiv (x^n) \cdot x$. The formula set is $\{\forall x (x^n = e \rightarrow x = e) : n \geq 1\}$.

> **Remark.** The notation x^n is only a shorthand for a certain term which depends on what the number n is.

(c) Torsion-free was handled in (b). Abelian: $\forall x \forall y\, (x \cdot y = y \cdot x)$. Divisible: for every $n \geq 2$ add $\forall x \exists y\, (y^n = x)$. No single (or finitely many) formula can express divisibility, see Problem 705.

(d) If p is a prime then the formula $\forall x\, (x^p = e)$ works, as in this case the order of x divides p, and for $x \neq e$ the order is not 1.

334. The shifts form the centralizer of π, thus the formula $x \cdot \pi = \pi \cdot x$ works.

335. $(x < y \wedge y < z \rightarrow x < z)$, $(x < y) \rightarrow \neg(y < x)$, $(x < y) \vee (x = y) \vee (y < x)$.

336. (a) $\forall x \forall y \exists z \big(x \leq z \wedge y \leq z \wedge \forall w((x \leq w \wedge y \leq w) \rightarrow z \leq w)\big)$.
(b) $\forall x \forall y \exists z \big(z \leq x \wedge z \leq y \wedge \forall w((w \leq x \wedge w \leq y) \rightarrow w \leq z)\big)$.
(c) $\exists x \exists y \forall z \big((z \leq x \rightarrow z = x) \wedge (y \leq z \rightarrow y = z)\big)$.

337. Use the formula $\forall x(\varphi_{\text{left}}(x) \vee \varphi_{\text{right}}(x))$, where $\varphi_{\text{left}}(x)$ is

$$\exists y\, (y \leq x \wedge y \neq x) \wedge$$
$$\forall y\big((y \leq x \wedge y \neq x) \rightarrow \exists z(z \neq y \wedge z \neq x \wedge y \leq z \wedge z \leq x)\big) \wedge$$
$$\exists y\big((x \leq y \wedge x \neq y) \wedge \forall z((x \leq z \wedge z \leq y) \rightarrow (x = z \vee y = z))\big).$$

$\varphi_{\text{right}}(x)$ is analogous. The formula is simpler when using $<$ instead of $\leq$.

Such an ordering does exist: take $\sum_\eta\{0, 1\}$, that is, replace each rational number by two consecutive points. However, this side cannot be the same for all points in the ordering. To see this, suppose, seeking a contradiction, that each point has an immediate right successor but on the left side there are elements arbitrarily close. Let x be arbitrary and y be the closest to x on the right hand side. Then x is on the left hand side of y, hence, as there must be elements arbitrary close to y on the left hand side, there must exist some $x < z < y$ which contradicts the choice of y.

338. $\exists x(\exists y(y < x \wedge \exists x(x < y \wedge \exists y(y < x \wedge \ldots))))$.

339. (a) For $n \geq 0$ let $\pi(n)$ denote the term $((((0 + 1) + 1) + \ldots) + 1$ (n times, bracketed from left to right); for negative $n \in \mathbb{Z}$ let $\pi(n) = -\pi(-n)$. Similarly, x^n denotes the term $x \cdots x$ (n times). Put

$$\Gamma = \big\{\exists x\big(\pi(a_n)x^n + \cdots + \pi(a_0) = 0\big) : n \geq 1,\ a_i \in \mathbb{Z},\ a_n \neq 0\big\}.$$

(b) $1 + 1 = 0$

(c) For every prime p write $1 + \cdots + 1 \neq 0$ using p ones.

(d) $\Gamma = \{\forall x_1 \cdots \forall x_n(x_1^2 + \ldots + x_n^2 \neq -1) : n \geq 1\}$. A field satisfies Γ if and only if an ordering can be defined on the field which has properties similar to that of the real numbers. Such fields are called *formally real*, and must be of characteristic zero.

(e) A field is algebraically closed if every polynomial *with coefficients from the field* has a root. This can be expressed by the following formula set:

$$\Gamma = \big\{\forall a_n \cdots \forall a_0 \exists x\big(a_n \neq 0 \rightarrow a_n x^n + \cdots + a_0 = 0\big) : n \geq 1\big\}.$$

(f) If (d) also holds, the field is called *real closed*.

$$\Gamma = \big\{\forall a_n \cdots \forall a_0 \exists x\big(a_n \neq 0 \rightarrow a_n x^n + \cdots + a_0 = 0\big) : n \geq 1,\ n \text{ is odd}\big\}$$

340. (a) $x \approx x$, $\quad (x \approx y) \rightarrow (y \approx x)$, $\quad (x \approx y \wedge y \approx z) \rightarrow (x \approx z)$. For an n-variable function symbol f and relation symbol r add

$$\bigwedge_i x_i \approx y_i \rightarrow f(x_1, \ldots, x_n) \approx f(y_1, \ldots, y_n), \text{ and}$$

$$\bigwedge_i x_i \approx y_j \rightarrow (r(x_1, \ldots, x_n) \leftrightarrow r(y_1, \ldots, y_n)).$$

(b) Suppose first that there is a model $\mathfrak{A} \vDash \Gamma$ and an evaluation e over $\mathfrak{A}$ such that $\mathfrak{A} \nvDash \varphi[e]$. Interpret $\approx$ as the equality relation on A. Then $\mathfrak{A}^{\approx} \vDash \Gamma^{\approx} \cup \Delta$, and with the same evaluation e, $\mathfrak{A}^{\approx} \nvDash \varphi^{\approx}$.

For the other direction let $\mathfrak{A}^{\approx} \vDash \Gamma^{\approx} \cup \Delta$. Define $\mathfrak{A}$ as the factor $\mathfrak{A}^{\approx}/\approx$. Since the formulas in Δ are true in $\mathfrak{A}^{\approx}$, this is a sound definition. Let $e^{\approx}$ be an evaluation over $\mathfrak{A}^{\approx}$, and $e(x) = e^{\approx}(x)/\approx$ be the corresponding evaluation over $\mathfrak{A}$. It is easy to check by induction on the complexity of formulas that for every τ-type formula φ, $\mathfrak{A}^{\approx} \vDash \varphi^{\approx}[e^{\approx}]$ if and only if $\mathfrak{A} \vDash \varphi[e]$. Thus $\mathfrak{A} \vDash \Gamma$, and if $\mathfrak{A}^{\approx} \nvDash \varphi^{\approx}[e^{\approx}]$ for some $e^{\approx}$, then $\mathfrak{A} \nvDash \varphi[e]$, giving the inverse implication.

341. Solution 1. We will construct formulas $\varphi_k(x, y)$ by induction which express that $x \geq y + 2^k$ and use the only additional variable symbol z. Let $\varphi_0(x, y) \equiv y < x$, and suppose $\varphi_k(x, y)$ has already been defined. Then put

$$\varphi_{k+1}(x, y) = \exists z \big(\exists y (y = z \wedge \varphi_k(x, y)) \wedge \exists x (x = z \wedge \varphi_k(x, y)) \big).$$

Solution 2. Similarly to Solution 338,

$$\exists z (x < z \wedge z < y \wedge \exists x (z < x \wedge x < y \wedge \exists z (\ldots))).$$

342. Formulas (a)–(c) hold in $\mathfrak{A}$, (d) does not.

(a) This formula says that in the binary representation of x no digit is 1, and $x = 0$ is such a number.

(b) According to this formula the u-th digit of z must be 1 if and only if the u-th digit of either x or y is 1. Take z to be the "bitwise or" of x and y.

(c) $\forall v (v \in u \rightarrow v \in x)$ holds if the binary representation of u can be got from that of x after changing some of the 1's to 0. y is the number which has 1 in exactly these positions. As there are finitely many such u, there is a required y.

(d) The formula says that x is not zero (has 1 at some position), and if x has a 1 at the y-th position, then it also has a 1 at the v-th position, where v is bigger than y. Clearly no such an x exists.

343. For each x there are finitely many u such that $u \in^{\mathfrak{A}} x$. For each such u take the smallest v such that $\mathfrak{A} \vDash \varphi[u, v]$, if such a v exists. Let us say the collections of such v's is the set $\{v_1, \ldots, v_n\}$. Then let y be the natural number whose binary digits up to $\max\{v_i\}$ are 1. This y fulfills the requirements.

344. (a) $\varphi_a(n) \equiv (1 + 1 \leq n) \wedge \forall x (x \leq 1 \vee n \leq x \vee \neg \exists z (x \cdot z = n))$.

(b) $\varphi_b(n) \equiv (n \geq 1) \wedge \forall x (\exists z (x \cdot z = n) \wedge \varphi_a(x) \rightarrow x = 1 + 1)$.

(c) $\varphi_c(n) \equiv (n \geq 1) \wedge \exists y \forall x (\varphi_a(x) \wedge \exists z (x \cdot z = n) \rightarrow x = y)$.

(d) $\varphi_d(n) \equiv \exists x_1 \ldots \exists x_9 (n = x_1 \cdot x_1 \cdot x_1 + \cdots + x_9 \cdot x_9 \cdot x_9)$.

Remark. Waring's theorem states that every natural number can be written as the sum of nine perfect cubes. Thus $\mathfrak{A} \vDash \varphi_d[a]$ for each natural number $a \in \omega$.

345. The twin prime conjecture states that there are infinitely many primes p such that $p + 2$ is also prime. The formula $\varphi_a(n)$ from Solution 344 is true for prime numbers only. The property "there are infinitely many" can be expressed by referring to unboundedness:

$$\forall n \exists p(n \leq p \wedge \varphi_a(p) \wedge \varphi_a(p + 1 + 1)).$$

346. (a) The formula $x + x = x$ is satisfied by 0 only.

(b) Only 1 satisfies $x \neq 0 \wedge \forall y(y \neq 0 \rightarrow x \mid y)$.

(c) By (a) and (b) 0 and 1 can be used in our formulas as shorthands for the unique elements determined by formulas φ_0 and φ_1. Express the least common multiple of two positive numbers using the divisibility relation as follows:

$$\text{lcm}(x, u, v) \equiv u \mid x \wedge v \mid x \wedge \forall y((u \mid y \wedge v \mid y) \rightarrow x \mid y).$$

For $a, b \geq 1$ $\mathfrak{A} \vDash \text{lcm}[c, a, b]$ iff c is the least common multiple of a and b.

Note that b and $b + 1$ are always coprime, thus $b(b + 1)$ is their lcm. Therefore $\varphi_3(a, b)$ can be the formula

$$(a = 0 \wedge b = 0) \vee (b \neq 0 \wedge \exists u(\text{lcm}(u, b, b + 1) \wedge u = a + b).$$

347. Observe that $(1 + az)(1 + bz) = 1 + z^2(1 + ab)$ if and only if $z = 0$ or $z = a + b$. Also, $z = 0$ iff $\forall u(z \cdot u = z)$, so the following formula works:

$$z \neq 0 \wedge S(x \cdot z) \cdot S(y \cdot z) = S(z \cdot z \cdot S(x \cdot y)) \ \vee$$
$$z = 0 \wedge \forall z'\big(S(x \cdot z') \cdot S(y \cdot z') = S(z' \cdot z' \cdot S(x \cdot y)) \rightarrow z' = 0\big).$$

348. (a) $z + z = z$ iff $z = 0$, thus $x \neq 0 \wedge x + 1 \neq 0 \wedge 0 + 1 \neq x$ works.

(b) $|x|$ is a prime if $|x| > 1$ and if y is a divisor of x and $|y| > 1$, then x is a divisor of y:

$$\varphi_2(x) \equiv |x| > 1 \wedge \forall y\big(|y| > 1 \wedge y \mid x \rightarrow x \mid y\big).$$

(c) $|u|$ is a power of the prime $|x|$ if $u \neq 0$ and all prime divisors of u are divisors of x (allowing $|u| = 1$):

$$\varphi_3(x, u) \equiv u \neq 0 \wedge \varphi_2(x) \wedge \forall z(z \mid u \wedge \varphi_2(z) \rightarrow z \div x).$$

(d) Note that for a prime p, p^n has exactly $2n + 2$ divisors: $\pm 1, \pm p, \ldots, \pm p^n$. Thus the formula expressing that u is a power of the prime $|x|$ and u has exactly 6 divisors will do:

$$\varphi_4(x, u) \equiv \varphi_3(x, u) \wedge \exists y_1 \ldots y_6 (\bigwedge_i y_i \mid u \wedge \bigwedge_{i \neq j} y_i \neq y_j \wedge \forall y(y \mid u \rightarrow \bigvee_i y = y_i)).$$

349. The map $\pi : i \mapsto -i$ is an automorphism of the structure $\mathfrak{A} = \langle \mathbb{Z}, +, | \rangle$, thus $\mathfrak{A} \vDash \varphi[a]$ iff $\mathfrak{A} \vDash \varphi[-a]$. As a consequence, no formula distinguishes 1 and -1.

350. (a) Following the hint, let $\mathsf{lcm}(z, x, y)$ be the formula

$$\mathsf{lcm}(z, x, y) \equiv x \,|\, z \wedge y \,|\, z \wedge \forall u\,((x \,|\, u \wedge y \,|\, u) \to z \,|\, u).$$

If none of a and b is zero, then $\mathfrak{A} \vDash \psi[c, a, b]$ iff c or $-c$ is $\mathsf{lcm}(a, b)$.

For an integer $i > 1$, $\mathsf{lcm}(i, i+1) = \pm i(i+1)$ and $\mathsf{lcm}(i, i-1) = i(i-1)$. Observe that

$$\pm i(i+1) \pm i(i-1) = \begin{cases} \pm 2i^2 \\ \pm 2i \end{cases}$$

If $|i| > 1$ then the four values are all different, thus for $|i| > 1$ the following formula is satisfied by $j = \pm i^2$ only:

$$\varphi_1(i, j) \equiv \exists x \exists y \big(\mathsf{lcm}(x, i, i+1) \wedge \mathsf{lcm}(y, i, i-1) \wedge$$
$$\wedge\, j + j = x + y \wedge j \neq i \wedge i + j \neq 0 \big).$$

When $|i| \leq 1$ simply check what j is.

(b) The problem is to choose between i^2 and $-i^2$. Suppose $|x| > 1$. Then $|x| + 1$ and $|x| - 1$ both divide $x^2 - 1$, while $|x| + 1$ does not divide $-x^2 - 1$. (Otherwise we would have $(|x| + 1) \,|\, (|x| + 1)^2 - x^2 - 1$, hence $|x| + 1 \,|\, 2|x|$, which implies $x = 0$ or $|x| = 1$). Thus the *positive* square j of $|i| > 1$ can be singled out by

$$\varphi_2(i, j) \equiv \varphi_1(i, j) \wedge i + 1 \,|\, j - 1 \wedge i - 1 \,|\, j - 1.$$

(c) An integer is non-negative iff it is the sum of four squares, thus $\varphi_3(i)$ can be

$$\varphi_3(i) \equiv \exists u \exists v \exists w \exists y \big(\varphi_2(u) \wedge \cdots \wedge \varphi_2(y) \wedge i = u + v + w + y \big).$$

351. We make use of the following abbreviation. If $\vartheta(x)$ is a formula with one free variable symbol, then $(\forall x \in \vartheta)\varphi$ abbreviates $\forall x(\vartheta(x) \to \varphi)$, and similarly for $\exists$. Let $\vartheta(x)$ be $R(x) \wedge 0 \leq x \wedge 0 \neq x$.

The required formula is $\varphi(f)$ below, where f is a free variable.

$$F(f) \to (\forall x \in R)(\forall \varepsilon \in \vartheta)(\exists \delta \in \vartheta)$$
$$\big((\forall y \in R)(|x - y| \leq \delta \to |H(f, x) - H(f, y)| \leq \varepsilon) \big)$$

Here $|a - b| \leq c$ is, for example, the abbreviation of the formula

$$\big((a - b \leq 0) \to -(a - b) \leq c \big) \wedge \big((0 \leq a - b) \to (a - b) \leq c \big)$$

352. There is no such λ. Let $\mathfrak{A} = \langle \lambda, c_\xi \rangle$ with $c_\xi^{\mathfrak{A}} = \xi$ for $\xi < \lambda$ and $\mathfrak{B} = \langle \lambda \cup \{a\}, c_\xi \rangle$, where $a \notin \lambda$ is arbitrary. That is, $\mathfrak{B}$ is obtained from $\mathfrak{A}$ by adding a new element to its universe. Then $\mathfrak{A}, \mathfrak{B} \vDash \Gamma$, $|\mathfrak{A}| = |\mathfrak{B}| = \lambda$ and $\mathfrak{A} \not\cong \mathfrak{B}$.

353. (a) Two structures of empty type are isomorphic iff they have the same cardinality. Hence the number of countable τ-structures is $\sum_{\alpha \leq \omega} \alpha = \omega$ countable.

(b) Let the only unary relation symbol in the language be R. $\mathfrak{A} = \langle A, R^{\mathfrak{A}} \rangle$ and $\mathfrak{B} = \langle B, R^{\mathfrak{B}} \rangle$ are isomorphic iff $|R^{\mathfrak{A}}| = |R^{\mathfrak{B}}|$, and $|A \smallsetminus R^{\mathfrak{A}}| = |B \smallsetminus R^{\mathfrak{B}}|$. Hence, there are countably many countable τ-structures.

(c) Each of $R_1^{\varepsilon_1} \cap \cdots \cap R_n^{\varepsilon_n}$ can be chosen countably many ways (here $\varepsilon_i \in \{0,1\}$ and $R^1 = R$, $R^0 = A \smallsetminus R$). Thus the answer is countable in this case, too.

354. We can assume that the universe of our structure $\mathfrak{A}$ is κ. n-ary relations and functions of $\mathfrak{A}$ are respectively subsets of κ^n and κ^{n+1}. The number of such subsets is 2^κ. Therefore the number of non-isomorphic τ-structures of cardinality κ is at most $|t| \cdot 2^\kappa = 2^\kappa$.

Let τ consists of λ many constant symbols $\{c_i : i < \lambda\}$. For $X \subseteq \lambda$ write $\Gamma_X = \{c_i = c_j : i, j \in X\} \cup \{c_i \neq c_j : i \in X, j \notin X\}$. For distinct $X, Y \subseteq \lambda$ no $\mathfrak{A} \vDash \Gamma_X$ and $\mathfrak{B} \vDash \Gamma_Y$ can be isomorphic. For each $X \subseteq \lambda$ there is a two-element model $\mathfrak{A}_X \vDash \Gamma_X$, therefore the number of non-isomorphic countable τ-structures is at least 2^λ.

355. By Problem 354 there can be no more than 2^κ non-isomorphic structures. Suppose τ contains a binary relation $\leq$, and consider only those structures where $\leq^{\mathfrak{A}}$ is a κ-type well-ordering of the ground set A. If two such structures are isomorphic, then the isomorphism is unique, as it preserves the well-ordering. Add another unary relation symbol R to the type. Interpreting R as different subsets (relative to the ordering), one gets 2^κ pairwise non-isomorphic structures.

> **Remark.** Only one binary relation symbol is enough, see Problem 364, and also Problem 356 (b).

356. (a) There are continuum many. The idea is that from the structure one can recover an infinite subset $X \subseteq \omega$ uniquely, and there are continuum many such subsets. The structure will contain a single element with $f(a) = a$. For each $i \in X$ add $i + 1$ new elements $a_0^i, a_1^i, \ldots, a_i^i$ so that $f(a_i^i) = a$, and $f(a_j^i) = a_{j+1}^i$ otherwise.

(b) Such a structure can be considered as a directed graph where each vertex has out-degree at most one, and $f(v) = v$ iff the out-degree is zero (v is a *sink*). We will consider only connected graphs with a single sink. Suppose we have κ many pairwise non-isomorphic such graphs of cardinality $< \kappa$ arranged as $\langle G_\alpha : \alpha < \kappa \rangle$. It is easy to construct 2^κ many non-isomorphic graphs of cardinality κ. For a subset $X \subseteq \kappa$ of size κ let a be the sink of the graph G_X, and for every $\alpha \in X$ add an edge from the sink of a copy of G_α to a. For different subsets X these graphs are different: removing a from G_X the connected components of the remaining graph are copies of G_α, thus one can recover the subset X from G_X.

It remained to show that for each κ there is a graph sequence $\langle G_\alpha \rangle$. It is clear if $\kappa \leq 2^\lambda$ for some $\lambda < \kappa$. Otherwise there is an increasing sequence of cardinals $\lambda_\xi < \kappa$ converging to κ, and the union of the non-isomorphic graphs for λ_ξ works.

357. The similarity type contains exactly one unary relation symbol R, and Γ expresses that in any model having more than three elements the relation R is full. Γ can be the single formula

$$\exists x_1 \exists x_2 \exists x_3 \exists x_4 \Big(\bigwedge_{i \neq j} x_i \neq x_j \Big) \;\to\; \forall x R(x).$$

In a model $\mathfrak{A}$ having three elements, we have $|R^{\mathfrak{A}}| \leq 3$, thus there are exactly four non-isomorphic models of Γ on three elements.

358. For a natural number $n \geq 1$ let $p(n)$ be the smallest prime number larger or equal than n (e.g. $p(3) = 3$, $p(4) = 5$). Let the formula φ_n express that if there are at least n distinct elements, then there are at least $p(n)$ distinct elements:

$$\Big(\exists x_1 \cdots \exists x_n \bigwedge_{i \neq j} x_i \neq x_j \Big) \;\longrightarrow\; \Big(\exists x_1 \cdots \exists x_{p(n)} \bigwedge_{i \neq j} x_i \neq x_j \Big)$$

Then $\Gamma = \{\varphi_n : n \in \omega\}$ works.

359. Let the similarity type contain a single unary relation symbol R and let Γ be the (infinite) set of formulas expressing that R and its complement are infinite. Each countable models of Γ are isomorphic, while for $\mathfrak{A} \vDash \Gamma$ with $|\mathfrak{A}| = \omega_1$ we can have $|R^{\mathfrak{A}}| = \omega$ or $|R^{\mathfrak{A}}| = \omega_1$, leading to two non-isomorphic models.

360. The formula φ_n below says that *there is* a cycle of length n:

$$\varphi_n \equiv \exists x_1 \cdots \exists x_n \Big(\bigwedge_{i \neq j} x_i \neq x_j \wedge \bigwedge_{i=1}^{n-1} E(x_i, x_{i+1}) \wedge E(x_n, x_1) \Big).$$

The degree of x is three if

$$\psi(x) \equiv \exists y_1 \exists y_2 \exists y_3 \Big(\bigwedge_{i \neq j} y_i \neq y_j \wedge \bigwedge_i E(x, y_i) \wedge \forall z (E(x, z) \to \bigvee_i z = y_i) \Big).$$

Then

$$\Gamma = \big\{ \forall x \psi(x), \neg \varphi_n : n \geq 3 \big\}.$$

Let G be the countable 3-regular, cycle free, connected graph. G is an infinite tree with degrees three: such a graph is unique. Any countable model of Γ is the disjoint union of copies of G. There are countably many non-isomorphic countable models of Γ. If $\mathfrak{A} \vDash \Gamma$ is an uncountable model, then $\mathfrak{A}$ is the disjoint union of $|\mathfrak{A}|$ many copies of G, hence there is exactly one model of Γ of cardinality $\kappa > \omega$. (Cf. Problem 632).

361. E is an equivalence relation:

$$\forall x E(x,x) \qquad\qquad\qquad\qquad \text{reflexive,}$$
$$\forall x \forall y (E(x,y) \to E(y,x)) \qquad\qquad \text{symmetric,}$$
$$\forall x \forall y \forall z (E(x,y) \wedge E(y,z) \to E(x,z)) \quad \text{transitive.}$$

E has at least n equivalence classes; each equivalence class has at least n members:

$$\varphi_n \equiv \exists x_1 \cdots \exists x_n \bigwedge_{i \neq j} \neg E(x_i, x_j),$$
$$\psi_n \equiv \forall x \exists y_1 \cdots \exists y_n \Big(\bigwedge_{i \neq j} y_i \neq y_j \wedge \bigwedge_i E(x,y_j) \Big).$$

Then $\Gamma = \{E \text{ is equivalence}\} \cup \{\varphi_n, \psi_n : n \geq 2\}$.

For a model $\mathfrak{A} \vDash \Gamma$ let $n_\lambda^{\mathfrak{A}}$ denote the number of equivalence classes having cardinality λ. Two models $\mathfrak{A}, \mathfrak{B}$ of Γ are isomorphic if and only if $n_\lambda^{\mathfrak{A}} = n_\lambda^{\mathfrak{B}}$ for all cardinals $\lambda \geq \omega$ (as in this case there is a bijection between the equivalence classes of the same cardinality).

(a) In the countable case $n_\omega^{\mathfrak{A}} = \omega$ and $n_\lambda^{\mathfrak{A}} = 0$ for any $\lambda > \omega$. Therefore, up to isomorphism, there is a unique countable model of Γ.

(b) In this case either $\lambda = \omega$ or $\lambda = \omega_1$, and n_λ can be $n \in \omega$, ω, or ω_1. Thus there are countable different models.

(c) The possibilities for λ are ω, ω_1, ω_2, and, as above, there are countably many models.

(d) In this case there are countably many possible λ, and each can take countably many possibilities. The number of non-isomorphic models is 2^ω.

362. Let the relation symbols be R_i. For a τ-type model $\mathfrak{A}$ and $\vec{\epsilon} \in \{0,1\}^k$ write $R_{\vec{\epsilon}}^{\mathfrak{A}}$ for $\bigcap_i \epsilon_i R_i^{\mathfrak{A}}$ where $\epsilon_i R_i^{\mathfrak{A}}$ means $R_i^{\mathfrak{A}}$ for $\epsilon_i = 1$ and its complement for $\epsilon_i = 0$. For and integer $s > 0$ the models $\mathfrak{A}$ and $\mathfrak{B}$ are s-*close*, if

$$\min\{s, |R_{\vec{\epsilon}}^{\mathfrak{A}}|\} = \min\{s, |R_{\vec{\epsilon}}^{\mathfrak{B}}|\} \quad \text{for all } \vec{\epsilon} \in \{0,1\}^k.$$

The vectors $\vec{a} \in A^n$ and $\vec{b} \in B^n$ are *similar*, if for every atomic formula ψ, $\mathfrak{A} \vDash \varphi[\vec{a}]$ iff $\mathfrak{B} \vDash [\vec{b}]$. Observe that if $\mathfrak{A}$ and $\mathfrak{B}$ are s-close, $\vec{a}$ and $\vec{b}$ are similar and $n < s$, then for any $a' \in A$ one can find a $b' \in B$ such that $\langle \vec{a}, a' \rangle$ and $\langle \vec{b}, b' \rangle$ are similar. The *weight* of a formula $\varphi \in F(\tau)$ is the number of its free variables plus the number of quantifiers in it; this is clearly smaller than the length. The following lemma can be proved by induction on the complexity of φ.

Lemma. Suppose $\mathfrak{A}$ and $\mathfrak{B}$ are s-close, the weight of $\varphi(x_1, \ldots, x_n)$ is less than s, and $\langle a_1, \ldots, a_n \rangle \in A^n$ and $\langle b_1, \ldots, b_n \rangle \in B^n$ are similar. Then $\mathfrak{A} \vDash \varphi[\vec{a}]$ iff $\mathfrak{B} \vDash \varphi[\vec{b}]$.

Indeed, the lemma clearly holds for atomic formulas, and the only nontrivial step is to check it for $\varphi(\vec{x}) \equiv \exists y \psi(\vec{x}, y)$. This, however, follows from the remark above on extension of similar sequences.

For any τ-type structure $\mathfrak{A}$ there is a $\mathfrak{B}$ which is s-close to $\mathfrak{A}$ and has at most $s \cdot 2^k$ elements: for each $\vec{e}$ pick elements from $R_{\vec{e}}^{\mathfrak{A}}$ and stop when either there are no more elements, or s elements have been picked. According to the Lemma, if φ has length $< s$, then $\mathfrak{A} \vDash \varphi$ iff $\mathfrak{B} \vDash \varphi$, as was claimed in the problem.

> **Remark.** The followed method is a special case of the Ehrenfeucht-Fraïssé game, see Problem 520.

363. From the solution of Problem 362 it follows that if the structures $\mathfrak{A}$ and $\mathfrak{B}$ are s-close for every $s \in \omega$, then formulas true in $\mathfrak{A}$ are the same as formulas true in $\mathfrak{B}$. For two relation symbols P and R this means that if in the two structures the four–four disjoint sets defined by $P(x) \wedge R(x)$, $\neg P(x) \wedge R(x)$, $P(x) \wedge \neg R(x)$, $\neg P(x) \wedge \neg R(x)$ have pairwise the same finite number of elements, or both are infinite, then $\mathfrak{A}$ and $\mathfrak{B}$ satisfy the same set of formulas.

By assumption Γ has a unique model of size ω_1. In this case exactly one of the four sets must be infinite, as otherwise choosing one infinite set to have cardinality ω_1, and the other to be countable would give two non-isomorphic models of size ω_1. Thus three of the above sets have a fixed finite number of elements, and the fourth one is infinite. But then a countable model of Γ must have the same structure, thus it is unique.

The reverse implication is not true, see also Problem 359. There is a unique countable model when all four sets are infinite, but there are 15 non-isomorphic models of size ω_1 (at least one partition must be of size ω_1).

364. Solution 1. Each ordering is a binary relation and the number of binary relations on κ is $2^{(\kappa^2)} = 2^\kappa$. Therefore there cannot exist more than 2^κ linear orderings on κ. (See also Problem 354).

Take κ as a well-ordered set of ordinals, and replace each $\alpha < \kappa$ by a copy of either ω or ω^* (the reverse order of ω. From this ordered set one can recover the places where ω and ω^* was used. Put two elements into a group if one is an immediate successor of the other one. Elements in a group form a subset isomorphic to either ω or ω^* – in which case it was the replaced sequence –, or $\omega^* + \omega$, in which case a replacement by ω^* was followed by another one by ω (but we don't know where it was split). Consequently no two of these orderings are isomorphic, giving 2^κ many different orderings.

Solution 2. For a subset $X \subseteq \kappa$ consider the linear ordering that results from replacing every $\alpha \in X$ with a copy of $\omega^* + \omega$ (the order type of the integers). See also Problem 368.

365. We show first, by a back and forth argument, that any two countable dense orderings without endpoints are isomorphic. Let $(P, <)$ and $(Q, <)$ be two such orderings and fix enumerations $P = \{p_n : n \in \omega\}$ and $Q = \{q_n : n \in \omega\}$. We construct, by induction on n, an increasing sequence f_n of finite order-preserving functions such that $(\star)$

$$p_n \in \text{dom}(f_n) \text{ and } q_n \in \text{ran}(f_n). \qquad (\star)$$

Set $f_0 = \emptyset$ and suppose f_k has already been defined for each $k < n$.

(Forth) If $p_n \in \text{dom}(f_{n-1})$ then we let $h = f_{n-1}$, otherwise using that Q is dense and that $|f_{n-1}|$ is finite, one can find an element $q \in Q$ such that the finite suborderings $(\text{dom}(f_{n-1}) \cup \{p_n\}, <)$ and $(\text{ran}(f_{n-1}) \cup \{q\}, <)$ are isomorphic. Then, set $h = f_{n-1} \cup \{\langle p_n, q \rangle\}$.

(Back) If $q_n \in \text{ran}(h)$ then put $f_n = h$, else using denseness of P there exists an element $p \in P$ such that $(\text{dom}(h) \cup \{p\}, <)$ and $(\text{ran}(h) \cup \{q_n\}, <)$ are isomorphic. Let $f_n = h \cup \{\langle p, q_n \rangle\}$.

Obviously $(\star)$ remains true for f_n. Letting $f = \bigcup_{n \in \omega} f_n$ we get an isomorphism f between P and Q.

It can be proved similarly that any countable dense ordering is isomorphic to one the following:

$$\{0\}, \quad \mathbb{Q} \cap (0,1), \quad \mathbb{Q} \cap [0,1), \quad \mathbb{Q} \cap (0,1], \quad \mathbb{Q} \cap [0,1].$$

Remark. The proof technique used here is called *back and forth method*. Similar arguments will be used many times later on in Chapter 8, see also Definition 8.14.

366. It is not hard to see that the following coloring works:

$$c(x) = \begin{cases} p & \text{if } x = \frac{r}{p^n}, \text{ where } p \text{ is a prime and } p, r \text{ are coprime,} \\ 0 & \text{otherwise.} \end{cases}$$

As for the isomorphism, suppose the colorings $c, d : \mathbb{Q} \to \omega$ are such that between any two rationals all colors occur. Enumerate $\mathbb{Q}$ as $\{q_i : i \in \omega\}$. Similarly to Problem 365 we build an isomorphism $f = \bigcup_{n \in \omega} f_n$ between $(\mathbb{Q}, <, c)$ and $(\mathbb{Q}, <, d)$ such that the increasing sequence f_n of finite order-preserving functions satisfies $(\star)$ below for each n.

$$q_n \in \text{dom}(f_n) \cap \text{ran}(f_n), \quad c(x) = d(f_n(x)) \text{ for all } x \in \text{dom}(f_n). \qquad (\star)$$

Let $f_0 = \emptyset$ and suppose f_k has already been defined for each $k < n$.

(Forth) If $q_n \in \text{dom}(f_{n-1})$ then we let $h = f_{n-1}$, otherwise using that $\mathbb{Q}$ is dense, that $|f_{n-1}|$ is finite and that between any two rationals all colors occur, one can find an element $q \in \mathbb{Q}$ such that the finite suborderings $(\text{dom}(f_{n-1}) \cup \{q_n\}, <)$ and $(\text{ran}(f_{n-1}) \cup \{q\}, <)$ are isomorphic and $d(q) = c(q_n)$. Then, set $h = f_{n-1} \cup \{\langle q_n, q \rangle\}$.

(Back) If $q_n \in \text{ran}(h)$ then put $f_n = h$, else using denseness of $\mathbb{Q}$ and that between any two rationals all colors occur, there exists an element $p \in \mathbb{Q}$ such that $(\text{dom}(h) \cup \{p\}, <)$ and $(\text{ran}(h) \cup \{q_n\}, <)$ are isomorphic and $c(p) = d(q_n)$. Let $f_n = h \cup \{\langle p, q_n \rangle\}$.

Obviously $(\star)$ remains true for f_n.

367. Let $(P, <)$ be the countable ordered set. Enumerate P as $\{p_n : n \in \omega\}$. The required order-preserving embedding can be created using the *forth* step of Solution 365.

368. Similarly to Solution 364 one can construct continuum many different linear orders on ω by replacing every $i \in \omega$ either by ω or by ω^*.

The problem is that these orderings are not discrete: some elements might not have immediate predecessors. To solve this problem, replace each point with a copy of $\omega^* + \omega$. After the replacement the ordering will be discrete. As an isomorphism must map an $\omega^* + \omega$ group to another one, it also gives an isomorphism between the underlying orderings, thus not two of them are isomorphic.

369. Consider the orderings $\kappa \times \mathbb{Q}$ and $\kappa^* \times \mathbb{Q}$ (each point in the first ordering is replaced by a copy of the second one). Both are dense without endpoints, and they are not isomorphic: the first contains an increasing sequence of type κ, while every increasing sequence in the second one is countable.

In general, for a subset $A \subseteq \kappa$ let $D(A)$ be the ordering $\sum_{\alpha<\kappa} D(A, \alpha)$, where

$$D(A, \alpha) = \begin{cases} (\omega_1 + \omega_1^*) \times \mathbb{Q} & \text{if } \alpha \notin A \\ (\omega_1 + 1 + \omega_1^*) \times \mathbb{Q} & \text{otherwise.} \end{cases}$$

Each $D(A)$ is a dense ordering without endpoints of cardinality κ. We claim that for different subsets A and B the orderings $D(A)$ and $D(B)$ are not isomorphic.

Define $a \sim b$ if the interval $[a, b]$ contains countably many elements. In $D(A)/\sim$ there are two types of equivalence classes: those of size ω_1 corresponding to elements of κ ($\omega_1^* \times \mathbb{Q}$ from α and $\omega_1 \times \mathbb{Q}$ from $\alpha + 1$ are in the same class), and those of size ω corresponding to $1 \times \mathbb{Q}$ when $\alpha \in A$. The first type of classes are well ordered in type κ, thus are isomorphic to κ, and from the second type classes one can recover the subset A. Consequently $D(A)$ and $D(B)$ cannot be isomorphic.

370. The order $\omega_1 \times \mathbb{Q} + \{a\} + \mathbb{Q}$ works. The element $a \in A$ is distinguished by the property that for every $b < a$ the interval (b, a) has cardinality ω_1 (no other element has this property), thus a cannot be moved by an order-preserving permutation.

371. Endpoints have no weight as they do not belong to any open interval. If a is an internal point, then there is an open interval with $a \in (x, y)$, and a can be separated from any other point by an open interval. (Choose the other point as an endpoint of such an interval.) Thus the weight of a is at least 1 and as each non-empty set of cardinalities must have a least element, the weight must exist.

372. The intersection of finitely many open intervals containing a is an open interval containing a.

373. As every open interval of the reals contains infinitely many points, thus no point has weight one. Thus the weight of each point is infinite, and since $\mathbb{Q}$ is dense in $\mathbb{R}$, the weight of each point of $\mathbb{R}$ is ω.

374. Solution 1. Take all functions $f:\omega_1 \to \mathbb{Z}$ and order them lexicographically: $f < g$ if $f(\alpha) < g(\alpha)$ and $f(\beta) = g(\beta)$ for all $\beta < \alpha$. In this ordering the intersection of countably many open intervals contains an open interval, thus no weight is ω.

Solution 2. Any σ-saturated, dense ordering (e.g. a non-trivial ultrapower of the rationals) is suitable by Problems 641 and 681.

375. All points of the rationals have countable weight, hence the ordering $\kappa \times \mathbb{Q}$ is as required.

376. The top element of the linear ordering $\omega + \omega^*$ bounds all increasing sequences, still the first ω block does not have a least upper bound.

377. Consider the set of countable subsets of an uncountable set, ordered by inclusion. As the union of countably many countable sets is itself countable, it follows that all countable increasing chains have an upper bound. Still, there is no maximal countable subset (therefore the conclusion of Zorn's lemma 1.3 does not hold).

378. Using Zorn's lemma it does follow. Let $A \subseteq P$ be totally ordered, and let S be the set of increasing sequences with elements from A ordered by $s < s'$ if s' is an extension of s. If $C \subseteq S$ is a chain, then $\cup C$ is a sequence extending all elements of C, thus the assumption of Zorn's lemma holds, consequently there is a maximal sequence $s \in S$. Now for every $a \in A$ there is an element s_α such that $s_\alpha < a$, as otherwise extending s by this a the sequence $s' = s {}^\frown a \in S$, $s < s'$, contradicting the maximality of s.

By assumption the maximal s has an upper bound in P, every element of A is bounded by some element of s, thus A is bounded as well. If Zorn's lemma – or equivalently, the axiom of choice – fails, then there might exist an infinite, unbounded set of reals, such that each of its well-ordered subset is finite, thus bounded.

379. That finite partial orderings can be extended to total orderings can be done by a fairly easy inductive argument, so we concentrate on the infinite case. Let $(P, \leq)$ be a partial order and consider

$$Q = \{(X, \leq_X) : X \subset P, \text{ and } \leq_X \text{ is a linear order extending } \leq\}$$

For $(X, \leq_X)$, $(Y, \leq_Y)$ in Q define $(X, \leq_X) < (Y, \leq_Y)$ if $X \subset Y$ and $\leq_Y \restriction (X \times X) = \leq_X$. Then $(Q, <)$ is a partial ordering that satisfies the conditions of Zorn's lemma, thus it has a maximal element. That maximal element must be a total order $(P, \leq_P)$ extending $(P, \leq)$.

380. Every propositional tautology evaluates to $\top$ for every value of the propositional variables. As the evaluation is fixed, the subformulas which replace the propositional variables evaluate to the same value at every place they are inserted.

381. Using $(u)_i < u$, namely that the elements of a sequence are smaller than the (code of the) sequence, all properties can be defined by a lengthy, but otherwise trivial course-of-values recursion. As an example, the recursive definition of the relation $R \subset \omega$ in (f) which consists of (the codes of) those triplets $\langle \alpha(\varphi), \alpha(x), \alpha(t) \rangle$ where the substitution $\varphi[x/t]$ is admissible, can go as follows. The triplet $\langle u, x, t \rangle$ is in R iff

- u is a formula code, x is a variable code, t is a term code;
- if u is an atomic formula, then yes;
- if u is a negation of u' (then $u' < u$), the result is yes if $\langle u', x, t \rangle \in R$, otherwise it is no;
- if u is an "or" of u' and u'', then both $\langle u', x, t \rangle \in R$ and $\langle u'', x, t \rangle \in R$ are required;
- if u is the code of the formula $\exists y \psi$, then
  - if $\alpha(y) = x$ then yes;
  - if $\alpha(y)$ does not occur in t, then yes;
  - otherwise no.

We used that the relation "x occurs t" for codes of variables and terms is recursive.

382. The problem is that there is no control over the function which assigns the arity n_i to the i-th function symbol. Checking whether $\alpha(f_i(x_0))$ is a code of a valid term one can recover the set $\{i \in \omega : n_i = 1\}$. If this set is not recursive, the code of (valid) τ-type terms cannot be recursive either.

383. By primitive recursion. Let $f(0) = \alpha(\pi_0)$, and

$$f(n+1) = \alpha(\pi_n + 1) = \langle 4, \alpha(+), \alpha(\pi_n), \alpha(1) \rangle$$
$$= \langle 4, \alpha(+), f(n), \alpha(1) \rangle.$$

where $\alpha(\pi_0)$, $\alpha(+)$, and $\alpha(1)$ are fixed natural numbers.

384. By a brute force method. φ is a tautology iff there is a propositional formula ψ such that φ is the result of replacing propositional variables by certain subformulas of φ. Using a reasonable coding of propositional formulas the code of ψ will be smaller than the code of φ (when the propositional variables have small codes). Suppose we have a sequence s of subformulas of φ defining the first-order formulas the propositional variables are to be replaced with. The function which, given the code of a propositional formula and the sequence s returns the code of the replaced first-order formula, is clearly recursive. The possible values of the replacement sequence s can be bounded by a recursive function of the code of φ, see Problem 142.

Thus deciding whether φ is a tautology can be done by using bounded quantifiers: Check every propositional formula with code $< \alpha(\varphi)$ whether it is a tautology or not (this is recursive by Problem 304); if yes, then for each sequence s (bounded by a recursive function) check if using this replacement for the propositional tautology gives φ or not.

385. Similarly to "justified computations" (Definition 4.15) let the (code of the) sequence $u = \langle d_0, \ldots, d_{n-1} \rangle$ be a justified evaluation, if each d_i is a triplet of a formula, evaluation, and true/false expressing that the formula under the evaluation is true or not. As the structure is finite, the condition that such a triplet is always justified by earlier triplets in the sequence is a recursive relation. $\mathfrak{A} \vDash \varphi$ if for all evaluations e of variables in φ (recoverable by a recursive function) the shortest justified evaluation, ending with a triplet with $\alpha(\varphi)$ and $\alpha(e)$ as the first two elements, has $\alpha(\top)$ as the last element. This is a recursive relation as such a justified evaluation always exists.

12.7   FUNDAMENTAL THEOREMS

386. Every propositional tautology has a derivation using MP and instances of Ax_1–Ax_8. Replacing propositional variables with first-order formulas the sequence becomes a valid first-order derivation.

387. By induction on the length of the derivation, as both MP and G preserves this property.

388. The ($\Leftarrow$) part is straightforward. In reference to Solution 291 it is enough to prove $\psi \to \varphi \vdash \psi \to (\forall x \varphi)$ for closed ψ. But this is immediate applying first G and then Ax_9.

389. ($\Rightarrow$) Apply G repeatedly until all necessary universal quantifiers appear in the front.
($\Leftarrow$) Apply Ax_{11} repeatedly for the obviously admissible substitution $\varphi[x_i / x_i]$ to get rid of the universal quantifiers at the beginning of the formula.

390. Use the tautology $(\neg A \to B) \to (\neg B \to A)$ where $A \equiv \varphi[x/t]$ and $B \equiv \exists \neg \varphi$ and Ax_{10} for $\neg \varphi$ to get

$$\vdash (\neg \exists \neg \varphi) \to \varphi[x/t].$$

391. (a) follows from the maximality of Σ.
   (b) We cannot have both φ and $\neg \varphi$ in Σ, otherwise (by propositional logic) $\bot$ would be derivable from Σ (even if φ is not closed). If φ is not in Σ, then by maximality $\Sigma, \varphi \vdash \bot$. Using that φ is closed, the deduction theorem 7.1 gives $\Sigma \vdash \varphi \to \bot$, from where $\Sigma \vdash \neg \varphi$ by propositional logic, and then $\neg \varphi \in \Sigma$ by maximality.
   (c) can be proved similarly.

392. Let Γ be the union, and suppose by contradiction that $\Gamma \vdash \bot$. Take a derivation $\varphi_1, \ldots, \varphi_n$ of $\bot$ from Γ. By definition of a union there is a smallest $\alpha < \kappa$ such that all $\varphi_i \in \Gamma_\alpha$. But then the sequence $\varphi_1, \ldots, \varphi_n$ is a derivation of $\bot$ from Γ_α too, contradicting the consistency of Γ_α.

393. Enumerate the formulas of $F(\tau)$ as $\{\varphi_\alpha : \alpha < \kappa\}$. Define by induction an increasing chain of s-consistent theories Γ_α. Let Γ_0 be the given theory. Suppose Γ_β has already been defined for all $\beta < \alpha$. If α is a limit, let

$\Gamma_\alpha = \bigcup\{\Gamma_\beta : \beta < \alpha\}$. By Problem 392, Γ_α is s-consistent. If α is successor, say $\alpha = \beta + 1$, then let $\Gamma_\alpha = \Gamma_\beta \cup \{\varphi_\beta\}$ if it is s-consistent, and $\Gamma_\alpha = \Gamma_\beta$ otherwise. By induction, Γ_α is s-consistent.

Put $\Gamma = \bigcup\{\Gamma_\alpha : \alpha < \kappa\}$. It is s-consistent and maximal: if $\varphi_\beta \notin \Gamma$, then $\Gamma_\beta \cup \{\varphi_\beta\}$ is not s-consistent, so neither is $\Gamma \cup \{\varphi_\beta\}$.

394. (a) Consider the derivation:

1. $y = y$ (axiom Ex_1)
2. $(y = y) \to \exists x(x = x)$ (axiom Ax_{10})
3. $\exists x(x = x)$ (MP)

Therefore the formula $\exists x(x = x)$ is derivable from Γ. As Γ is Henkin, there must be a constant $c \in \tau$ with $\Gamma \vdash c = c$.

(b) Let t be the term $f(\bar{c})$ and $\varphi(x)$ be the atomic formula $x = t$. It suffices to show that $\exists x(t = x)$ can be derived from Γ which is shown by

1. $x = x$ (axiom Ex_1)
2. $\forall x(x = x)$ (G)
3. $\forall x(x = x) \to (x = x)[x/t]$ (Ax_{11})
4. $\varphi[x/t]$ (same as $t = t$)
5. $\varphi[x/t] \to \exists x\varphi$ (Ax_{10})
6. $\exists x(t = x)$ (MP: 5,4)

395. If $c = d$ for distinct constants is in Γ, then it cannot be satisfied in the model defined on the set of constant symbols.

396. (a) We need to check that A is not empty, and that A is closed under $f^{\mathfrak{B}}$ for each function symbol $f \in \tau$. As $\mathfrak{B} \vDash \exists x(x = x)$, there must be at least one constant symbol in τ. For showing that A is closed for $f^{\mathfrak{B}}$ take elements $c_1, \ldots, c_n \in A$. We have $\mathfrak{B} \vDash \exists y\, f(c_1, \ldots, c_n) = y$, thus $\mathfrak{B} \vDash f(c_1, \ldots, c_n) = c$ for some constant symbol c, and then $f^{\mathfrak{B}}(c_1, \ldots, c_n) \in A$, as required.

(b) Assume ψ is closed and proceed by induction on the complexity of ψ. For atomic and quantifier-free formulas the statement holds as $\mathfrak{A}$ is a substructure. The induction step is trivial for $\vee$ and $\neg$. For closed formulas of the form $\exists x\varphi(x)$ we have, by assumption, $\mathfrak{B} \vDash \exists x\varphi(x)$ iff there is a $c \in A$ such that $\mathfrak{B} \vDash \varphi[x/c]$. By the inductive hypothesis this happens iff $\mathfrak{A} \vDash \varphi[x/c]$, thus iff $\mathfrak{A} \vDash \exists x\varphi(x)$.

397. By Problem 388 $\Gamma \vdash \varphi$ and $\Gamma \vdash \bar{\varphi}$ are equivalent, thus φ can be assumed to be closed. Work by induction on the complexity of φ. Equality-free closed atomic formulas are of the form $r(c_1, \ldots, c_n)$, and the claim follows from the definition of $r^{\mathfrak{A}}$.

Using maximality of Γ for $\varphi \vee \psi \in \Gamma$ we have either $\varphi \in \Gamma$ or $\psi \in \Gamma$ (by Problem 391). Then, by induction, we get either $\mathfrak{A} \vDash \varphi$ or $\mathfrak{A} \vDash \psi$, therefore $\mathfrak{A} \vDash \varphi \vee \psi$.

$\mathfrak{A} \vDash \neg\varphi$ iff $\mathfrak{A} \nvDash \varphi$, which, by the inductive hypothesis, implies $\varphi \notin \Gamma$. By maximality of Γ we have then $\neg\varphi \in \Gamma$.

Finally, assume $\exists x\varphi \in \Gamma$ and let c be the Henkin witness of $\exists x\varphi$. Then $\mathfrak{A} \vDash \varphi[c]$. For the other direction, if $\mathfrak{A} \vDash \exists x\varphi$, then $\mathfrak{A} \vDash \varphi[x/c]$ for some constant symbol $c \in \tau$ by the Substitution lemma 6.7. As $\varphi[x/c]$ is closed, the induction hypothesis gives $\Gamma \vdash \varphi[x/c]$, from where Ax_{10} gives $\Gamma \vdash \exists x\varphi$ as required.

398. We show that $\sim$ is reflexive, i.e. $\Sigma \vdash c = c$, others are similar. Here is the derivation:

1. $x = x$ (axiom Ex_1)
2. $\forall x(x = x)$ (generalization)
3. $\forall x(x = x) \rightarrow (c = c)$ (instance of Ax_{11})
4. $c = c$ (MP: 3,2)

399. (a) $F(\tau) \subseteq F(\tau')$, thus every τ-type axiom (and inference rule) is also a τ'-type axiom (inference rule). If $\Sigma \vdash \bot$ in type τ, then the same formula sequence is a correct τ'-type derivation, thus $\Sigma \vdash \bot$ in type τ'.

(b) Suppose $D' = \langle \varphi'_1, \ldots, \varphi'_n \rangle$ is a τ'-type derivation of $\bot$ from $\Sigma \subseteq F(\tau)$. We need to show that then there is a τ-type derivation of $\bot$ as well. Each φ'_i in D' will be replaced by some τ-type formula φ_i such that the sequence $\varphi_1, \ldots, \varphi_n$ is a valid τ-type derivation of $\bot$.

There are only constant symbols in $\tau' \smallsetminus \tau$, and only finitely many among these symbols occur in the derivation D'. For each of them pick a variable symbol not occurring in D', and replace that constant symbol by the corresponding variable symbol. The following facts are clear. (i) Formulas in $F(\tau)$ (in particular, elements of Σ) do not change. (ii) After replacement each formula will be an element of $F(\tau)$; finally (iii) applications of the derivation rules MP and G remain valid. To finish the proof we only need to check that (iv) the image of a τ'-type axiom is a τ-type axiom. This is clear for tautologies and for equality axioms (as the latter ones do not contain constant symbols), and the axioms Ax_{12} which will be introduced later. The schemes Ax_9, Ax_{10} and Ax_{11} were carefully tailored to satisfy this requirement.

For a more general statement see Problem 407.

400. Let $\varphi_1, \ldots, \varphi_n$ be the derivation of $\varphi[x/c]$ from Σ. Choose a variable symbol z which does not occur in the derivation, and replace every occurrence of c by z. Elements of Σ remain the same, axioms remain axioms (see Problem 399), and applications of MP and G remain correct. The sequence ends with the formula $\varphi[x/z]$, thus $\Sigma \vdash \varphi[x/z]$. In the formula $\varphi[x/z]$ replacing z by x is an admissible substitution, thus $\forall z\varphi[x/z] \rightarrow \varphi$ is an instance of Ax_{11}. Using G, this axiom and MP we have $\Sigma \vdash \varphi$, from where another application of G gives the result.

401. Define $c_0 \sim c_1$ iff $\Sigma \vdash c_0 = c_1$. By Problem 398 this is an equivalence relation, let $\tilde{c}$ be the equivalence class of c. The ground set is $\mathfrak{A}$ is the collection of equivalence classes – this is not empty by Problem 394(a). The interpretation of c is $c^{\mathfrak{A}} = \tilde{c}$; for the other symbols

$$f^{\mathfrak{A}}(\tilde{c}_1, \ldots, \tilde{c}_n) = \tilde{c} \quad \text{iff} \quad \Sigma \vdash f(c_1, \ldots, c_n) = c,$$

$$\langle \tilde{c}_1, \ldots, \tilde{c}_n \rangle \in r^{\mathfrak{A}} \quad \text{iff} \quad \Sigma \vdash r(c_1, \ldots, c_n).$$

The equality axioms Ex_2 and Ex_3 ensure that these definitions are sound, namely they do not depend on the representative elements of the equivalence classes. By Problem 394(b) $f^{\mathfrak{A}}$ is defined for all possible arguments.

Next we show that for closed formulas $\varphi \in \Sigma$ iff $\mathfrak{A} \vDash \varphi$. By Problem 389 a formula is in Σ if and only if its universal closure is in Σ, moreover a formula is true in $\mathfrak{A}$ just in case its universal closure is true, thus from this it follows that $\mathfrak{A} \vDash \Sigma$.

By definition, and by the equality axioms, the required equivalence holds for atomic formulas. For $\neg\varphi$ it follows from Problem 391(b), for $\varphi \vee \psi$ it follows from Problem 391(c). The remaining case is $\exists x \varphi(x)$ where this formula is closed. Since Σ is Henkin, $\Sigma \vdash \exists x \varphi$ implies $\Sigma \vdash \varphi[x/c]$ for some constant symbol $c \in \tau$. Then $\mathfrak{A} \vDash \varphi[x/c]$ by induction, therefore $\mathfrak{A} \vDash \exists x \varphi$.

Conversely, if $\mathfrak{A} \vDash \exists x \varphi$, then for some evaluation e and constant symbol $c \in \tau$, $\mathfrak{A} \vDash \varphi[e(x/c^{\mathfrak{A}})]$. Using the Substitution Lemma 6.7 for the term c, we get $\mathfrak{A} \vDash (\varphi[x/c])[e]$, and since $\varphi[x/c]$ has no free variables, this means $\mathfrak{A} \vDash \varphi[x/c]$, and then $\Sigma \vdash \varphi[x/c]$ by induction. From here Ax_{10} gives $\Sigma \vdash \exists x \varphi(x)$ as required.

402. Let $\kappa = |F(\tau)|$, C be a set of κ many new constant symbols, and let $\tau' = \tau \cup C$ adding these symbols. Clearly, $|F(\tau')| = \kappa$, enumerate it as $\{\varphi_\alpha : \alpha < \kappa\}$.

By Problem 399(b) Σ is τ'-consistent. From this point on work in type τ'. Problem 393 explained how to extend Σ to a maximal s-consistent theory. We follow the same method with some extra work which guarantees that at the end we get a Henkin theory as well. So let $\Sigma_0 = \Sigma$, and define the increasing τ'-consistent theories Σ_α as follows. For limit α take the union. For $\alpha = \beta + 1$ distinguish three cases.

Case 1. If $\Sigma_\beta \cup \{\varphi_\beta\}$ is not τ'-consistent, then let $\Sigma_\alpha = \Sigma_\beta$.

Case 2. If $\Sigma_\beta \cup \{\varphi_\beta\}$ is τ'-consistent, but either φ_β is not closed, or it is not of the form $\exists x \psi_\beta(x)$, then $\Sigma_\alpha = \Sigma_\beta \cup \{\varphi_\beta\}$.

Case 3. $\Sigma_\beta \cup \{\varphi_\beta\}$ is τ'-consistent, and φ_β is closed and is of the form $\exists x \psi_\beta(x)$. Take a constant symbol $c_\beta \in C$ which does not occur in $\Sigma_\beta \cup \{\varphi_\beta\}$. As this set contains less than κ many new constant symbols, such a c_β exists. In this case let $\Sigma_\alpha = \Sigma_\beta \cup \{\psi_\beta[x/c_\beta]\}$.

We claim that Σ_α is τ'-consistent. It is clear in Cases 1 and 2. In Case 3 suppose $\Sigma_\alpha \vdash \bot$. As $\psi_\beta[x/c_\beta]$ is closed, the Syntactic deduction lemma 7.1 and propositional logic gives $\Sigma_\beta \vdash \neg\psi_\beta[x/c_\beta]$. The constant symbol c_β does not occur neither in Σ_β nor in ψ_β, thus we have $\Sigma_\beta \vdash \neg\psi_\beta$ by Problem 400. Now G and Ax_{12} gives $\Sigma_\beta \vdash \neg\exists x \psi_\beta(x)$, contradicting that $\Sigma_\beta \cup \{\varphi_\beta\}$ is s-consistent.

The union $\Sigma' = \bigcup\{\Sigma_\alpha\}$ is maximal syntactically consistent and Henkin.

403. By Problem 402 the type can be extended by new constant symbols such that $\Sigma \subseteq \Sigma' \subset F(\tau')$ is a maximal s-consistent Henkin theory. Problem 401 says that Σ' has a model. The τ-type reduct of this model (drop symbols which do not occur in τ) is a model of Σ.

404. The derivation is sound because all the axioms Ax_1–Ax_{12} and Ex_1–Ex_3 are true in every structure.

As for the converse, we may assume that φ is closed by Problem 389. Now if $\Sigma \nvdash \varphi$ then, by the Deduction Lemma 7.1, $\Sigma \cup \{\neg\varphi\}$ is syntactically consistent. Therefore, by the first completeness theorem 10.6 it has a model $\mathfrak{A}$. But then $\mathfrak{A} \vDash \Sigma$ and $\mathfrak{A} \nvDash \varphi$, showing $\Sigma \nvDash \varphi$.

405. In Problems 399 and 400. When limiting the available variable symbols the proof system becomes sensitive to what other (constant, function, relation) symbols are available for formulas in the derivation $\Sigma \vdash \varphi$, but which appear neither in Σ, nor in φ. See also Problem 407.

406. Check that during the proof of the completeness theorem we used the equality axioms Ex_1–Ex_3 only when the equality symbol appeared in one of the formulas in Γ or in φ. The only exception is Problem 394(a) when the existence of a constant symbol was proved. Here one can use $\top$ instead of $x = x$.

407. (a) Not necessarily. Let c, d, and e be constant symbols in τ' and suppose τ contains c and d only. Let φ be the formula $c = d$, and $\Gamma = \{c = e, e = d\}$. Then $\Gamma \vdash \varphi$, but this derivation needs to refer to e.

(b) If $\Gamma \subset F(\tau)$, then there is a derivation of φ from Γ that uses τ-formulas and axioms only. For we have $\Sigma \vDash \varphi$ for τ-structures, thus by Gödel's second completeness theorem we have $\Sigma \vdash \varphi$ in type τ.

> **Remark.** This general property of Hilbert-type derivation follows easily from the completeness theorem. The special case discussed in Problem 399 was needed in the proof of the theorem, thus had to be established before. The approach used there (replacing formulas in a τ' derivation by formulas from $F(\tau)$) works in general, but the details are quite messy.

408. Apply the following rules until they are applicable, where Q is any of the quantifiers $\exists$ and $\forall$:

$$\left.\begin{aligned}(Qx\varphi) \wedge \psi &\Rightarrow Qx'(\varphi[x/x'] \wedge \psi) \\ (Qx\varphi) \vee \psi &\Rightarrow Qx'(\varphi[x/x'] \vee \psi)\end{aligned}\right\} \text{ where } x' \text{ does not occur in } \varphi, \psi,$$

$$\begin{aligned}\neg(\exists x\varphi) &\Rightarrow \forall x\neg\varphi, \\ \neg(\forall x\varphi) &\Rightarrow \exists x\neg\varphi.\end{aligned}$$

In each step $\vDash \varphi \leftrightarrow \varphi^*$, thus all formulas are equivalent to the original one. If the procedure halts, the last formula is in prenex form. In each step the total depth of the quantifiers decreases, thus the procedure must halt after finitely many steps.

409. ($\Rightarrow$) Suppose $\mathfrak{A} \vDash \Sigma$. Take a well-ordering $<$ of the universe A, and define, for all $\vec{a} \in A^n$,

$$f_\psi^{\mathfrak{A}}(\vec{a}) = \begin{cases} \min_< \{b \in A : \mathfrak{A} \vDash \varphi[b, \vec{a}]\} & \text{if this set is non-empty,} \\ \min_< A & \text{otherwise.} \end{cases}$$

If $\mathfrak{A} \vDash \exists y \varphi(y, \vec{x})$ then $f_\psi^{\mathfrak{A}}$ always takes a value which satisfies φ.

($\Leftarrow$) The τ-type reduct of any model of Σ^* is a model of Σ as the function $f_\psi^{\mathfrak{A}}$ provides the element required by the existential quantifier.

410. Given the formula set Γ, make the following transformations which do not change whether the set has a model or not.

1. Convert every element of Γ into prenex normal form (Problem 408).
2. From the front of each formula delete all universal quantifiers (see Problem 311).
3. If some of the formulas starts with an existential quantifier as $\psi \equiv \exists y \varphi(y, \vec{x})$, then expand the type with the Skolem function f_ψ and replace ψ by $\varphi[y/f_\psi(\vec{x})]$ (Problem 409). Continue at Step 2.

When Step 3 is not more applicable, the new formula set Γ^* contains no quantifies, and has a model if and only if the original set does. Actually, if $\mathfrak{A} \vDash \Gamma$, then the ground set of a model of Γ^* can be chosen to be $\mathfrak{A}$, and the interpretation of the Skolem functions as in Problem 409.

411. (a) If there are no constant symbols then the free term-structure is empty.

(b) If there are no functions symbols in τ, then $|\mathfrak{M}|$ equals the cardinality of constant symbols. If there is at least one function symbol, then $|\mathfrak{M}|$ is $\omega \cdot |C_\tau \cup F_\tau|$.

(c) By induction on the complexity of the term t.

(d) There are two constant symbols c and d in τ, and Γ contains the formula $c = d$.

412. Let $\mathfrak{A} \vDash \Gamma$ be a model, $\mathfrak{M}$ be the Herbrand structure of signature τ, and for an n-place relation symbol $r \in \tau$ and variable-free terms $t_1, \ldots, t_n \in K(\tau) = M$ define

$$\langle t_1, \ldots, t_n \rangle \in r^{\mathfrak{M}} \quad \text{iff} \quad \mathfrak{A} \vDash r(t_1, \ldots, t_n).$$

We claim first that for each $\varphi \in F(\tau)$ which contains no variables, quantifiers or equality symbols, $\mathfrak{A} \vDash \varphi$ if and only if $\mathfrak{M} \vDash \varphi$. This holds for atomic formulas by definition, and easily follows for $\neg\varphi$ and $\varphi_1 \vee \varphi_2$.

Let $\varphi \in \Gamma$ and $e^{\mathfrak{M}}$ be an evaluation over $\mathfrak{M}$, that is, $e^{\mathfrak{M}}(x_i) = t_i$ is a variable-free term. $\mathfrak{M} \vDash \varphi[e^{\mathfrak{M}}]$ iff $\mathfrak{M} \vDash (\varphi[x_1/t_1, \ldots x_n/t_n])$ by the substitution lemma 6.7. Let $e^{\mathfrak{A}}$ be the evaluation over $\mathfrak{A}$ where $e(x_i) = t_i^{\mathfrak{A}}$. As $\mathfrak{A} \vDash \varphi[e^{\mathfrak{A}}]$ by assumption, $\mathfrak{A} \vDash (\varphi[x_1/t_1, \ldots, x_n/t_n])$ by the substitution lemma again. As $\varphi[x/1/t_1, \ldots, x_n/t_n]$ is closed, contains no quantifiers or equality symbols, this implies $\mathfrak{M} \vDash \varphi[x_1/t_1, \ldots, x_n/t_n]$, as required.

As for why both conditions are needed, observe first that in the presence of equality we might have $\mathfrak{A} \vDash c = d$ for distinct constant symbols c and d. However, in the Herbrand structure $c^{\mathfrak{M}}$ and $d^{\mathfrak{M}}$ are always different.

Secondly, let τ contain the unary function symbol f, constant symbol c, and binary relation R; and Γ be the theory which says that $R(x, y) \to \neg R(y, x)$, $R(c, f^k(c))$, for all $k \geq 1$, and $\exists x\, R(x, c)$. The Herbrand universe consists of the terms $c, f(c), f^2(c), \ldots$, and none of them can satisfy $R(x, c)$.

413. As suggested by the hint, replace the equality symbol by the binary relation symbol $\approx$ to get the formula set $\Gamma^\approx$, and put the equality axioms Ex_1, Ex_2, Ex_3 written for $\approx$ into Δ. By Problem 340, Γ has a model iff $\Gamma^\approx \cup \Delta$ has one; and the factor or a model of $\Gamma^\approx \cup \Delta$ is a model of Γ.

Use the procedure in Problem 410 of adding Skolem functions to the type to get the formula set Γ^* from $\Gamma^\approx \cup \Delta$ so that Γ^* has neither equality symbol, nor quantifiers and the $\tau \cup \{\approx\}$-type reduct of any model of Γ^* is a model of $\Gamma^\approx \cup \Delta$. Observe that the number of Skolem functions is at most $\max(\omega, |\tau|)$, which upper bounds the cardinality of the similarity type of Γ^*. Apply Problem 412 which claims that Γ^* has a Herbrand model. The cardinality of the Herbrand structure is at most $\max(\omega, |\tau|)$ by Problem 411(b), from where the claim follows.

To get the universe of a model of the original formula set Γ, the Herbrand structure should be factored by (the interpretation of) the relation $\approx$.

414. This follows from Problem 412.

415. Use the procedure outlined in Problem 413 to get rid of the equality symbol and quantifiers. Add a constant symbol to τ if it has none. Convert the remaining formulas to conjunctive normal form (Problem 276).

The only point which adds potentially infinitely many formulas is the one which adds the equality axioms for each function and relation symbol in τ. If Σ is finite, then it suffices to handle only those symbols which appear in Σ, thus keeping the formula set finite.

416. If $\mathcal{C}$ has a model, then by Problem 414 it has a Herbrand model $\mathfrak{M}$. Let the propositional variable $r(t_1, \ldots, t_n)$ be true if this formula holds in $\mathfrak{M}$. As $\mathfrak{M} \vDash \mathcal{C}$, each clause in $\mathcal{C}^o$ must contain a true literal.

To see the converse, if $\mathcal{C}^o$ is satisfiable, then define the interpretation of the relation $r \in \tau$ on the Herbrand structure $\mathfrak{M}$ as $\langle t_1, \ldots, t_n \rangle \in r^{\mathfrak{M}}$ iff the propositional variable $r(t_1, \ldots, t_n)$ is true. As this truth assignment satisfies every clause in $\mathcal{C}^o$, for clause $c \in \mathcal{C}$ and every evaluation e over $\mathfrak{M}$ some literal in $c[e]$ will be true in $\mathfrak{M}$. Thus $\mathfrak{M}$ is a model of $\mathcal{C}$ as was required.

417. By the completeness theorem 10.14 we have $\Sigma \vDash \varphi$ iff $\Sigma \vdash \varphi$. $\vdash$ has the compactness property: there is a finite $\Gamma \subset \Sigma$ with $\Gamma \vdash \varphi$, hence $\Gamma \vDash \varphi$ again by the completeness theorem.

418. Apply the compactness theorem with $\varphi \equiv \bot$.

419. Let φ_n be the formula which is true if there are at least n different elements:

$$\varphi_n \equiv \exists x_1 \ldots \exists x_n \bigwedge_{1 \leq i < j \leq n} x_i \neq x_j.$$

By assumption, every finite subset of $\Delta = \Sigma \cup \{\varphi_n : n \geq 2\}$ has a model, thus Δ has a model, which must be infinite.

420. To show that there is a model of Σ of size at least κ add κ many new constant symbols to τ, and consider the theory $\Delta = \Sigma \cup \{c_\alpha \neq c_\beta : \alpha < \beta < \kappa\}$. Any infinite model $\mathfrak{A}$ of Σ can be turned into a model of a finite subset of Δ: interpret the finitely many new constant symbols as different elements of A. By the compactness theorem Δ has a model, which must have cardinality at least κ.

421. Add ω_1 many constant symbols to the type τ, and take the formula set $\Delta = \Gamma \cup \{c_\beta < c_\alpha : \alpha < \beta < \omega_1\}$. As Γ has an infinite model, it can be made a model for every finite subset of Δ (as it requires the ordering for finitely many of the new constant symbols, the rest can be interpreted arbitrarily). A model of Δ works.

422. Suppose the claim to be false. Let c_i be new constant symbols for $i \in \omega$, and consider $\Delta = \Sigma \cup \{c_{i+1} \, E \, c_i : i \in \omega\}$. As each finite subset of Δ has a model, so has Δ, but in that model $\mathfrak{A}$ the subset $\{c_i^{\mathfrak{A}} : i \in \omega\}$ has no $E^{\mathfrak{A}}$-minimal element.

423. Suppose by contradiction that Σ has models of arbitrary large diameter. Add two constant symbols c_1, c_2 to τ, and let φ_n be the formula which says that there is no path of length $\leq n$ between c_1 and c_2. Every finite subset of $\Delta = \Sigma \cup \{C(c_1, c_2)\} \cup \{\varphi_n : n \in \omega\}$ has a model, thus Δ has a model, too. In that model $C(c_1, c_2)$ holds, but there is no path of any length between c_1 and c_2.

424. Add countably new constants c_i to the language, and add the formulas to Γ which say that $c_i \, R \, c_{i+1}$ and $\neg(c_i \, R \, c_j)$ for $i + 1 < j$. We need to check that every finite subset of the extended set has a model.

By assumption, there is a model of Γ, where $S_n = R^0 \cup R^1 \cup \cdots \cup R^n$ is not transitive. Treat R as the set of the edges of a directed graph: there is a directed edge from a to b if $(a, b) \in R$. Then $(a, b) \in S_n$ if there is a directed path of length at most n from a to b. If S_n is not transitive, then there must be two points, a and b, such that the shortest path from a to b has length n. If not, then any two points connected by a directed path are connected by such a path shorter than n, thus S_n would be transitive. Let $a = a_0, a_1, \ldots, a_n = b$ be the points on this path, then there is a directed edge from a_i to a_{i+1}, but no edge goes from a_i to any later a_j (as in this case there would be a shorter path connecting a and b). It means that $a_0, \ldots, a_n$ satisfy the requirements required from the constants $c_0, \ldots, c_n$, providing the required consistency.

425. (i)$\Rightarrow$(ii) Let $S = R^0 \cup \cdots \cup R^n$. Define by induction the formula $\varphi_k(x, y)$ as follows. $\varphi_0(x, y) \equiv x = y$, and $\varphi_{k+1}(x, y) \equiv \exists z_{k+1}(R(x, z_{k+1}) \wedge \varphi_k(z_{k+1}, y))$. Write

$$\varphi(x, y) \equiv \varphi_0(x, y) \vee \varphi_1(x, y) \vee \cdots \vee \varphi_n(x, y).$$

Then $\mathfrak{A} \vDash \varphi[a,b]$ iff $\langle a,b \rangle \in S^{\mathfrak{A}}$.

(ii)$\Rightarrow$(i) By way of contradiction suppose $\varphi(x,y)$ is the transitive closure of R in every model of Γ, but there is no n such that $R^0 \cup \cdots \cup R^n$ is always transitive. It means that for each $n \in \omega$ there is a model $\mathfrak{A}_n \vDash \Gamma$ and a pair $\langle a_n, b_n \rangle$ such that $\mathfrak{A} \vDash \varphi[a_n, b_n]$ but $\langle a_n, b_n \rangle \notin R^0 \cup \cdots \cup R^n$ (interpreted in $\mathfrak{A}_n$), that is, $\mathfrak{A}_n \nvDash \varphi_j[a_n, b_n]$ for all $j \leq n$. Consequently every finite subset of

$$\Gamma \cup \{\varphi(c_1, c_2), \neg \varphi_j(c_1, c_2) : j \in \omega\}$$

has a model, and then by compactness, this theory has a model $\mathfrak{B}$. As $\mathfrak{B} \vDash \neg \varphi_j(c_1, c_2)$ there is no directed $R^{\mathfrak{B}}$-path from $c_1^{\mathfrak{B}}$ to $c_2^{\mathfrak{B}}$, thus $c_2^{\mathfrak{B}}$ is not in the transitive closure of $c_1^{\mathfrak{B}}$. However, $\mathfrak{B} \vDash \varphi(c_1, c_2)$, a contradiction.

426. By way of contradiction assume that for all $n \in \omega$ there is a model $\mathfrak{A}_n$ and $a_n \in A_n$ such that $\mathfrak{A}_n \vDash \neg \varphi_n[a_n]$. But the assumption gives that $\mathfrak{A}_n \vDash \neg \varphi_j[a_n]$ for all $j < n$ as well. It means that every finite subset of

$$\Gamma \cup \{\neg \varphi_i(c) : i < \omega\}$$

has a model. By compactness, this set also has a model $\mathfrak{B}$, however in this model $c^{\mathfrak{B}}$ does not satisfy any of the formulas φ_j.

427. Using the hint, let $\Delta' = \{\delta \in \Delta : \Gamma \vDash \delta\}$. First, $\Gamma \vDash \Delta'$ is clear. To show that $\Delta' \vDash \Gamma$ pick any model $\mathfrak{B} \vDash \Delta'$ and define

$$\Delta'' = \{\neg \delta \in \Delta : \mathfrak{B} \vDash \neg \delta\}.$$

If we can find any $\mathfrak{A} \vDash \Gamma \cup \Delta''$ then we are done: condition $(\star)$ applied to $\mathfrak{A}$ and $\mathfrak{B}$ gives $\mathfrak{B} \vDash \Gamma$ (as every element of Δ true in $\mathfrak{A}$ is also true in $\mathfrak{B}$), consequently every model of Δ' also satisfies Γ, i.e., $\Delta' \vDash \Gamma$.

Thus it suffices to show that $\Gamma \cup \Delta''$ is consistent. Suppose not, then the compactness theorem and the deduction lemma says that there are finitely many $\neg \delta_i \in \Delta''$ such that

$$\Gamma \vDash \neg(\neg \delta_1 \wedge \cdots \wedge \neg \delta_n).$$

By assumption on Δ, $\delta \equiv \bigvee_i \delta_i \in \Delta$, and $\Gamma \vDash \delta$, thus $\delta \in \Delta'$, and then $\mathfrak{B} \vDash \delta$. But this contradicts the fact that $\mathfrak{B} \vDash \neg \delta_i$ for all i.

428. The similarity type τ will consist of a constant symbol c and unary relation symbols R_i for $i < \omega$. The formula set Γ says that no four different R_i is true for c:

$$\Gamma = \left\{ \neg \exists x \big(R_{i_1}(x) \wedge R_{i_2}(x) \wedge R_{i_3}(x) \wedge R_{i_4}(x) \big) : i_1 < i_2 < i_3 < i_4 < \omega \right\}$$

and for $i < \omega$ set $\Gamma_i = \Gamma \cup \{R_i(c)\}$.

429. (a) Assume Robinson's consistency theorem and suppose $\varphi \vDash \psi$. Let $\bar{\varphi}$ and $\bar{\psi}$ be the universal closure of φ and ψ, respectively. If the conditions in

Robinson's theorem does not hold for the theories $\Gamma_1 = \{\bar{\varphi}\}$ and $\Gamma_2 = \{\neg\bar{\psi}\}$, then there is a closed $\vartheta \in F(\tau_1 \cap \tau_2)$ such that $\Gamma_1 \vDash \vartheta$ and $\Gamma_2 \vDash \neg\vartheta$. But in this case we would have $\varphi \vDash \vartheta$ and $\vartheta \vDash \psi$ providing the required interpolant. If the condition in Robinson's theorem holds, then $\Gamma_1 \cup \Gamma_2 = \{\bar{\varphi}, \neg\bar{\psi}\}$ is consistent, contradicting the assumption that $\varphi \vDash \psi$.

(b) Assume Craig's interpolation theorem, and let $\Gamma_1 \subset F(\tau_1)$, $\Gamma_2 \subset F(\tau_2)$. If $\Gamma_1 \cup \Gamma_2$ has no model, then by the compactness theorem 7.10 there are finitely many formulas $\varphi_0, \ldots, \varphi_{n-1} \in \Gamma_1$ and $\psi_0, \ldots, \psi_{k-1} \in \Gamma_2$ such that $\{\varphi_i, \psi_j : i < n, j < k\}$ is inconsistent. Note that the inconsistency means $\bigwedge_{i<n} \bar{\varphi}_i \vDash \neg\bigwedge_{j<k} \bar{\psi}_j$ for the closure of the formulas. By Craig's theorem there is an interpolant formula $\vartheta \in F(\tau_1 \cap \tau_2)$ for them. But then ϑ is a formula such that $\Gamma_1 \vDash \vartheta$ and $\Gamma_2 \vDash \neg\vartheta$.

430. Any formula $\psi \in F(\tau_2)$ is of the form $\psi \equiv \psi'[\vec{x}/\vec{c}]$ where $\psi' \in F(\tau)$ and $\vec{c}$ are constant symbols from $\tau_2 \smallsetminus \tau$. ($\varphi$ may contain additional free variables.) If $\varphi \vDash \psi$, then $\varphi \vdash \psi$ by completeness theorem (the derivation is in type $\tau_1 \cup \tau_2$). As the constant symbols $\vec{c}$ do not occur in φ, Problem 400 gives $\varphi \vdash \forall \vec{x} \psi'(\vec{x})$. As this formula is in $F(\tau)$, and clearly implies $\psi'[\vec{x}/\vec{c}]$, this is the interpolant.

431. No, even $\Gamma_i \cup \Gamma_2 \cup \Gamma_3$ need not be consistent. Let τ contain three constant symbols, and Γ_i consist of two formulas: a formula which says that there are exactly two elements in the structure, and either $c_1 \neq c_2$, or $c_2 \neq c_3$, or $c_3 \neq c_1$, respectively. The union of any two theories is consistent, but $\Gamma_1 \cup \Gamma_2 \cup \Gamma_3$ is not.

432. Let $I = \{1\}$ and $J = \{2, \ldots, n\}$. If $\Gamma_J = \bigcup_{j \in J} \Gamma_j$ were not consistent, then the condition for the pair (I, J) would not hold, simply choose $\varphi \equiv \top$. Then apply Robinson's consistency theorem for Γ_1 and Γ_J.

433. According to the hint, $\Sigma(P) \cup \Sigma(P') \vDash P(\vec{c}) \leftrightarrow P'(\vec{c})$. By the compactness we may assume that $\Sigma(P)$ consists of finitely many closed formulas only, and then by the deduction lemma 6.9 $\Sigma(P), P(\vec{c}) \vDash \Sigma(P'), P'(\vec{c})$. By Craig's interpolation theorem there is a $\vartheta(\vec{x}) \in F(\tau)$ such that $\vartheta(\vec{c})$ is an interpolant:

$$\Sigma(P), P(\vec{c}) \vDash \vartheta(\vec{c}),$$
$$\vartheta(\vec{c}) \vDash \Sigma(P'), P'(\vec{c}).$$

From there the deduction lemma gives $\Sigma(P) \vDash \vartheta(\vec{c}) \leftrightarrow P(\vec{c})$. As the constant symbols $\vec{c}$ do not occur in $\Sigma(P)$, this shows that ϑ defines P explicitly.

434. By assumption there is $\vartheta(\vec{x}) \in \tau$ such that $\mathfrak{A} \vDash \vartheta(\vec{x}) \leftrightarrow P(\vec{x})$. As f is an automorphism with respect to symbols in τ, f preserves ϑ. But then, by the equivalence of ϑ and P, f should preserve P as well.

435. Let π be an automorphism of $\mathfrak{A}$. By induction on the complexity of the formulas, for every evaluation e over $\mathfrak{A}$, $\mathfrak{A} \vDash \varphi[e]$ iff $\mathfrak{A} \vDash \varphi[\pi \circ e]$. Thus if $\mathfrak{A} \vDash \varphi[\vec{a}, \vec{p}]$ then $\mathfrak{A} \vDash \varphi[\pi(\vec{a}), \pi(\vec{p})]$. By assumption $\pi(\vec{p}) = \vec{p}$, thus $\pi(\vec{a}) \in X$ for all $\vec{a} \in X$. As pi is an automorphism (and thus one-to-one), we must have $\pi(X) = X$.

436. Create a transcendence basis T of $\mathbb{C}$ over the field of rationals $\mathbb{Q}$ such that it has both a real number $r \in T$ and a complex number $z \in T$. Take the permutation of T which swaps r and z. This permutation extends uniquely to an automorphism of $\mathbb{Q}(T)$ which extends to an automorphism of $\mathbb{C}$. This automorphism does not keep the real line fixed, hence Problem 435 applies.

437. Every permutation of prime numbers induces an automorphism of $\mathfrak{A}$, and such an automorphism does not preserve addition.

438. The map $q \mapsto 2q$ is an automorphism of $\langle \mathbb{Q}, \le, + \rangle$ which does not preserve $\{1\}$, thus Problem 435 applies.

439. Assume, by way of contradiction, that $\Phi(x)$ defines the set of even numbers in $\mathfrak{A}$. In particular,

$$\mathfrak{A} \models \forall x(\Phi(x) \leftrightarrow \neg\Phi(S(x))). \qquad (\star)$$

Let Σ be the set of all formulas true in $\mathfrak{A}$. By Problem 420, Σ has a model $\mathfrak{B}$ of cardinality ω_1, thus there is an element $b^* \in B$ which differs from all of 0, $S(0)$, $S(S(0))$, etc. We claim that the function

$$\pi(b) = \begin{cases} b & \text{if } b = S^k(0) \text{ for some } k \ge 0, \\ S^{\mathfrak{B}}(b) & \text{otherwise,} \end{cases}$$

is an automorphism of $\mathfrak{B}$. Indeed, π preserves the interpretation of 0 and S, thus we only need to check that it is one-to-one. But this follows easily from the fact that in $\mathfrak{B}$ the following formulas are true (as they are true on $\mathfrak{A}$):

$$\forall x(S(x) \ne 0),$$
$$\forall x \forall y(S(x) = S(y) \rightarrow x = y),$$
$$\forall x(x \ne 0 \rightarrow \exists y(x = S(y)).$$

Therefore $\mathfrak{B} \models \Phi[b]$ iff $\mathfrak{B} \models \Phi[\pi(b)]$ for every $b \in B$. But this contradicts $(\star)$ for the choice $b = b^*$. See also Problem 552.

440. If $\Phi(x, y, z)$ defines the addition, that is, $\mathfrak{A} \models \Phi[a, b, c]$ iff $a + b = c$, then $\exists y\Phi(y, y, x)$ defines the even numbers in $\mathfrak{A} = \langle \omega, 0, \le, S \rangle$. The reasoning of Solution 439 shows that it is impossible as the function π defined there preserves the ordering as well.

441. If there is no such a model (not necessarily countable), then the function f is defined implicitly, thus, by Beth theorem 7.14, it is defined explicitly. But in the model $\langle \omega, 0, \le, S \rangle$ only the addition satisfies these formulas, so the addition would be definable in $\langle \omega, 0, \le, S \rangle$ contradicting Problem 440.

So let $\mathfrak{A}$ be a discrete ordering with different functions $f_1^{\mathfrak{A}}$ and $f_2^{\mathfrak{A}}$ satisfying the required formulas, namely $f_1^{\mathfrak{A}}(a, b) \ne f_2^{\mathfrak{A}}(a, b)$ for some $a, b \in A$. Take a countable elementary submodel of $\mathfrak{A}$ generated by $\{a, b\}$ (in the type extended by f_1 and f_2) to get the required countable structure.

442. $x \le y$ iff $\mathfrak{A} \models \exists z(z + x = y)$. 0 is the only element satisfying $x + x = x$, and 1 is the successor of 0 in the ordering $\le$: $x \ne 0 \wedge \forall y(y = 0 \vee x \le y)$.

443. The claim follows from the fact that if $\mathfrak{B}$ is a semantical substructure of $\mathfrak{A}$, and $X \subseteq B^n$ is $\mathfrak{B}$-definable, then it is also $\mathfrak{A}$-definable. This can be seen as follows. Every symbol in τ' is definable by some formula from $F(\tau)$, and the ground set of $\mathfrak{B}$ is also definable, thus for each $\varphi(\vec{x}) \in F(\tau')$ there is a $\varphi^*(\vec{x}) \in F(\tau)$ such that

$$\{\vec{b} \in B^n : \mathfrak{B} \models \varphi[\vec{b}]\} = \{\vec{a} \in A^n : \mathfrak{A} \models \varphi^*[\vec{a}]\}.$$

Such a φ^* can be created by recursion on the complexity of φ.

444. Let $\mathfrak{A} = \langle \mathbb{Z}, +, \cdot \rangle$. ω is a subset of $\mathbb{Z}$ and the interpretation of addition and multiplication in ω and in $\mathbb{Z}$ are the same. Thus it suffices to define the set of non-negative integers in $\mathfrak{A}$. As every such number is the sum of four squares, and only non-negative integers are of this form, the following formula defines ω in $\mathbb{Z}$:

$$\varphi(x) \equiv \exists y_1 \exists y_2 \exists y_2 \exists y_4 (x = y_1 \cdot y_1 + y_2 \cdot y_2 + y_3 \cdot y_3 + y_4 \cdot y_4).$$

445. Since $x \cdot y = z$ iff $(w^x)^y = w^z$ for all w, the following formula defines the multiplication:

$$\varphi(x, y, z) \equiv \forall w(e(e(w, x), y) = e(w, z)).$$

Addition can be defined using multiplication and exponentiation as follows

$$\psi(x, y, z) \equiv \forall w \, e(w, x) \cdot e(w, y) = e(w, z)).$$

446. According to Problem 442, $0^{\mathfrak{A}}$, $1^{\mathfrak{A}}$ and $\le^{\mathfrak{A}}$ can be defined by formulas, so they can be used as if they were present in the type. The function $q(x)$ defining the square of x can be defined as the only element y which satisfies $\mathsf{sq}(y) \wedge \mathsf{sq}(y + x + x + 1)$ and no element between y and $y + x + x + 1$ satisfies sq. Now $a \cdot b = c$ iff $(a + b)^2 = a^2 + 2c + b^2$. Thus $\varphi(x, z, y)$ can be

$$q(x + y) = q(x) + q(y) + z + z$$

447. (a) The ordinal numbers in $\mathfrak{A}$ endowed with the ordinal addition and ordinal multiplication form a model isomorphic to $\mathfrak{N}$. All of them are formula-definable. For example, $x \in H$ is an ordinal, if x is transitive: $(\forall y \in x)(\forall z \in y)$ $(z \in x)$, and the relation $\in$ restricted to x is an ordering: $(\forall y_1, y_2 \in x)(y_1 = y_2 \vee y_1 \in y_2 \vee y_2 \in y_1)$. (Transitivity of $\in \restriction x$ follows from the transitivity of x.) Ordinal addition and multiplication is defined by transfinite recursion, which translates to a defining formula via the transfinite recursion theorem.

(b) By Problem 342 the relation $i \in^{\mathfrak{N}} j$ iff the $(i + 1)$-st digit (counting from the right) in the binary representation of j is 1, makes a model of the hereditarily finite sets with ground set ω, the Ackermann model. Thus we are done if this relation is definable. But $i \in^M j$ iff $j = a + 2^i + 2^{i+1}b$ for some $a < 2^i$ and b which gives the defining formula

$$\varphi(i, j) \equiv \exists a \exists b(a < f(i) \wedge j = a + f(i) + f(i + 1) \cdot b).$$

Remark. By Problem 833 the function $f(i) = 2^i$ is definable in $\mathfrak{N}$, thus $\langle H, \in \rangle$ can be semantically interpreted in $\mathfrak{N} = \langle \omega, +, \cdot \rangle$. When two structures can be mutually semantically interpreted in each other, they are called *definitionally equivalent.*

448. Treat models of the binary relation symbol ρ as directed graphs: there is a directed edge from v to w if $(v, w) \in \rho^{\mathfrak{A}}$. Let $\mathfrak{B}$ be a structure with the binary function symbol f. We need to create a directed graph $\mathfrak{A}$ so that from this graph we can recover the value of the function $f^{\mathfrak{B}}$. The structure of such a graph is depicted on Figure 12.3. There is a single node v with in-degree

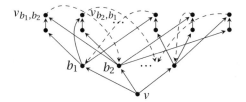

Figure 12.3: Recovering a function from a directed graph

zero, thus it can be defined by a formula. Points corresponding to the base set of the structure $\mathfrak{B}$ are the out-neighbors of v – again, a definable set. For any pair of them, say b_1 and b_2, there is exactly one vertex v_{b_1, b_2} which can be reached from b_1 by a path of length two, and directly from b_2. The only outgoing edge from v_{b_1, b_2} goes to the point corresponding to the value of the function $f(b_1, b_2)$. From this description it is clear that f can be recovered by a formula: $f(b_1, b_2) = c$ iff $\forall v \forall w\, (\rho(b_1, w) \wedge \rho(w, v) \wedge \rho(b_2, v) \rightarrow \rho(v, c))$.

449. Similarly to Problem 448 we will define a graph from which the binary relation ρ can be retrieved by some formula. Such a graph is sketched in Figure 12.4. Vertex a is the only one which has a single neighbor of degree

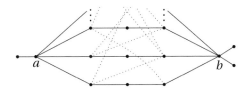

Figure 12.4: Embedding a relation into a graph

one, and b is the only vertex which has two such neighbors, thus they can be recovered by some formula. Other vertices adjacent to a are called A-vertices, and similarly for B vertices. Vertices in the middle – the C vertices – are not adjacent to either a or b; they have degree exactly 2 with one A and one B neighbor. Every A-vertex and every B-vertex has a unique C-neighbor. All further edges of the graph go between A and B vertices.

It is clear that there is a single graph formula which, if true, forces the graph to have the structure described above. The base of the substructure will be formed by the A-vertices. Say that the A-vertices u and v are in relation ρ (in this order), if there is an edge from u to v', where v' is the unique B vertex connected to v through a C-vertex. It is clear that any binary relation can be encoded this way.

450. Using the transitivity of the semantical definition (Problem 443). $\mathfrak{N}$ can be defined in $\mathfrak{A} = \langle \omega, e \rangle$ for a binary function e (Problem 445). Any structure with a single binary function can be defined in a structure with a single binary relation symbol (Problem 448), and any such structure can be defined in a graph (Problem 449).

451. Let G be a graph. The lattice will have four levels: the minimal element 0, all vertices of the graph, all edges of the graph, and the maximal element 1. A vertex is below an edge if it is adjacent to it.

Using the lattice, vertices of the graph correspond to those lattice elements which are directly above 0, and two such points are connected by an edge if their join is not the maximal element 1.

452. By Problem 350(c) the set of non-negative integers in $\mathfrak{A}$ is definable. By point (b) of the same problem, the set of perfect squares, denoted by sq, is also definable. Thus $\langle \omega, +, \mathsf{sq} \rangle$ is semantically definable in $\mathfrak{A}$, and then Problem 446 so is $\langle \omega, +, \cdot \rangle$.

453. By Problem 334 the set of shifts is defined by the formula $x \circ \pi = \pi \circ x$. For $j \in \mathbb{Z}$ let π_j be the shift $x \mapsto x + j$. Then $\pi_1 = \pi$, and $\pi_{j_1+j_2} = \pi_{j_1} \circ \pi_{j_2}$, thus both 1 and addition is definable in G. According to the hint we seek for a definition for the divisibility relation. For a permutation σ of $\mathbb{Z}$, $\pi_i \circ \sigma = \sigma \circ \pi_i$ is equivalent to $\sigma(x+i) = \sigma(x)+i$ for all $x \in \mathbb{Z}$, and then $\sigma(x+j) = \sigma(x)+j$ also holds for each multiple of i. From here it is easy to check that the formula

$$\forall \sigma \, (\pi_i \circ \sigma = \sigma \circ \pi_i \rightarrow \pi_j \circ \sigma = \sigma \circ \pi_j)$$

holds in G if and only if i is a divisor of j. It means that the structure $\langle \mathbb{Z}, 1, +, | \rangle$ is semantically definable in G, and then $\mathfrak{N}$ is also definable in G by Problem 451.

454. (a) A finite subset of $T(x)$ is realizable in a model $\mathfrak{A}$ if there is an x different from the finitely many $c_i^{\mathfrak{A}}$ mentioned in that subset. As Σ has an infinite model, such an element always exists.

(b) The type $T(x)$ is realized by any element of the structure which differs from all $c_i^{\mathfrak{A}}$. Thus $\mathfrak{A}$ omits $T(x)$ exactly in the case every element of A is an interpretation of some c_i.

455. The ordered structure $\mathfrak{A}$ realizes $T(x, y)$, if there are two elements $a, b \in A$ with infinitely many elements between them. $\mathfrak{A}$ will omit T if $\mathfrak{A}$ is finite, and $\mathfrak{A}$ cannot omit T if the cardinality of $\mathfrak{A}$ is at least ω_1. So suppose $\mathfrak{A}$ is countable and omits T. Between any two elements there are only finitely many others, thus every element has an immediate successor and an immediate predecessor, thus the ordering is discrete. There are three discrete orderings omitting T, namely ω, ω^* (the reverse of ω), and $\omega^* + \omega$.

456. Let the type τ consist of the constant symbols c_i for $i < \omega$, and let $\Gamma = \{c_i \neq c_j : i \neq j\}$. Take $\mathfrak{B}_k = \langle \omega \cup \{a_i : i < k\}, c_i^{\mathfrak{B}_k}\rangle$, where $c_i^{\mathfrak{B}_k} = i$ and the a_i's are new elements. $\mathfrak{B}_k \models \Gamma$, and the type

$$T(x) \equiv \{x \neq c_i : i \in \omega\}$$

is realized in $\mathfrak{B}_k$ by exactly k elements.

457. Let τ consists of two disjoint sets C and D of constant symbols, where C is uncountable and D is countably infinite. Write $\Sigma = \{a \neq b : a, b \in C, a \neq b\}$ and let $T(x)$ be the type $\{x \neq d : d \in D\}$. Every model of Σ is uncountable, thus every model of Σ realizes $T(x)$, as there is always an element which differs from the interpretation of all constant symbols in C, see Problem 454. It is straightforward that $T(x)$ is not isolated.

458. The structure $\mathfrak{A}$ omits the type $T(x)$.

459. (a) $\mathfrak{A}$ is an ω-model iff $\mathfrak{A}$ omits the type $T(x) = \{x \neq \pi_i : i \in \omega\}$. We use Theorem 7.19 to show that Γ has a model omitting T. First, T is a type as in any model of Γ the interpretations of π_i are different. Thus we only need to check that T is not isolated. To this end suppose $\varphi(x)$ is a formula such that $\Gamma \models \varphi(x) \rightarrow (x \neq \pi_i)$ for all $i \in \omega$. In this case ω-completeness gives $\Gamma \models \forall y(\varphi(x) \rightarrow (x \neq y))$, that is $\Gamma \models \neg\varphi(x)$. It means that $\Gamma \cup \{\exists x \varphi(x)\}$ is not consistent, showing that T is not isolated indeed.

(b) If $\mathfrak{A} \models \Gamma$ is an ω-model and $\mathfrak{A} \models \varphi(\pi_0)$, $\mathfrak{A} \models \varphi(\pi_1)$, ..., then surely $\mathfrak{A} \not\models \exists x \neg\varphi(x)$, thus the formula $\exists x \neg\varphi(x)$ cannot be a consequence of Γ.

460. Let Γ' be the set of all closed formulas provable from Γ in the extended inference system. Clearly Γ' is syntactically consistent in the traditional sense, thus it has a model. As Γ' is ω-complete, by Problem 459 it has an ω-model as well.

If Γ has an ω-model, then $\mathrm{Th}(\mathfrak{A})$, the set of formulas true in $\mathfrak{A}$, extends Γ. They are closed for the ω-rule, thus only formulas in $\mathrm{Th}(\mathfrak{A})$ can be derived from Γ (even using the ω-rule). As $\bot$ is not in $\mathrm{Th}(\mathfrak{A})$, Γ is syntactically consistent.

461. A syntactically ω-consistent theory is ω-complete, thus by Problem 460 such a theory has an ω-model. All formulas true in that model form a maximal syntactically ω-consistent set extending the given theory.

462. Let the similarity type τ contain 0, S and a constant symbol c and put $\Sigma = \{\pi_n \neq c : n \in \omega\}$. With $\varphi(x) \equiv x \neq c$ the ω-rule implies

$$\frac{\varphi(\pi_0),\ \varphi(\pi_1),\ \varphi(\pi_2),\ \varphi(\pi_3),\ \dots}{\forall x \varphi(x)}$$

Thus $\Sigma \vdash \forall x(x \neq c)$. But clearly no finite $\Gamma \subset \Sigma$ gives $\Gamma \vdash \forall x(x \neq c)$. (Note: Σ is ω-inconsistent.)

12.8   ELEMENTARY EQUIVALENCE

463. Let $j:\mathfrak{A}\to\mathfrak{B}$ be an isomorphism and e be an evaluation over $\mathfrak{A}$. In this case $j\circ e$ is an evaluation over $\mathfrak{B}$. As $j(c^{\mathfrak{A}})=c^{\mathfrak{B}}$ for each constant symbol, and j maps $f^{\mathfrak{A}}$ to $f^{\mathfrak{B}}$, an easy induction gives $j(t^{\mathfrak{A}}[e])=t^{\mathfrak{B}}[j\circ e]$ for each τ-term t. This means that the equivalence $\mathfrak{A}\vDash\varphi[e]$ iff $\mathfrak{B}\vDash\varphi[j\circ e]$ holds for atomic formulas, and another induction show it for arbitrary formulas. Finally, the claim follows from the fact that every evaluation over $\mathfrak{B}$ is of the form $j\circ e$ for some evaluation e over $\mathfrak{A}$.

464. Let $\mathfrak{A}$ be infinite. By Problem 420, $\mathrm{Th}(\mathfrak{A})$ has models of arbitrary large cardinality (as it has an infinite model $\mathfrak{A}$). If $\mathfrak{A}$ and $\mathfrak{B}$ are isomorphic, they must have the same cardinality.

465. Suppose first that the similarity type τ is finite and $\mathfrak{A}$ has n elements. Fix an evaluation e over $\mathfrak{A}$ which assigns different elements to $x_1,\dots x_n$. Construct the formula

$$\varphi \equiv \exists x_1\dots\exists x_n\big(\psi(\vec{x}) \wedge \forall y\bigvee_i y = x_i\big)$$

where $\psi(\vec{x})$ is the conjunction of all atomic formulas and their negations with unnested terms that hold in $\mathfrak{A}$ under the evaluation e. As $\mathfrak{A}$ and $\mathfrak{B}$ are elementarily equivalent, $\mathfrak{B}\vDash\varphi$, thus there are elements $b_1,\dots,b_n\in B$ in place of the x_i's, and the map $a_i\mapsto b_i$ is an isomorphism.

For the general case suppose τ is infinite and for each finite $\sigma\subseteq\tau$ fix an isomorphism $f_\sigma:\mathfrak{A}\!\restriction\!\sigma\to\mathfrak{B}\!\restriction\!\sigma$. As $\mathfrak{A}$ (and thus $\mathfrak{B}$) has n elements, there are finitely many functions between $\mathfrak{A}$ and $\mathfrak{B}$ only, say $g_0,\dots,g_k$. If $\mathfrak{A}$ and $\mathfrak{B}$ are not isomorphic then g_ℓ is not an isomorphism, thus there exists either a function or a relation symbol which is not mapped properly by g_ℓ. Pick one of these symbols, and let σ be the collection of these violated symbols. As there are finitely many possible functions, σ is finite. But then f_σ cannot be an isomorphism, a contradiction.

466. Here is a counterexample. The type τ has countably many unary relations R_n. For $n\in\omega$ let $R_n^{\mathfrak{A}}=(-\infty,-n)$ and $R_n^{\mathfrak{B}}=(-\infty,-n)\cup\{1\}$. Then $\mathfrak{A}=\langle\mathbb{R},R_n^{\mathfrak{A}}\rangle_{n\in\omega}$ and $\mathfrak{B}=\langle\mathbb{R},R_n^{\mathfrak{B}}\rangle_{n\in\omega}$ are not isomorphic because $\bigcap_{n\in\omega}R_n^{\mathfrak{A}}=\emptyset$ while $\bigcap_{n\in\omega}R_n^{\mathfrak{B}}=\{1\}$. But for any finite $\sigma=\{R_0,\dots,R_k\}$ it is easy to see that $\mathfrak{A}\!\restriction\!\sigma$ and $\mathfrak{B}\!\restriction\!\sigma$ are isomorphic.

467. (a) There are continuum many subsets of $F(\tau)$ (as it is countable), thus among more than continuum many subsets of $F(\tau)$ there must be two equal.

(b) Let $a_1,a_2,\dots$ be a (countable) sequence of ≥ 2 natural numbers, and consider the ordering $a_1+\eta+a_2+\eta+\cdots$ where η is the order type of rationals. From this ordering one can recover the sequence $a_1,a_2,\dots$ by formulas: there are exactly a_1 elements which are smaller than any other element. The first element of the next block of a_2 elements is the smallest element which is bigger than the first a_1 and has an immediate successor. There are exactly

a_2 discrete elements in this block, etc. Consequently orderings created from different $a_1, a_2, \ldots$ sequences are not elementarily equivalent, and there are continuum many such sequences.

468. Consider the linear orderings $2 + \eta + 3 + \eta + 2 + \cdots$ and $3 + \eta + 2 + \eta + 3 + \cdots$, where η is the order type of the rationals. These orderings are not elementarily equivalent by Solution 467, but clearly can be embedded one to the other. A modified construction can give continuum many non-equivalent orderings such that any two of them are mutually embeddable.

469. Let $\mathfrak{A}_n$ be the closed interval $[1/n, 1 - 1/n]$ with the usual ordering. As they are isomorphic, Problem 463 implies $\mathfrak{A}_n \equiv \mathfrak{A}_m$. However $\bigcup_n \mathfrak{A}_n = (0, 1)$ has no maximal element, thus it is not elementarily equivalent to any $\mathfrak{A}_n$.

470. Not true. Let $\{G_i : i < \omega\}$ be the set of all finite graphs and let G be the structure which is their disjoint union. The structure H contains a copy of G and a new vertex v which is connected to every other vertex. The set of spanned finite subgraphs of G and H coincide (both are the set of all finite graphs), and G and H are not elementarily equivalent shown by the formula $\exists v \forall x\, E(v, x)$.

471. No. By the Erdős-DeBruijn theorem a graph has chromatic number $k < \omega$ if and only if each of its finite subgraphs can be colored with k colors and there is a finite subgraph with chromatic number k. Suppose $G_1 \equiv G_2$ and $\chi(G_1) < \chi(G_2) = k < \omega$. There is a finite subgraph $H \subseteq G_2$ with $\chi(H) = k$ and a formula φ expressing "there are $|H|$ points which induce a subgraph isomorphic to H". As H is fixed, this can be expressed with a first-order formula. Now, $G_2 \vDash \varphi$ implies $G_1 \vDash \varphi$ and therefore there is a finite subgraph of G_1 isomorphic to H. But then the chromatic number of G_1 cannot be less than $\chi(H) = k$.

472. By induction on the complexity of the formulas using that $t^{\mathfrak{A}}[e] = t^{\mathfrak{B}}[e]$ for all terms t (by definition of the substructure and induction on the terms).

473. Choose τ with no function or constant symbols, and the τ-structure $\mathfrak{A}$ which has at least two elements. Let $\mathfrak{B}$ be a substructure that consists of a single point of $\mathfrak{A}$. If φ is the formula $\forall y(x = y)$, then $\mathfrak{B} \vDash \exists x \forall y(x = y)$ but $\mathfrak{A} \nvDash \exists x \forall y(x = y)$.

The formula φ cannot be quantifier-free. Suppose it is, then $\mathfrak{B} \vDash \exists x \varphi[e]$ iff $\mathfrak{B} \vDash \varphi[e(x/b)]$ for some $b \in B$. By Problem 472 this is equivalent to $\mathfrak{A} \vDash \varphi[e(x/b)]$ which holds iff $\mathfrak{A} \vDash \exists x \varphi[e]$.

474. This is just another way of saying that for all evaluations e over $\mathfrak{B}$, $\mathfrak{B} \vDash \varphi[e]$ iff $\mathfrak{A} \vDash \varphi[e]$.

475. A proper substructure $\mathfrak{B}$ of $\mathfrak{A} = \langle \mathbb{Z}, \leq \rangle$ either has an endpoint (an thus $\mathfrak{B}$ is not even elementarily equivalent to $\mathfrak{A}$), or contains two consecutive points $a, b \in B$ which are not consecutive in $\mathfrak{A}$. If $\varphi(x, y)$ denotes the formula which expresses that there are no elements between x and y, then we get $\mathfrak{B} \vDash \varphi[a, b]$ but $\mathfrak{A} \nvDash \varphi[a, b]$.

476. Every property, expressible by a formula, of a node in the elementary subgraph H is true in H iff it is true in G. As every node has degree 2 in G, the same must be true in H, thus the only elementary subgraph of G is itself.

477. Let $\mathfrak{B} = \langle \omega, \leq \rangle$ and $\mathfrak{A} = \langle \omega - \{0\}, \leq \rangle$. As $f(n) = n + 1$ is an isomorphism between $\mathfrak{A}$ and $\mathfrak{B}$, the two structures are elementarily equivalent (see Problem 463). Let $\varphi(v)$ be the formula which expresses that v is the least element. Then $\mathfrak{A} \models \varphi[v = 1]$ but $\mathfrak{B} \not\models \varphi[v = 1]$, consequently $\mathfrak{A}$ is not an elementary substructure of $\mathfrak{B}$.

478. (a) Problems 472 and 473 imply that for quantifier-free φ we have $\mathfrak{B} \models \exists \vec{x}\varphi[e]$ iff $\mathfrak{A} \models \exists \vec{x}\varphi[e]$. Therefore it is enough to give $\mathfrak{B} \subseteq \mathfrak{A}$ such that $\mathfrak{B}$ is not an elementary substructure of $\mathfrak{A}$. Take, for example, $\mathfrak{A}$ to be finite with the empty language, and $|B| < |A|$.

(b) Let $\mathfrak{B} = \langle \omega, < \rangle$, and let $\mathfrak{A}$ has one more point which is bigger than any element in $\mathfrak{B}$. The truth value of a quantifier-free formula $\varphi(\vec{x}, y_1, \ldots, y_n)$ depends only on the order of the interpretation of the variables. Thus if $\mathfrak{B} \models \forall \vec{x}\varphi[e]$ for an evaluation e over $\mathfrak{B}$, then the same formula is true in $\mathfrak{A}$. And $\mathfrak{B}$ is not an elementary substructure of $\mathfrak{A}$ as $\mathfrak{A}$ has a biggest element, while $\mathfrak{B}$ has not.

479. Both are true. To check $\mathfrak{B} \prec \mathfrak{A}$ observe that for an evaluation e over $\mathfrak{B}$ we have

$$\mathfrak{B} \models \varphi[e] \quad \Leftrightarrow \quad \mathfrak{C} \models \varphi[e] \quad \Leftrightarrow \quad \mathfrak{A} \models \varphi[e].$$

This also implies $\mathfrak{A} \equiv \mathfrak{B}$.

480. No finite structure can have a proper elementary substructure (e.g., because it can be expressed by a first-order formula that a structure has n elements for some fixed $n \in \omega$). Regarding infinite groups: $\mathbb{Z}_2^\infty \triangleleft H \times \mathbb{Z}_2^\infty$ for any group H. Clearly, $\mathbb{Z}_2^\infty \models \forall x(x + x = 0)$ while H can be chosen so that $H \times \mathbb{Z}_2^\infty \models \exists x(x + x \neq 0)$ (e.g. take $H = \mathbb{Z}_3$).

481. No. $G = \bigoplus_\omega \mathbb{Z}_4$ has a subgroup $H = \bigoplus_\omega \{0, 2\}$, but G is not even elementarily equivalent to H as $H \models \forall x(x + x = 0)$, while $G \models 3 + 3 \neq 0$.

482. ($\Rightarrow$) If $\mathfrak{B} \prec \mathfrak{A}$, then $\mathfrak{A} \models \exists x\varphi[\vec{b}]$ iff $\mathfrak{B} \models (\exists x\varphi)[\vec{b}]$ iff there is a $c \in A$ with $\mathfrak{B} \models \varphi[c, \vec{b}]$ iff $\mathfrak{A} \models \varphi[c, \vec{b}]$.

($\Leftarrow$) We need to show $\mathfrak{B} \models \varphi[\vec{b}] \Leftrightarrow \mathfrak{A} \models \varphi[\vec{b}]$ for all $\vec{b} \in B$ and formula $\varphi(\vec{v})$. As $\mathfrak{B}$ is a substructure of $\mathfrak{A}$, this holds for quantifier-free formulas (see Problem 472). Suppose (inductive hypothesis) that it holds for φ and consider the formula $\exists x\varphi$. (For the other cases the induction is easy.) Then $\mathfrak{A} \models (\exists x\varphi)[\vec{b}]$ implies (by assumption of the theorem) that there is $c \in B$ such that $\mathfrak{A} \models \varphi[c, \vec{b}]$. By the inductive hypothesis this is equivalent to $\mathfrak{B} \models \varphi[c, \vec{b}]$, which is equivalent to $\mathfrak{B} \models (\exists x\varphi)[\vec{b}]$. The reverse implication is immediate.

483. If f is a function symbol and $\vec{b} \in B$, then there is an automorphism j which fixes $\vec{b}$ and moves $a = f^{\mathfrak{B}}(\vec{b})$ to B. As j is an automorphism, $j(a) = a$, which implies $a \in B$. Thus $\mathfrak{B}$ is the ground set of a substructure.

To prove that $\mathfrak{B} \prec \mathfrak{A}$ use Theorem 8.3. Take $\vec{b} \in B$ and formula $\varphi(y, \vec{x})$ and assume $\mathfrak{A} \vDash (\exists x \varphi)[\vec{b}]$. We have to find a good $x \in B$. There is an $a \in A$ such that $\mathfrak{A} \vDash \varphi[a, \vec{b}]$, and, by assumption, there is an automorphism j which moves a into B while keeping $\vec{b}$ fixed. As automorphisms preserve formulas, we have

$$\mathfrak{A} \vDash \varphi[a, \vec{b}] \quad \Leftrightarrow \quad \mathfrak{A} \vDash \varphi[j(a), j(\vec{b})] \quad \Leftrightarrow \quad \mathfrak{B} \vDash \varphi[j(a), \vec{b}].$$

As $j(a) \in B$, we are done.

484. We use the method of Problem 483. Fix the real numbers $a_1, \ldots, a_n \in (b, c)$. For any real number $r \in (a, d)$ there is an automorphism (order-preserving map) of (a, d) which keeps all a_i and moves r into the open interval (b, c).

485. For each finite subset $F \subseteq \mathbb{Q}$ and $a \in \mathbb{R}$ there is an order-preserving permutation $j : \mathbb{R} \to \mathbb{R}$ that fixes F pointwise and moves a into $\mathbb{Q}$ (such a permutation can be chosen to be a piecewise linear function). Therefore Problem 483 applies.

486. Since A is countable, the construction in Problem 365 shows that for any finite subset $F \subset B$ there is an order-preserving permutation of $\mathfrak{A}$ which fixes F, and moves $a \in A$ into a different element. Thus Problem 483 applies.

> **Remark.** By Problem 370 the same idea does not work for uncountable structures. The statement, however, is true in general, see Problem 545.

487. The elementary substructure $\mathfrak{B}$ must contain the point of the broom (the only point which has degree ≥ 3), all points of the handle (they are the points with degree 2, and if any of them is in B, then both neighbors must be in B as well), and infinitely many edges from the brush (as $\mathfrak{B}$ must contain at least n one-degree nodes for each n). That is, remove edges from the brush ensuring infinitely many remain. By Problem 483 all of these substructures are elementary.

488. We can remove paths of length two that are joined to v taking care that infinitely many of such paths should remain joined to v. The rest is similar to Solution 487.

489. (a) $\mathfrak{A}_n$ is the usual ordering of the open interval $(0, 1/n)$. They are elementary submodels of each other by Problem 484.
   (b) $\mathfrak{A}_n$ is the usual ordering of the open interval $(-1/n, 1/n)$.

490. The Skolem function of the formula $\exists y (f(\vec{x}) = y)$ is the interpretation of f, thus B is closed for all functions. The second claim is immediate from the Tarski–Vaught test 8.3 as the Skolem function f_ψ of the formula $\psi \equiv \exists y \varphi(y, \vec{x})$ satisfies, for all $\vec{a} \in \mathfrak{A}$,

$$\mathfrak{A} \vDash (\exists y \varphi)[\vec{a}] \quad \text{implies} \quad \mathfrak{A} \vDash \varphi[f_\psi(\vec{a}), \vec{a}].$$

491. If $\mathfrak{B} \subseteq \mathfrak{A}$ is a substructure, $t(\vec{x})$ is a term and $\vec{b} \in B$, then clearly $t[\vec{b}] \in B$. From here the claim follows from the Tarski–Vaught test and the definition of built-in Skolem functions.

492. If X is empty, add a single point to it. Observe that there are at most $\max(|\tau|, \omega)$ Skolem functions as there are that many formulas. Let B be the closure of X for all Skolem functions. The cardinality of B is clearly at most $\max(|X|, |F(\tau)|)$. Finally, $\mathfrak{B}$ is an elementary substructure of $\mathfrak{A}$ by Problem 490.

493. (a) By Problem 420, Γ has arbitrary large models. Apply the Downward Löwenheim–Skolem Theorem 8.5 for a $X \subseteq A$ of cardinality κ. (b) Apply (a) to the theory $\mathrm{Th}(\mathfrak{A})$.

494. Use the idea of Problem 358. The similarity type is empty, and for $i \geq 1$ let φ_i be the formula which says that if there are i different elements, then there are $i + 1$ different elements. Let $\Gamma = \{\varphi_i : i \notin K\}$.

495. Let $\mathfrak{A} \models \Gamma$ of size κ. There are 2^κ many different constants, relations and functions in the structure, thus among the interpretation of the symbols in τ there are at most 2^κ many different ones. Replace different symbols with the same interpretation by the same symbol in Γ. The new theory has models for each $\lambda \geq 2^\kappa$, and interpreting the missing symbols equal to their representative ones makes it a model of Γ.

> **Remark.** Problem 614 constructs directly a model of size 2^κ from any model of size κ. That construction, however, falls short to prove that there are models for all $\lambda \geq 2^\kappa$.

496. (ii)$\Rightarrow$(i)$\Rightarrow$complete is clear. The implication complete$\Rightarrow$(ii) follows from the fact that if two structures $\mathfrak{A}$ and $\mathfrak{B}$ are not elementarily equivalent, then there must be closed formula φ such that $\mathfrak{A} \models \varphi$ and $\mathfrak{B} \models \neg\varphi$.

497. For $\mathfrak{A}, \mathfrak{B} \models \Gamma$, using Problem 493 one can find models $\mathfrak{A}', \mathfrak{B}'$ of cardinality κ such that $\mathfrak{A}' \equiv \mathfrak{A}$ and $\mathfrak{B}' \equiv \mathfrak{B}$. By assumption we have $\mathfrak{A}' \cong \mathfrak{B}'$, thus $\mathfrak{A} \equiv \mathfrak{A}' \cong \mathfrak{B}' \equiv \mathfrak{B}$. Combining this with Problem 463 we get $\mathfrak{A} \equiv \mathfrak{B}$.

498. Let Γ be the empty theory of the empty language (models on this language are pure sets). Each two sets of the same cardinality are isomorphic, however no finite set can be elementarily equivalent to any other set that has different cardinality.

499. Yes. Infinite complete graphs are elementarily equivalent (because complete graphs of the same cardinality are isomorphic and the theory of an infinite complete graph has no finite models; see Problem 497). Consequently, for different infinite cardinals κ and λ we have $\kappa = \chi(K_\kappa) \neq \chi(K_\lambda) = \lambda$, while K_κ and K_λ are elementarily equivalent.

500. Any uncountable model $\mathfrak{A} \models \Gamma$ consists of $|A|$ many $\mathbb{Z}$-chains, hence uncountable models of Γ are isomorphic. The type is finite, thus by Theorem 8.7, it is complete.

501. In a model $\mathfrak{A} \vDash \Gamma$ draw a directed edge from x to $S(x)$. From the formulas it follows that every node, except for $0^{\mathfrak{A}}$ has in-degree exactly one; and there are no loops. Thus $\mathfrak{A}$ consists of an infinite half-line starting from $0^{\mathfrak{A}}$, and lines isomorphic to $\mathbb{Z}$. Thus any two models of cardinality ω_1 are isomorphic, and Theorem 8.7 applies. See also Problem 552.

502. By Problem 365 it has a unique countable model. Use Theorem 8.7.

503. Torsion-free divisible Abelian groups are isomorphic to the additive groups of vector spaces over the field of rational numbers $\mathbb{Q}$. Any two such vector spaces are isomorphic if and only if their dimensions are the same. If the cardinality is uncountable, then the dimension equals the cardinality, thus there is only one such vector space for all $\kappa > \aleph_0$.

The theory is not $\aleph_0$-categorical; actually there are $\aleph_0$ non-isomorphic models corresponding to the dimension which can be $n \geq 1$ or countably infinite.

504. An algebraically closed field is uniquely determined by its transcendence degree over the base field, which is $\mathbb{Q}$ for characteristic zero fields. If the degree is uncountable, then the field has the same cardinality.

505. Expand the similarity type with constant symbols c, d and let

$$\varphi_n(c, d) \equiv \exists x_1 \ldots \exists x_n \big(c < x_1 < \cdots < x_n < d \,\wedge\, Z(x_1) \wedge \cdots \wedge Z(x_n) \big)$$

Let $\Gamma' = \Gamma \cup \{\varphi_n(c, d) : n \in \omega\}$. Then Γ' is finitely satisfiable, so by compactness there is a model $\mathfrak{A} \vDash \Gamma'$. By the Löwenheim–Skolem theorem we can assume $\mathfrak{A}$ is countable. Let $\mathfrak{B}$ be a countable elementary substructure of $\langle \mathbb{R}, \leq, Z \rangle$. Then the reduct of $\mathfrak{A}$ to the original similarity type and $\mathfrak{B}$ are countable models of Γ but they are not isomorphic.

506. The similarity type consists of a single unary relation symbol R. Let Γ be the theory which says that infinitely many elements are in R, and infinitely many elements are not in R. All countable models of Γ are isomorphic, thus the theory is complete (Theorem 8.7). Let $\mathfrak{A} = \langle A, R^{\mathfrak{A}} \rangle$ be such that $R^{\mathfrak{A}}$ is countable, $A \smallsetminus R^{\mathfrak{A}}$ is uncountable, and let $\mathfrak{B} = \langle B, R^{\mathfrak{B}} \rangle$ be such that $R^{\mathfrak{B}}$ is uncountable, $B \smallsetminus R^{\mathfrak{B}}$ is countable. Then $\mathfrak{A}, \mathfrak{B} \vDash \Gamma$, and so $\mathfrak{A} \equiv \mathfrak{B}$, but there is no embedding from either one to the other.

507. (a) Let $j : A \to B$ be defined by $j(a) = c_a^{\mathfrak{B}}$. If $a, b \in A$ are distinct elements, then $c_a \neq c_b \in \Delta_{\mathfrak{A}}^0$, thus $j(a) \neq j(b)$. If $f \in \tau$ is a function symbol and $f^{\mathfrak{A}}(a_1, \ldots, a_n) = b$, then the formula $f(c_{a_1}, \ldots, c_{a_n}) = c_b$ belongs to $\Delta_{\mathfrak{A}}^0$ and $f^{\mathfrak{B}}(j(a_1), \ldots, j(a_n)) = j(b)$. For a relation symbol $R \in \tau$ we have $\langle a_1, \ldots, a_n \rangle \in R^{\mathfrak{A}}$ iff $R(c_{a_1}, \ldots, c_{a_n}) \in \Delta_{\mathfrak{A}}^0$, thus $\langle j(a_1), \ldots, j(a_n) \rangle \in R^{\mathfrak{B}}$. Therefore the image of j is a substructure.

(b) Let $j : A \to B$ be defined by $j(a) = c_a^{\mathfrak{B}}$. For a formula φ and elements $a_1, \ldots, a_n \in A$ we have $\mathfrak{A} \vDash \varphi[a_1, \ldots, a_n]$ iff $\varphi(c_{a_1}, \ldots, c_{a_n}) \in \Delta_{\mathfrak{A}}$ iff $\mathfrak{B} \vDash \varphi[c_{a_1}^{\mathfrak{B}}, \ldots, c_{a_n}^{\mathfrak{B}}]$ iff $\mathfrak{B} \vDash \varphi[j(a_1), \ldots, j(a_n)]$.

508. $\Delta_{\mathfrak{A}}$ has arbitrary large models by Problem 420, and $\mathfrak{A}$ can be embedded elementarily to every model of $\Delta_{\mathfrak{A}}$ by Problem 507(b).

509. Any model of $\Delta_{\mathfrak{A}} \cup \Sigma$ is an extension of $\mathfrak{A}$ that models Σ (see Problem 507). Using compactness we show $\Delta_{\mathfrak{A}} \cup \Sigma$ is consistent. Take a finite subset $\Gamma \subset \Delta_{\mathfrak{A}} \cup \Sigma$. Then $\mathfrak{A}$ has a finitely generated substructure generated by the constants appearing in $\Gamma \cap \Delta_{\mathfrak{A}}$. By the hypothesis this structure can be extended to be a model of Σ, thus Γ is consistent.

510. (a) Let τ_1 and τ_2 be the similarity type of $\Delta_{\mathfrak{A}}$ and $\Delta_{\mathfrak{B}}$, respectively. Then $\tau_1 \cap \tau_2 = \tau$. By the elementary equivalence we have that $\varphi \in \Delta_{\mathfrak{A}}$ iff $\varphi \in \Delta_{\mathfrak{B}}$ for any closed formula $\varphi \in F(\tau)$. Thus Robinson's consistency theorem 7.11 gives that $\Delta_{\mathfrak{A}} \cup \Delta_{\mathfrak{B}}$ is consistent.

> **Remark.** As τ_1 and τ_2 has additional constant symbols only, a direct approach similar to the one used in Problem 430 gives that $\Delta_{\mathfrak{A}} \cup \Delta_{\mathfrak{B}}$ is consistent.

(b) By Problem 507(b) both $\mathfrak{A}$ and $\mathfrak{B}$ can be embedded into a model of $\Delta_{\mathfrak{A}} \cup \Delta_{\mathfrak{B}}$.

511. Similarly to Solution 510, it suffices to show that $\bigcup_{i \in I} \Delta_{\mathfrak{A}_i}$ is consistent. For finite I proceed by induction on $|I|$. For $|I| = n$ embed $n-1$ structures into $\mathfrak{B}$. As $\mathfrak{B}$ and $\mathfrak{A}_n$ (as τ-structures) are elementarily equivalent, they can be embedded elementarily into a τ-structure $\mathfrak{C}$. By the transitivity of embedding (Problem 479) each $\mathfrak{A}_i$ embeds to $\mathfrak{C}$.

If $|I|$ is infinite, then any finite subset of $\Sigma = \bigcup_{i \in I} \Delta_{\mathfrak{A}_i}$ is in a union of finitely many Δ_i's, thus Σ is consistent.

512. ($\Rightarrow$) Using Robinson's consistency theorem $\mathrm{Th}(\mathfrak{A}_1) \cup \mathrm{Th}(\mathfrak{A}_2)$ is consistent, and both $\mathfrak{A}_1$ and $\mathfrak{A}_2$ can be embedded to its models.

($\Leftarrow$) Suppose Γ_1 and Γ_2 satisfies the condition of Robinson's theorem. For every closed $\varphi \in F(\tau)$ either φ or $\neg\varphi$ can be added to both Γ_1 and Γ_2, otherwise the condition would be violated. Thus Γ_1 and Γ_2 can be extended so that their τ-type consequences is the same maximal consistent theory. Take $\mathfrak{A}_i \vDash \Gamma_i$, the τ-type reducts are elementarily equivalent, thus they can be embedded into a $\tau_1 \cup \tau_2$-structure $\mathfrak{B}$. But then $\Gamma_1 \cup \Gamma_2$ is consistent.

513. Not true even for $|I| = 3$. There are three binary relation symbols R_1, R_2, R_3 (counted modulo 3), and in $\mathfrak{A}_i$ $R_i^{\mathfrak{A}_i}$ is a dense linear order without endpoints, and $R_{i+1}^{\mathfrak{A}_i}$ is the reverse order. By Problem 502 the pairwise reducts are elementarily equivalent, thus $\mathfrak{A}_i$ and $\mathfrak{A}_{i+1}$ embeds into a joint structure, where R_i and R_{i+2} is the same relation (both are the reverse of R_{i+1}). But not all three embeds elementarily into the same structure.

514. By Problem 507 it is enough to show that $\Delta_{\mathfrak{A}}^0 \cup \Gamma$ is consistent. If not, then by compactness and as $\Delta_{\mathfrak{A}}^0$ is closed for $\wedge$, there is a single $\delta(\bar{a}) \in \Delta_{\mathfrak{A}}^0$ such that $\Gamma \vDash \neg\delta(\bar{a})$ where $\delta(\bar{x}) \in F(\tau)$ is quantifier-free and $\bar{a}$ are constant symbols denoting elements of $\mathfrak{A}$. As $\bar{a}$ does not occur in the similarity type

τ, we have $\Gamma \vDash \forall \vec{x} \neg \delta(\vec{x})$. As this universal formula is a consequence of Γ, it must hold in $\mathfrak{A}$, contradicting that $\delta(\vec{a}) \in \Delta^0_{\mathfrak{A}}$.

515. Suppose $\varphi(\vec{x})$ is not equivalent to any quantifier-free formula. Let $\mathfrak{A} \vDash \Gamma$, $\mathfrak{A} \vDash \varphi[\vec{b}]$, and let $\mathfrak{B}$ be the substructure generated by $\vec{b}$. (If no such an $\mathfrak{A}$ exists, then $\varphi(\vec{x})$ is equivalent to the quantifier-free formula $\bot$.) If $\Gamma' = \Gamma \cup \Delta^0_{\mathfrak{B}} \cup \{\neg\varphi(\vec{b})\}$ has a model $\mathfrak{A}'$, then $\mathfrak{A}$, $\mathfrak{A}'$ and $\mathfrak{B}$ show that the conclusion in problem holds. If Γ' were not consistent, then by compactness and as $\Delta^0_{\mathfrak{B}}$ is closed for $\wedge$, there is a single $\vartheta \in \Delta^0_{\mathfrak{B}}$ such that $\Gamma, \vartheta \vDash \varphi(\vec{b})$. As $\mathfrak{B}$ is the substructure generated from $\vec{b}$, every element in $\Delta^0_{\mathfrak{B}}$ is of the form $\delta(\vec{b})$ where $\delta(\vec{x}) \in F(\tau)$ is quantifier-free. Consequently in this case $\Gamma \vDash \delta(\vec{b}) \to \varphi(\vec{b})$ for some quantifier-free $\delta(\vec{x})$ with $\delta(\vec{b}) \in \Delta^0_{\mathfrak{B}}$. We get the required contradiction if none of such $\delta(\vec{b})$ formulas are true in $\mathfrak{A}$. Thus we need $\mathfrak{A}$ to be a model of this set:

$$\Gamma \cup \{\varphi(\vec{b})\} \cup \{\neg\delta(\vec{b}) : \delta \text{ is quantifier-free and } \Gamma \vDash \delta(\vec{b}) \to \varphi(\vec{b})\}.$$

So we are done if this set is consistent. Suppose again it is not, then there are finitely many $\delta_i(\vec{b})$ such that

$$\Gamma \vDash \varphi(\vec{b}) \to \bigvee_i \delta_i(\vec{b}).$$

As $\Gamma \vDash \delta_i(\vec{b}) \to \varphi(\vec{b})$ for all i, this shows that $\Gamma \vDash \varphi(\vec{b}) \leftrightarrow \bigvee_i \delta_i(\vec{b})$, contradicting that φ is not equivalent to any quantifier-free formula.

516. (a) Any structure with a $\Subset$-maximal element.

(b) The standard model of arithmetic $\langle \omega, 0, 1, +, \cdot, \rangle$ with $\Subset$ as the usual ordering, see Problem 784. More generally, every structure where $\Subset$ is an order of regular order type.

(c) Any countable structure $\mathfrak{A}$ where for each $a \in A$ there are only finitely many $b \in A$ with $b \Subset a$, and given finitely many $b_1, \ldots b_n \in A$ there is an $a \in A$ such that $b_i \Subset a$ for all i, see Problem 343.

(d) Models of set theory.

517. Denote the element a of $\mathfrak{A}$ by the constant symbol c_a, and let c be a brand new constant symbol. Take $\Sigma = \Delta_{\mathfrak{A}} \cup \{c_a \Subset c : a \in A\}$ (here $\Delta_{\mathfrak{A}}$ is the diagram of $\mathfrak{A}$, see Definition 8.8). For $a \in A$ let $T_a(x) = \{x \Subset c_a \wedge x \neq c_b : b \Subset^{\mathfrak{A}} a\}$ be a type.

If $\mathfrak{A}$ has no maximal element, then Σ is consistent, and any $\mathfrak{B} \vDash \Sigma$ is a proper elementary extension of $\mathfrak{A}$ (see Problem 507(b)). If $\mathfrak{B}$ omits each T_a, then the reduct of $\mathfrak{B}$ to the original signature is a proper end-extension of $\mathfrak{A}$. To apply the omitting types theorem 7.19 we need to show that T_a is non-isolated.

Assume that $\varphi(x, c)$ isolates T_a, that is, $\Sigma \cup \{\exists x \varphi(x, c)\}$ is consistent, and for all $b \Subset a$ we have $\Sigma \vDash \varphi(x, c) \to x \Subset c_a$ and

$$\Sigma \vDash \varphi(x, c) \to x \neq c_b,$$

that is, $\Sigma \vDash \neg\varphi(c_b, c)$. By compactness, there is a finite $A_0 \subset A$ so that

$$\Delta_{\mathfrak{A}} \vDash \bigwedge \{c_a \Subset c : a \in A_0\} \to \neg\varphi(c_b, c)$$

Let m be the $\Subset$-maximal element in A_0. Then $\Delta_{\mathfrak{A}} \vDash c_m \Subset c \to \neg\varphi(c_b, c)$. As the constant c does not occur in $\Delta_{\mathfrak{A}}$, we have

$$\Delta_{\mathfrak{A}} \vDash c_m \Subset y \to \neg\varphi(c_b, y).$$

This holds in $\mathfrak{A}$ for all $b \Subset a$, thus we have

$$\mathfrak{A} \vDash (\forall x \Subset c_a)(\exists m)\, \forall y\, (m \Subset y \to \neg\varphi(x, y)).$$

By the collection principle there is an $n \in A$ such that

$$\mathfrak{A} \vDash (\forall x \Subset c_a)(\exists m \Subset c_n)\, \forall y\, (m \Subset y \to \neg\varphi(x, y)),$$

thus $\mathfrak{A} \vDash (\forall x \Subset c_a)\, \forall y\, (c_n \Subset y \to \neg\varphi(x, y))$. But then $\Sigma \vDash (\forall x \Subset c_a)\, \neg\varphi(x, c)$, that is, $\Sigma \vDash \varphi(x, c) \to \neg(x \Subset c_a)$. However we have $\Sigma \vDash \varphi(x, c) \to (x \Subset c_a)$, hence $\Sigma \cup \{\exists x\varphi(x, c)\}$ cannot be consistent, a contradiction.

> **Remark.** The same solution works when $\Subset$ is a partial order where every pair has a strict upper bound.

518. After nodes a_i, b_i have been chosen, II wins if there is an edge between a_i and a_j iff there is an edge between b_i and b_j. So II wins if he can maintain the following invariant: If r more rounds are to come, then the (shorter) distance between a_i and a_j is exactly the same as the distance between b_i and b_j, except when both distances are larger than 2^r. When the game ends ($r = 0$), and this property holds, player II wins.

It is clear that II can play this way in the first two rounds. After that II can always respond properly, as if the new point is closer to both a_i and a_j than 2^{r-1}, then a_i and a_j must be closer to each other than 2^r, thus the corresponding point in the other structure exists.

> **Remark.** I can win the game in about $\log_2 |G_1|$ rounds. In the first three rounds I can secure corresponding nodes a, a' and b, b' such that their distances are different. Then always halving the smaller interval she can go down to distance 1.

519. (a) By assumption, $\mathfrak{A} \vDash \forall x \exists y\, \varphi(x, y)$ and $\mathfrak{B} \vDash \exists x \forall y\, \neg\varphi(x, y)$. In her 1st move I picks a $b_1 \in B$ $\mathfrak{B} \vDash \forall y\, \neg\varphi(b_1, y)$. Then II picks $a_1 \in A$. Then I picks $a_2 \in A$ so that $\mathfrak{A} \vDash \varphi(a_1, a_2)$ (such an a_2 exists by assumption). After that no matter what $b_2 \in B$ II picks, $\mathfrak{B} \vDash \neg\varphi(b_1, b_2)$.

As φ is quantifier-free, there is an atomic formula ψ such that the truth values of $\psi(a_1, a_2)$ and $\psi(b_1, b_2)$ are the opposite. Let us assume (for simplicity) that $\psi \equiv R(t_1, \ldots, t_n)$ or $\psi \equiv t_1 = t_2$. I can secure her winning by making sure that all arguments and all function values in the terms t_i are mentioned in structure $\mathfrak{A}$. The created map $a_i \mapsto b_i$ is a partial isomorphism only if

II responds with the corresponding values in $\mathfrak{B}$. Doing so, II will miss the correct value for the atomic formula ψ.

(b) Let N be fixed. The similarity type τ contains an N-place relation symbol R and binary function symbols f_i for $1 \le i \le N$. Let $\mathfrak{A} = \langle \omega, R^{\mathfrak{A}}, f_i^{\mathfrak{A}} \rangle$, where $f_i^{\mathfrak{A}}(x,y) = i$, and $R^{\mathfrak{A}}$ holds for an N-tuple if all arguments are different. Put $\mathfrak{B} = \langle \omega, R^{\mathfrak{B}}, f_i^{\mathfrak{B}} \rangle$, where $f_i^{\mathfrak{B}} = f_i^{\mathfrak{A}}$ for $i \le N$, but $R^{\mathfrak{B}}$ never holds.

Observe that any partial map $j : \mathfrak{A} \to \mathfrak{B}$ is a partial isomorphism as far as it keeps the integers $1 \le i \le N$, and its domain does not contain N different numbers. Thus II wins the N-1-round game.

Finally, let φ be the formula $R(f_1(x,y), \ldots, f_N(x,y))$. Clearly $\mathfrak{A} \models \forall x \exists y \varphi(x,y)$ and $\mathfrak{B} \models \neg \forall x \exists y \varphi(x,y)$, as required.

520. Suppose that $\mathfrak{A}$ and $\mathfrak{B}$ are not elementary equivalent witnessed by the closed formula φ. Write φ in prenex normal form (Problem 408) as

$$\varphi_1 \equiv Q_1 x_1 \ldots Q_n x_n \psi(x_1, \ldots, x_n)$$

where each Q_i is either existential or universal quantifier, and $\psi(\vec{x})$ is quantifier-free. Let $\varphi_i(x_1, \ldots, x_i)$ be the formula after the quantifier Q_i. I can play as follows. If $Q_1 = \exists$ then he picks $a_1 \in A$ such that $\mathfrak{A} \models \varphi_1[a_1]$ (and then $\mathfrak{B} \models \neg \varphi_1[b_1]$ whatever II's response is), and if $Q_1 = \forall$, then he picks b_1 such that $\mathfrak{B} \models \neg \varphi_1[b_1]$ (and then $\mathfrak{A} \models \varphi_1[a_1]$ for every response of II). In general, in the i-th round if $Q_j = \exists$, he picks $a_j \in A$ such that $\mathfrak{A} \models \varphi_j[a_1, \ldots, a_j]$, otherwise he picks $b_j \in B$ which satisfies $\mathfrak{B} \models \neg \varphi_j(b_1, \ldots, b_j)$.

After n rounds the picked elements satisfy

$$\mathfrak{A} \models \psi[a_1, \ldots, a_n] \text{ and } \mathfrak{B} \models \neg \psi[b_1, \ldots, b_n].$$

As ψ is quantifier-free, an argument similar to the one in Solution 519(a) shows that I wins the N-round game, where N depends only on the original formula φ.

521. (a) A discrete linear order without endpoints has the form $\omega + K \times \mathbb{Z}$ for some arbitrary (possibly empty) linear order K. Two elements a and b are *infinitely far apart* if they belong to different $\mathbb{Z}$-chains (that is, the distance between them is infinite).

The main problem that player II faces is that player I can pick elements infinitely far apart while forcing II to choose elements in the other structure which are finitely far apart only. The crux is that the number of rounds is fixed in advance and II can choose elements which are only "sufficiently far apart" securing her the win.

Similarly to Solution 518, Player II can maintain the following invariant. Add the initial points of the structures as a_0, b_0, they must correspond to each other. If r more rounds are to come, then a_i and b_j are ordered the same way, and the distance between a_i and a_j is exactly the same as the distance between b_i and b_j, *except* when both distances are larger than 2^r (possibly infinite).

To show that for each possible pick of player I, player II can choose a corresponding element in the other structure can be done similarly as in Solution 518.

(b) If N is not fixed in advance, then I can win the game, for example, for the structures $\mathfrak{A} = \langle \omega, < \rangle$ and $\mathfrak{B} = \langle \omega + \mathbb{Z}, < \rangle$. Player I will pick a decreasing sequence of elements from $\mathbb{Z}$. II is forced to pick a decreasing sequence from ω, and sooner or later she runs out of the available elements.

522. Color a point blue if it has an immediate successor, and green otherwise. II must keep both the order and the color of the chosen points. She can always do it (as between any two elements of the same color there are infinitely many other elements of both colors) except for points at the beginning or at the end of the orderings. Thus II wins the game if and only if the orders have minimal (maximal) elements at the same time.

523. If the element chosen by I is in some Boolean combination, II can respond by a new element from the same combination in the other structure up to the N-th round. Then she wins.

524. II wins the EF game for every N. The same strategy works as in Solution 521(a).

525. Proceed along the proof of Problem 520. If $\forall x_1 \exists x_2 \ldots \delta(\bar{x})$ is true in $\mathfrak{A}$ and false in $\mathfrak{B}$, then I can play so that after n moves $\mathfrak{A} \models \delta[a_1, \ldots, a_n]$, while $\mathfrak{B} \models \neg\delta[b_1, \ldots, b_n]$. As no constant or function symbols are in τ, it means that I won the game.

526. Suppose N is fixed. First, assume that in $\mathfrak{A}$ the path starting from v_0 has length at least 2^N. Then player II can play as follows. if I picks an element in $\mathfrak{A}$ not from the path starting at v_0, or picks an element in $\mathfrak{B}$ not from the infinite path, then II picks the same element in the other structure. On the paths starting from v_0 (both in $\mathfrak{A}$ and $\mathfrak{B}$), II plays by the strategy described in Solution 521(a) making sure that the pairwise distances of the chosen points in the corresponding paths starting from v_0 in $\mathfrak{A}$ and in $\mathfrak{B}$ are either equal or both are bigger than 2^r where r is the number of remaining rounds. Playing so II clearly wins the N-round game.

Now, if the path starting from v_0 is shorter than 2^N, then a modification of the previous strategy works. There is a point v_i in $\mathfrak{A}$ such that the path starting from v_i has length 2^N and otherwise the 2^N neighbourhood of v_i is isomorphic to that of $v_0^{\mathfrak{A}}$. If I plays any point not in this neighbourhood, then II replies with a point that has identical neighbourhood of size 2^r where r is the number of remaining rounds. See a more general statement in Problem 565.

527. The similarity type contains countably many constant symbols c_i for $i < \omega$. Let $\mathfrak{A} = \langle \omega, c_i^{\mathfrak{A}} \rangle$ be such that $c_i^{\mathfrak{A}} = i$, that is, each element of ω is denoted by a constant. Let $\mathfrak{B}$ be an arbitrary proper elementary extension of $\mathfrak{A}$, then there are elements in B which are not denoted by constants.

As partial isomorphisms should preserve constant symbols, player I can always win by choosing an element from $\mathfrak{B}$ which is not denoted by any of the constants.

528. If $f : \mathfrak{A} \to \mathfrak{B}$ is an isomorphism, then $I = \{f\}$ is a back and forth system. The converse statement is proved via the method in Solution 365. Suppose I is a back and forth system and let a_n, b_n be a one-to-one enumeration of A and B, respectively. Define, by induction on n, partial isomorphisms $f_n \in I$ so that f_0 is arbitrary and

$$
\begin{aligned}
f_{2n} &= \text{some } g \in I \text{ for which } f_{2n-1} \subseteq g \text{ and } a_n \in \mathrm{dom}(g), \\
f_{2n+1} &= \text{some } g \in I \text{ for which } f_{2n} \subseteq g \text{ and } b_n \in \mathrm{ran}(g).
\end{aligned}
$$

Then $\bigcup_{n \in \omega} f_n$ is an isomorphism between $\mathfrak{A}$ and $\mathfrak{B}$.

529. ($\Leftarrow$) Suppose $\mathfrak{A} \rightleftarrows \mathfrak{B}$. The $\equiv_{\infty,\omega}$-equivalence follows from the following fact: for every for $\varphi \in L_{\infty,\omega}$, $f \in I$ and $\vec{a} \in \mathrm{dom}(f)$ we have

$$
\mathfrak{A} \vDash \varphi(\vec{a}) \quad \text{if and only if} \quad \mathfrak{B} \vDash \varphi(f(\vec{a})).
$$

This can be proved by induction on the complexity of φ. For quantifier-free formulas this is straightforward from the definition of a partial isomorphism. For $\exists x \psi$ we have $\mathfrak{A} \vDash \exists x \psi(x, \vec{a})$ if and only if there is $a \in A$ so that $\mathfrak{A} \vDash \psi(a, \vec{a})$. Then there is $g \in I$ extending f such that $a \in \mathrm{dom}(g)$. By the induction hypothesis g preserves ψ, thus $\mathfrak{B} \vDash \psi(g(a), g(\vec{a}))$ and this ensures $\mathfrak{B} \vDash \exists x \psi(x, f(\vec{a}))$.

($\Rightarrow$) Suppose $\mathfrak{A} \equiv_{\infty,\omega} \mathfrak{B}$. We claim that the set

$$
I = \{f : A \to B : f \text{ is finite and preserves all } L_{\infty,\omega} - \text{formulas}\}
$$

is a back and forth system. Clearly $\emptyset \in I$ thus we have to verify the 'back' and 'forth' properties. Pick any $f \in I$ and $a \in A$. We shall find $g \in I$ with $f \subseteq g$, $a \in \mathrm{dom}(g)$. As $\mathrm{dom}(f)$ is finite, there is a tuple enumerating it: $\vec{a} = \mathrm{dom}(f)$. We have to find $b \in B$ such that $\langle a, \vec{a} \rangle$ satisfies the same formulas as $\langle b, f(\vec{a}) \rangle$ does. If there were no such b then for all $b \in B$ there exists an $L_{\infty,\omega}$-formula φ_b so that $\mathfrak{A} \vDash \varphi_b(a, \vec{a})$ while $\mathfrak{B} \nvDash \varphi_b(b, f(\vec{a}))$. But then

$$
\mathfrak{A} \vDash \exists x \bigwedge_{b \in B} \varphi_b(x, \vec{a}), \quad \text{while} \quad \mathfrak{B} \nvDash \exists x \bigwedge_{b \in B} \varphi_b(x, f(\vec{a})),
$$

which contradicts $\mathfrak{A} \equiv_{\infty,\omega} \mathfrak{B}$.

530. By Karp's theorem 8.15 we need to find $\mathfrak{A}$ and $\mathfrak{B}$ such that $\mathfrak{A} \equiv \mathfrak{B}$ but $\mathfrak{A} \not\equiv_{\infty,\omega} \mathfrak{B}$. Suppose the similarity type contains countably many constant symbols, and let $\mathfrak{A}$ be a countable structure where each element of $a \in A$ is denoted by a constant symbol $c_a^{\mathfrak{A}} = a$. Let $\mathfrak{B}$ its proper elementary extension of uncountable cardinality. Then $\mathfrak{A} \equiv \mathfrak{B}$. On the other hand with

$$
\varphi \equiv \forall x \bigvee_{a \in A} x = c_a
$$

we have $\mathfrak{A} \vDash \varphi$ but $\mathfrak{B} \nvDash \varphi$, showing $\mathfrak{A} \not\equiv_{\infty,\omega} \mathfrak{B}$.

531. (a), (b) Finite partial isomorphisms (that respect endpoints) form a back and forth system, see Problem 365.

(c) Let I be the set of all isomorphisms from finite subalgebras of $\mathfrak{A}$ into $\mathfrak{B}$.

532. $\mathfrak{A}$ is the Lebesgue measure algebra (measurable subsets of $\mathbb{R}$ modulo zero sets), and $\mathfrak{B}$ is the modulo finite quotient algebra $\wp(\omega)/[\omega]^{<\omega}$. Both are atomless of cardinality continuum. In $\mathfrak{A}$ every collection of pairwise disjoint sets (antichain) is countable, while in $\mathfrak{B}$ an almost disjoint family gives such a collection of size continuum. Thus $\mathfrak{A}$ and $\mathfrak{B}$ are not isomorphic.

533. The closed formula $\exists x P(x)$ is not equivalent (modulo Γ) to any quantifier-free formula as the only quantifier-free formulas are $\top$ and $\bot$. After adding the constant symbol c the theory $\Gamma \cup \{\exists x P(x) \leftrightarrow P(c)\}$ works.

534. Yes. In $\mathfrak{A}$ the formula $\exists x\,(ax^2 + bx + c = 0)$ is equivalent to

$$\big((a \neq 0 \wedge b^2 - 4ac \geq 0) \vee (a = 0 \wedge (b \neq 0 \vee c = 0))\big)$$

535. By induction on the complexity of the formulas. For atomic formulas and for the connectives $\wedge$ and $\neg$ the statement is trivial. For the quantifier suppose we have a formula $\exists x \varphi$. By induction, there is a quantifier-free δ' such that $\Gamma \vDash \varphi \leftrightarrow \delta'$, thus $\Gamma \vDash \exists x \varphi \leftrightarrow \exists x \delta'$. The rest can be done using disjunctive normal form, that $\exists$ distributes over $\vee$, and the assumption of the problem.

536. When using strict inequality $<$, atomic formulas can have only two forms: $u = v$ and $u < v$. Negated literals $\neg(u < v)$ and $\neg(u = v)$ can be replaced with $(u = v) \vee (u < v)$ and $(u < v) \vee (v < u)$, respectively, thus it suffices to check the condition in Problem 535 for the case when all literals are positive.

So let take a formula of the form $\exists x(\ell_1 \wedge \cdots \wedge \ell_n)$ where all ℓ_i is atomic. We can assume that each ℓ_i contains x since otherwise ℓ_i could be move out of the scope of $\exists x$. If ℓ_i is of the form $x = x$, then it can be deleted; if it is $x = u$ then replace x by u everywhere and delete it. The remaining literals have the form $x < x$, $u < x$ and $x < v$. If $x < x$ occurs, then φ is equivalent to $\bot$ (which is quantifier-free), otherwise the literals can be rearranged as $\bigwedge_i(u_i < x) \wedge \bigwedge_j(x < v_j)$. If bounds from any side are missing, then this is equivalent to $\top$, otherwise such an x exists iff $\bigwedge_{i,j}(u_i < v_j)$, a quantifier-free formula.

537. (a) Immediate from Problem 535.

(b) No. The dense linear order without endpoints has quantifier elimination. It has no terms at all, thus there is no Skolem function for the formula $\exists y(x < y)$.

538. (a) Using the Tarski–Vaught test 8.3. Suppose $\vec{b} \in B$ and $\mathfrak{A} \vDash \exists y \varphi(y, \vec{b})$. As Γ has built-in Skolem functions, there is a τ-term $t(\vec{x})$ such that $\mathfrak{A} \vDash \varphi[t^{\mathfrak{B}}(\vec{b}), \vec{b}]$. But $t^{\mathfrak{B}}(\vec{b}) \in B$ as $\mathfrak{B}$ is a substructure.

(b) By Problem 325, for a quantifier-free formula $\psi(\vec{x})$ and $\vec{b} \in B$ we have $\mathfrak{B} \vDash \psi(\vec{b})$ iff $\mathfrak{A} \vDash \psi(\vec{b})$. Let $\varphi(\vec{x}) \in F(\tau)$ be any formula. As Γ has quantifier elimination, $\Gamma \vDash \varphi(\vec{x}) \leftrightarrow \psi(\vec{x})$ for some quantifier-free $\psi(\vec{x})$. As both $\mathfrak{B}$ and $\mathfrak{A}$ are models of Γ, for any $\vec{b} \in \mathfrak{B}$,

$$\mathfrak{B} \vDash \varphi(\vec{b}) \quad \Leftrightarrow \quad \mathfrak{B} \vDash \psi(\vec{b}) \quad \Leftrightarrow \quad \mathfrak{A} \vDash \psi(\vec{b}) \quad \Leftrightarrow \quad \mathfrak{A} \vDash \varphi(\vec{b}),$$

so $\mathfrak{B}$ is an elementary submodel.

539. For each $\varphi(\vec{x}) \in F(\tau)$ let $R_\varphi(\vec{x})$ be a new relation symbol. Let $\tau' = \tau \cup \{R_\varphi : \varphi \in F(\tau)\}$ and define

$$\Gamma' = \Gamma \cup \{\forall \vec{x}(\varphi(\vec{x}) \leftrightarrow R_\varphi(\vec{x})) : \varphi \in F(\tau)\}.$$

The following claims can be checked easily.

1. Each model of Γ can uniquely be expanded to a model of Γ' on the same ground set.
2. For each $\varphi \in F(\tau)$ we have $\Gamma \vDash \varphi$ if and only if $\Gamma' \vDash \varphi$.
3. For each $\varphi \in F(\tau')$ there is $\psi \in F(\tau)$ such that $\Gamma' \vDash \varphi \leftrightarrow \psi$.
4. Γ' admits quantifier elimination.

> **Remark.** If Γ' satisfies the first three items above, then it is a *definitional expansion* of Γ.

540. No. Let Γ be a non-complete theory and take the conservative extension described in 539. The resulting theory is non-complete (being a conservative extension) but has quantifier elimination.

541. (a) The main observation is that whenever two tuples of elements $\vec{a}$ and $\vec{b}$ satisfy the same unnested atomic formulas (see Definition 8.11), then they satisfy the same formulas. This is because the former implies that the mapping $\vec{a} \mapsto \vec{b}$ is a partial isomorphism, which, by assumption, extends to an automorphism, and automorphisms preserve all formulas.

Take a formula $\varphi(\vec{x})$ with free variables $\vec{x}$. For $\vec{a} \in A$ having the same length as $\vec{x}$ write

$$\mathrm{tp}_{at}^{\mathfrak{A}}(\vec{a}) = \{\delta(\vec{x}) : \mathfrak{A} \vDash \delta[\vec{a}], \text{ and } \delta \text{ is an unnested atomic formula}\}.$$

As the similarity type is finite, $\vartheta_{\vec{a}}(\vec{x}) = \bigwedge\{\delta(\vec{x}) : \delta \in \mathrm{tp}_{at}^{\mathfrak{A}}(\vec{a})\}$ is a formula, and there is a *finite* list of such formulas as $\vec{a}$ runs over the vectors $\vec{a}$ with $\mathfrak{A} \vDash \varphi(\vec{a})$. Then $\mathfrak{A} \vDash \varphi(\vec{x}) \leftrightarrow \bigvee_i \vartheta_i(\vec{x})$, and this latter formula is quantifier-free.

(b) If $\mathfrak{A} \vDash \Gamma$ and Γ is complete, then $\mathfrak{A} \vDash \varphi \leftrightarrow \psi$ if and only if $\Gamma \vDash \varphi \leftrightarrow \psi$ which proves that Γ has quantifier elimination.

542. (i)$\Rightarrow$(ii) If δ is quantifier-free, then by Problem 472, $\mathfrak{A} \vDash \delta(\vec{b})$ iff $\mathfrak{B} \vDash \delta(\vec{b})$ iff $\mathfrak{A}' \vDash \delta(\vec{b})$, thus it cannot happen that $\varphi(\vec{b}) \leftrightarrow \delta(\vec{b})$ is true both in $\mathfrak{A}$ and $\mathfrak{A}'$.

(ii)$\Rightarrow$(i) See Problem 515.

543. By assumption, for all $\vec{b} \in B$ we have $\mathfrak{A} \vDash \exists x \delta(x, \vec{b})$ iff $\mathfrak{A}' \vDash \exists x \delta(x, \vec{b})$. By Problem 542, $\exists x \delta(x, \vec{y})$ is equivalent to a quantifier-free formula. According to Problem 535 this property implies that Γ has quantifier elimination.

544. Easily follows from Problem 472.

545. It is clear that the substructure on $B = A - \{a\}$ is also a dense linear order without endpoints. As the theory admits quantifier elimination (Problem 536), $\mathfrak{A}$ is model complete, thus every submodel is an elementary substructure by Problem 544.

546. Any two countable models of Γ are isomorphic, thus Γ is complete (Theorem 8.7). Also, every finite partial isomorphism of a countable model extends to an automorphism, thus Γ admits quantifier elimination (Problem 541). Therefore Γ is model complete (Problem 544).

547. Let $\mathfrak{A} = \langle \omega, S \rangle$ and $j : \mathfrak{A} \to \mathfrak{A}$ be the embedding defined by $j(n) = n + 1$. Put $\varphi(x) = \forall y (x \neq Sy)$ and notice $\mathfrak{A} \vDash \varphi[0]$ while $\mathfrak{A} \nvDash \varphi[j(0)]$, therefore j is not an elementary embedding.

548. It is enough to prove that $\Gamma \cup \Delta_{\mathfrak{B}}^0$ is consistent as, by Problem 507(a) $\mathfrak{B}$ embeds into any $\mathfrak{A} \vDash \Gamma \cup \Delta_{\mathfrak{B}}^0$ as a substructure.

Suppose that $\Gamma \cup \Delta_{\mathfrak{B}}^0$ is not consistent. By compactness, there is a finite $\Gamma' \subset \Gamma \cup \Delta_{\mathfrak{B}}^0$ which is inconsistent. The part of this finite subset coming from $\Delta_{\mathfrak{B}}^0$ can be considered as a single formula $\delta(\vec{c})$, where $\vec{c}$ are constants denoting elements of $\mathfrak{B}$, and $\delta(\vec{x})$ is quantifier-free. If $\Gamma \cup \{\delta(\vec{c})\}$ is inconsistent, then $\Gamma \vDash \forall \vec{x} \neg \delta(\vec{x})$, and, as this is a universal formula, $\mathfrak{B} \vDash \forall \vec{x} \neg \delta(\vec{x})$, contradicting that $\delta(\vec{c})$ is in the diagram $\Delta_{\mathfrak{B}}^0$.

549. That (ii) and (iii) are equivalent is easy: if $\neg \varphi$ is equivalent to an existential formula, then φ is equivalent to a universal formula, and vice versa.

(iii)$\Rightarrow$(i) If $\mathfrak{B}$ is a substructure of $\mathfrak{A}$, then every universal formula true in $\mathfrak{A}$ is true in $\mathfrak{B}$ (e.g. combine 311 and 325). In other words, substructures are "elementary substructures with respect to universal formulas". If every formula is equivalent to a universal formula, then every substructure of every model of Γ is an elementary substructure, thus Γ is model complete.

(i)$\Rightarrow$(iii) If $\mathfrak{B}$ is a model for every universal formula φ such that $\Gamma \vDash \varphi$, then by Problem 548, $\mathfrak{B}$ embeds into a model $\mathfrak{A}$ of Γ. But then $\mathfrak{B}$ is an elementary substructure of $\mathfrak{B}$ (by model completeness of Γ). Therefore models of Γ are the same structures as models of the universal formulas in Γ, thus every formula is equivalent to a universal formula modulo Γ.

550. Every closed formula $\varphi \in F(\tau')$ is equivalent modulo Γ' to a quantifier-free closed formula $\delta \in F(\tau')$. As for such formulas we have either $\Gamma' \vDash \delta$ or $\Gamma' \vDash \neg \delta$, the theory Γ' is complete. But Γ' is a conservative extension of Γ, thus for formulas $\vartheta \in F(\tau)$ we have $\Gamma \vDash \vartheta$ iff $\Gamma' \vDash \vartheta$. Therefore Γ is complete as well.

551. Let $\mathfrak{A} = \langle \omega, S \rangle$ and $j : \mathfrak{A} \to \mathfrak{A}$ be the embedding defined by $j(n) = n + 1$. Put $\varphi(x) = \forall y (x \neq Sy)$ and notice $\mathfrak{A} \vDash \varphi[0]$ while $\mathfrak{A} \nvDash \varphi[j(0)]$, therefore j is not an elementary embedding, hence by Problem 544 $\mathfrak{A}$ does not have quantifier elimination.

To prove that the theory of $\mathfrak{B} = \langle \mathbb{Z}, S \rangle$ has quantifier elimination apply Problem 543. If $\mathfrak{B}$ is a common substructure of $\mathfrak{A}$ and $\mathfrak{A}'$, then with any $b \in B$ it contains its successor (but not necessarily the unique predecessor). If there is an element $a \in A$ satisfying a quantifier-free formula $\delta[a, \vec{b}]$, then a should be equal to or differ from different successors / predecessors of elements in $\vec{b}$. Clearly, such an element can be found in $\mathfrak{A}'$.

552. (a) Atomic formulas are of the form $S^n(x) = S^k(y)$. By Problem 535 it is enough to check that every formula of the form $\exists x (\ell_1 \wedge \cdots \wedge \ell_n)$, where each ℓ_i is a literal (an atomic formula or its negation) is equivalent to a quantifier-free formula modulo Γ. But this is straightforward using the trick that $S^n(x) = S^k(y)$ and $x = S^{k-n}(y)$ are equivalent when $n \le k$, $x = S^k(x)$ is $\perp$ for $k > 0$, and $\exists x (S^k(x) = y)$ iff $\bigwedge_{j<k} y \ne S^j(0)$. Finally, the truth of quantifier-free atomic formulas are decided by the theory.

(b) Γ is complete and $\langle \omega, S, 0 \rangle$ is one of its models.

(c) The theory admits quantifier elimination, thus every definable set is definable by a quantifier-free formula. The claim is true for atomic formulas, as they define a single-element set. The collection of finite or co-finite sets is closed under complement, union, and intersection, thus quantifier-free formulas can define such sets only.

553. (a) The embedding of $\langle \mathbb{Z}, < \rangle$ to $\langle \mathbb{Z}, < \rangle$ defined by $f(n) = n$ for $n \le 0$ and $f(n) = n + 1$ otherwise is not elementary, therefore Γ is not model complete, thus it does not have quantifier elimination (see Problem 544).

(b) Recall that there are non-isomorphic countable discrete linear orderings without endpoints X_0 and X_1 (Problem 368). Take any discrete linear ordering without endpoints X of cardinality κ and consider $X_0 + X$ and $X_1 + X$.

(c) We prove that Γ has a finite conservative extension Γ' which admits quantifier elimination, and all variable-free formulas are decided by Γ'. By Problem 550 it implies that Γ is complete. Add a unary function symbol S to the language and complement Γ with the following axioms expressing S behaves like the successor function:

$$\forall x \forall y (Sx \le y \leftrightarrow (x \le y \wedge x \ne y)),$$
$$\forall x \exists y (x = Sy).$$

Similar to Problem 536 it is enough to consider formulas of the form

$$\exists x \left(\bigwedge_{i<l} t_i \le S^{p_i}(x) \wedge \bigwedge_{j<m} S^{q_j}(x) \le u_j \wedge \bigwedge_{k<n} S^{r_k}(x) = v_k \right),$$

where t_i, u_j and v_k are terms with no occurrence of x and p_i, q_j and r_k are natural numbers. (This is so as $S^k(x) = S^\ell(x)$ is either $\top$ or $\perp$.) Now $S^i(x) \le t$ if and only if $S^{i+j}(x) \le S^j(t)$, therefore one can replace all terms $S^i(x)$ with $S^N(x)$ and also replace $S^N(x)$ with a new variable. The resulting formula will be of the form

$$\varphi(x) \equiv \exists x \left(\bigwedge_{i<k} t_i \le x \wedge \bigwedge_{j<l} x \le u_j \right),$$

thus it suffices to show how to eliminate the quantifier in φ. If $k = 0$ or $l = 0$ then $\varphi(x) \leftrightarrow \top$; otherwise it is equivalent to $\bigwedge_{i,j} t_i \leq u_j$. As the only variable-free formulas are $\top$ and $\bot$, we are done.

554. Consider the model $\langle \omega, \leq \rangle$ of Γ. The function $j : n \mapsto n + 1$ is an embedding (order-preserving map), but not an elementary embedding (does not preserve the minimal element). Thus Γ does not have quantifier elimination by Problem 544.

For the conservative extension define 0 as the minimal element, and $S(x)$ as the successor of x (the smallest among those which are strictly bigger than x). The method indicated in solution 553(c) works as in this case as all facts used there are actually consequences of Γ.

(b) The truth of the quantifier-free formulas $S^k(0) \bar{\bar{\neq}} S^\ell(0)$ (and their Boolean combinations) in the conservative extension are decided by Γ, thus Γ is complete.

555. Countable dense linear orderings are isomorphic to the rationals (Problems 531, 365). Thus the universe of countable models of Γ can be identified with $\mathbb{Q}$. A standard back and forth argument shows that up to isomorphism Γ has exactly three countable models according to whether $\lim c_n$ is infinite, rational, or irrational (see Problem 366). Observe that reducts of countable models of Γ to a finite sublanguage are isomorphic (back and forth), consequently any two models of Γ are elementarily equivalent. This means that Γ is complete.

556. (a), (b) Countable models of Γ are isomorphic, therefore Γ is complete (see Problem 497). Take a countable model $\mathfrak{A} \vDash \Gamma$. It is straightforward to verify that a finite partial isomorphism of $\mathfrak{A}$ can be extended to an automorphism of $\mathfrak{A}$, thus Problem 541 applies.

(c) Countable models of Γ are isomorphic, nevertheless Problem 541 cannot be used directly because the partial isomorphism which maps an element from a 2-sized class into a 3-sized class does not extend to an automorphism. Indeed, Γ is not quantifier eliminable. For, consider two countable models $\mathfrak{A}, \mathfrak{B} \vDash \Gamma$ and the embedding $f : \mathfrak{A} \to \mathfrak{B}$ that maps equivalent elements of A into equivalence classes of $\mathfrak{B}$ having 3 members. Then f is not an elementary embedding, hence Γ is not model complete and thus Γ cannot have quantifier elimination (see Problem 544). However, there is a finite conservative extension $\Gamma' \supseteq \Gamma$ which has quantifier elimination: Extend the language with two unary relation symbols P and Q and add new axioms that stipulates that P consists of the 2-sized and Q consists of the 3-sized classes of E. Then countable models of Γ' are isomorphic and thus Γ' is complete, and it is straightforward to check that finite partial isomorphisms of countable models of Γ' extend to automorphisms. By Problem 541, Γ' has quantifier elimination.

(d) Γ has countably many countable models: each model contains one n-element equivalence class for all $n \in \omega$ and zero, one, $\ldots$, countably many infinite equivalence classes. Γ is not quantifier eliminable because it is not model complete. For, let $\mathfrak{A}$ be the model in which there is no infinite class and take the self-embedding which moves an n-sized class into the $n+1$-sized one. This embedding cannot be elementary. To construct a finite conservative extension $\Gamma' \supset \Gamma$ which has quantifier elimination, add a unary function symbol f to the language and add new axioms that express that f fixes equivalence classes setwise so that f visits all elements of the equivalence class on n elements in a circle (thus a substructure contains complete finite equivalence classes). Then Problem 543 applies.

(e) There are continuum many countable models of Γ: For each infinite $X \subseteq \omega$ there is a model $\mathfrak{A}_X$ that contains one n-element equivalence class for every $n \in X$. For $X \neq Y$ we have $\mathfrak{A}_X \not\equiv \mathfrak{A}_Y$. As for quantifier elimination use the argument in (d): add a unary function symbol f to the language and add new axioms that express that f makes a full circle in each n-element equivalence class.

Notice that this theory is not complete, and has a countable model with no finite equivalence class at all.

557. (a) Such a group is isomorphic to the additive group of an infinite dimensional vector space over $GF(p)$. Two such vector spaces of the same dimension are isomorphic, thus the theory is complete (Problem 497). As in this case a substructure is a linear subspace, Problem 543 applies.

(b) We employ Problem 543. Let $\mathfrak{A}$ and $\mathfrak{A}'$ be torsion-free divisible Abelian groups and $\mathfrak{B}$ be a common substructure (in this case, subgroup) of $\mathfrak{A}$ and $\mathfrak{A}'$. Pick $\vec{b} \in B$, $a \in A$ and suppose $\mathfrak{A} \vDash \varphi[a, \vec{b}]$ for some quantifier-free φ. Recall from algebra, that torsion-free Abelian groups have a divisible hull, in particular, there is a divisible torsion-free group $\mathfrak{B}'$ which extends $\mathfrak{B}$, contains $a \in A$, and can be embedded into $\mathfrak{A}$ and $\mathfrak{A}'$. Therefore $\mathfrak{B}' \vDash \exists x \varphi[x, \vec{b}]$ and thus there is an $a' \in A'$ with $\mathfrak{A}' \vDash \varphi[a', \vec{b}]$.

(c) We apply Problem 543 again. Let K, L be algebraically closed fields, and $\mathfrak{B}$ be a common substructure (a subfield). Suppose φ is quantifier-free and $\vec{b} \in B$, $a \in K$ are such that $K \vDash \varphi[a, \vec{b}]$. We shall show that there is $c \in L$ with $L \vDash \varphi[c, \vec{b}]$. The algebraic closure of B in K and in L are isomorphic, thus we may assume that B is also algebraically closed. It is enough to show that $\vec{b} \in B$, $a \in K$ and $K \vDash \varphi[a, \vec{b}]$ imply that there is a $c \in B$ such that $B \vDash \varphi[c, \vec{b}]$. By Problem 535, φ can be assumed to be a conjunction of literals, which are equivalent to polynomial equalities and non-equalities in the language of fields. Thus for some polynomials $p_1, \ldots, p_n$, and $q_1, \ldots, q_m \in B[x]$ our formula $\varphi(x, \vec{b})$ is equivalent to

$$\bigwedge_{i=1}^{n} p_i(x) = 0 \wedge \bigwedge_{j=1}^{m} q_j(x) \neq 0.$$

If any of the polynomials p_i are non-zero, then a is algebraic over B and as B is algebraically closed we have therefore $a \in B$. Hence we can assume that φ is equivalent to

$$\bigwedge_{i=1}^{m} q_i(x) \neq 0.$$

Each $q_i(x) = 0$ has finitely many solutions only, thus there can be only finitely many elements of B not satisfying φ. As algebraically closed fields are infinite, there is a $c \in B$ such that $B \vDash \varphi[c, \vec{b}]$, are required.

(d) Similar to (c) as the corresponding theorems are also true for real closed fields

(e) Problem 531(c) tells us that the theory of atomless Boolean algebras is complete, and finite partial isomorphisms between countable models extend to an automorphism. Then Problem 541 applies.

558. Yes. By Problem 557(c) algebraically closed fields admit quantifier elimination. As $\mathfrak{F}'$ is an algebraically closed substructure, Problem 538(b) claims that it is an elementary subfield.

559. Suppose $|t| \leq |\mathfrak{A}|$ is infinite and write $\Sigma = \Delta_{\mathfrak{A}} \cup \{c \neq c_a : a \in A\}$, where c is a new constant symbol and the constant c_a denotes the element $a \in A$. As each finite subset of Σ is consistent there is, by compactness, a model $\mathfrak{C} \vDash \Sigma$ which is an elementary extension of $\mathfrak{A}$ (Problem 507). Take $X = \{c\} \cup \{c_a : a \in A\}$ and note $|t| \leq |X| = |\mathfrak{A}|$. By the downward Löweinheim–Skolem theorem 8.5 $\mathfrak{C}$ has an elementary substructure $\mathfrak{B}$ containing X such that $|X| = |\mathfrak{B}|$.

560. The idea is that the sum of two odd numbers is even in $\mathbb{Z}$ while not in $\mathbb{Z} \oplus \mathbb{Z}$. Let $\mathrm{Even}(x)$ denote the formula $\exists w(w + w = x)$. Then

$$\mathbb{Z} \vDash \forall x \forall y \big((\neg \mathrm{Even}(x) \wedge \neg \mathrm{Even}(y)) \to \mathrm{Even}(x + y)\big),$$

while

$$\mathbb{Z} \oplus \mathbb{Z} \vDash \neg \mathrm{Even}[\langle 0, 1 \rangle] \wedge \neg \mathrm{Even}[\langle 1, 0 \rangle] \wedge \neg \mathrm{Even}[\langle 1, 1 \rangle].$$

561. Write $\Sigma = \mathrm{Th}(\mathbb{R}) \cup \{c \leq 1/n : n \in \omega\}$, where c is a new constant symbol. Note that each finite subset of Σ is consistent. By compactness there is a model $\mathfrak{A} \vDash \Sigma$. The field reduct $\mathfrak{B}$ of $\mathfrak{A}$ is elementarily equivalent to $\mathbb{R}$ as $\mathfrak{B} \vDash \mathrm{Th}(\mathbb{R})$, however there is an element $c \in B$ such that for all $n \in \omega$ we have $\mathfrak{B} \vDash c \leq 1/n$. This means that $\mathfrak{B}$ fails to satisfy the Archimedean property of fields.

562. Let $a \in A$ be an element not in $\mathbb{R}$. As either a or $-a$ is positive, we may assume that $a \geq 0$. If $r < a$ for all $r \in \mathbb{R}$, then $\mathfrak{A}$ is not Archimedean. Otherwise the set $\{r \in \mathbb{R} : r < a\}$ is bounded, and let $s \in \mathbb{R}$ be its lowest upper bound. If $a < s$, then $1/(s - a) \in A$ is larger than any number in $\mathbb{R}$; if $s < a$ then $1/(a - s)$ is larger than any real number in $\mathbb{R}$.

563. Use additive notation + for the group operation. Let G be a free group. Observe that for each $n \in \omega$ there is a non-zero group element $g \in G$ which is

2^n-divisible, meaning $G \models \exists h (g = 2^n \cdot h)$ (take a generator and add it to itself 2^n times). Let c be a new constant symbol and write $\Sigma = \mathrm{Th}(G) \cup \{\exists h (c = 2^n \cdot h) : n \in \omega\}$. Since each finite subset of Σ is consistent, by compactness there is a model $\mathfrak{A} \models \Sigma$. The group reduct $\mathfrak{B}$ of $\mathfrak{A}$ is elementarily equivalent to G as $\mathfrak{B} \models \mathrm{Th}(G)$, however $c \in B$ is infinitely divisible, thus $\mathfrak{B}$ cannot be a free group.

564. Let $H \subset G$ be a proper subgraph. As G is connected, there are connected vertices $v \in H$ and $w \in G \smallsetminus H$. If v has degree n in G then it satisfies the formula $\varphi_n(x)$ saying that x has exactly n different neighbours. The same formula is not true in H, thus H cannot be an elementary subgraph.

565. Suppose N is given and we play the N-round Ehrenfeucht–Fraïssé game. If II has a winning strategy, then by Problem 520 the two graphs are elementarily equivalent.

In the first round I picks a vertex, say a_1. Then let II pick b_1 such that $G_2(b_1, 2^N) = G_1(a_1, 2^N)$. If there are r rounds left, II plays as follows: if I picks, say, a_i, then II checks whether a_i belongs to any of the neighborhoods $G_1(a_j, 2^r)$ for $j < i$. If so, then II picks the corresponding point b_i from $G_2(b_j, 2^r)$. Otherwise II picks an arbitrary b_i far away from every point chosen earlier such that $G_2(b_i, 2^r) = G_1(a_i, 2^r)$.

566. (a) Take a vertex v and for each $n > 0$ attach a path of length n to v. This will be G_1. To get G_2 attach an additional path of infinite length to v. The graphs clearly satisfy the requirements, and are elementarily equivalent shown by the usual winning strategy of II in the N-round Ehrenfeucht–Fraïssé game.

(b) The graphs in Problem 526 are focally finite (every node has degree 2 or 3), non-isomorphic, elementarily equivalent, and clearly non-isomorphic.

567. For $\vartheta \geq \kappa + \lambda$ let $G = \bigcup_\vartheta K_\kappa$ and $H = \bigcup_\vartheta K_\lambda$.

568. Let $\mathfrak{A}$ be the usual ordering of the natural numbers and for each $n \in \omega$ let c_n denote a new constant symbol. Put $\Sigma = \mathrm{Th}(\mathfrak{A}) \cup \{c_0 < c_1 < \cdots < c_{n-1} : n \in \omega\}$. For each finite subset Γ of Σ there is an expansion $\mathfrak{A}_\Gamma$ of $\mathfrak{A}$ such that $\mathfrak{A}_\Gamma \models \Gamma$. By compactness it follows that Σ is consistent; the ordering-reduct of any model of Σ is not well ordered. See also Problem 421.

569. Use induction on the complexity of the formula. For details see the solution of Problem 536 where it is shown that the theory of $\mathfrak{A}$ admits elimination of quantifiers.

570. A standard back and forth argument, see Problem 365

571. Let $\mathfrak{A}$ be the ordering $\omega_2 \times \mathbb{Q}$, where each element of ω_2 is replaced by a copy of the rationals. Let $a_1 < a_2 \in A$ be two points such that the interval (a_1, a_2) contains ω_1 many different elements, and $b_1 < b_2$ such that there are countably many elements between $(b_1$ and b_2. The map $f : a_i \mapsto b_i$ is a partial isomorphism which does not extend to an isomorphism.

572. If a theory Γ eliminates quantifiers and $\mathfrak{A}, \mathfrak{B} \models \Gamma$ are such that $\mathfrak{A}$ is a submodel of $\mathfrak{B}$, then $\mathfrak{A} \prec \mathfrak{B}$, see Problem 538(b). Note that the theory of dense linear orderings without endpoints eliminates quantifiers (see Problem 536).

573. The conservative extension with the unary function S denoting the successor has quantifier elimination, see Problem 554. $\mathfrak{B}$ is a substructure and also a discrete order in that type, thus Problem 538(b) gives that it is an elementary substructure.

574. Take $\mathfrak{A} = \omega + \mathbb{Z} \times \mathbb{R}$. That $\langle \mathbb{R}, \leq \rangle$ can be embedded into $\mathfrak{A}$ is clear: map each $r \in \mathbb{R}$ into any element of the corresponding $\mathbb{Z}$-chain.

As this is a discrete order with initial element and $\langle \omega, \leq \rangle$ is its initial segment, Problem 573 gives that $\langle \omega, \leq \rangle$ is an elementary submodel of $\mathfrak{A}$.

575. (a) The back and forth argument in Solution 570 implies that $\langle \mathbb{Q}, < \rangle$ can be properly embedded into itself. By Problem 572 this embedding is elementary.

(b) Any proper embedding $\alpha : \langle \mathbb{Q}, < \rangle \to \langle \mathbb{Q}, < \rangle$ induces a proper elementary embedding $\alpha^* : \mathbb{Q} \times \mathbb{Z} \to \mathbb{Q} \times \mathbb{Z}$ by Problem 553(c).

(c) Similarly to (b), the embedding $\alpha^* : (\omega + \mathbb{Q} \times \mathbb{Z}) \to (\omega + \mathbb{Q} \times \mathbb{Z})$ which keeps ω fixed, is elementary by Problem 573.

576. Take the dense linear ordering $\mathfrak{A} = \langle \mathbb{Q}, \leq \rangle$, and $X = \emptyset$. An elementary substructure of $\mathfrak{A}$ is an infinite dense linear ordering (in fact, all such substructures are elementary, see Problem 572) but clearly there is no minimal one.

577. Replacing each point in a linear order I by $\mathbb{Z}$ gives a discrete linear ordering without endpoints. Take the two orderings in Solution 468, make the above substitution. The resulting structures are elementarily equivalent, not isomorphic, and each can be embedded into the other.

578. Denote the set of function symbols of $\mathfrak{A}$ by F, then $|F| \leq \omega$.

(a) Define, by induction on n, an increasing sequence $\alpha_0 \leq \alpha_1 \leq \ldots$ of countable ordinals. Choose α_0 so that it contains the interpretation of all constant symbols. (As there are only countably many of them, there is such an $\alpha_0 < \omega_1$.) If α_n has been chosen, let $\alpha_{n+1} < \omega_1$ be an upper bound for the countable set $\{f^{\mathfrak{A}}[\alpha_n] : f \in F\}$. (This set is countable as there are at most countably many function symbols, and there are countably many finite tuples from α_n.) Put $\alpha = \sup\{\alpha_n : n \in \omega\} < \omega_1$.

Now, for each $\vec{x} \in \alpha$ and $f \in F$ there exists $n \in \omega$ so that $\vec{x} \in \alpha_n$, and thus $f^{\mathfrak{A}}(\vec{x}) \in \alpha_{n+1} \subseteq \alpha$. Consequently we have $f^{\mathfrak{A}}[\alpha] \subseteq \alpha$ for all $f \in F$, that is, α is the universe of a substructure of $\mathfrak{A}$.

(b) As τ is countable, there are countably many Skolem functions, see Definition 7.6. Add these symbols to the type, and let $\mathfrak{A}^*$ be the expanded structure (adding the interpretation of the new function symbols). Now $\mathfrak{A}^*$ has built-in Skolem functions (Definition 8.4, has the ground set ω_1, and has a countable substructure (part (a)). But every substructure of $\mathfrak{A}^*$ is elementary, by Problem 538(a), which is an elementary submodel of $\mathfrak{A}$ as well.

579. (a) $\omega + 4$ has a maximal element while ω does not have any, thus they are not even elementarily equivalent.

(b) There is exactly one element in $\omega + \omega$ which does not have an immediate predecessor, while there are infinitely many such elements in $\omega \cdot \omega$. This property can be expressed by a first-order formula, hence the two orderings are not even elementarily equivalent.

(c) Let $\varphi(x)$ be the formula which expresses that there is no immediate predecessor of x. In ω_1 it is true that there is an x (this is just $\omega \cdot \omega$) such that the elements below x satisfying the formula φ are not bounded: $(\forall u < x)\,(\exists v < x)\,(u < v \wedge \varphi(v))$, while this is not true for any element in $\omega \cdot \omega$.

580. Regard ordinals as structures with the usual ordering. Problem 578 gives a countable ordinal α such that $\langle \alpha, < \rangle$ is an elementary submodel of $\langle \omega_1, < \rangle$. In fact, Solution 578 gives more: this α can be arbitrarily large (as α_0 can be any ordinal in the proof of 578). Therefore there are countable ordinals $\alpha < \beta$ such that both $\langle \alpha, < \rangle$ and $\langle \beta, < \rangle$ are elementary submodels of $\langle \omega_1, < \rangle$. Problem 479 implies $\alpha < \beta$.

Remark. An example for such countable ordinals is $\omega^\omega < \omega^\omega + \omega^\omega$.

581. (a) Take $\Gamma = \emptyset$ with the empty language. Countable models (=sets) are isomorphic.
- There is a unique countable complete graph as well.
- The theory of a single unary relation R such that there are infinitely many elements both in R and in the complement of R, see Solution 359.
- Dense linear order without endpoints by Problem 502.
- Atomless Boolean algebras by Problem 531(c).

(b) Let Γ be the theory of simple graphs in which each vertex has degree two and there are no cycles (this latter can be expressed by a countable set of formulas each stating that there is no cycle of length n). Models of Γ consist of disjoint paths isomorphic to $\mathbb{Z}$. It has countable many countable models, and κ-categorical for every $\kappa > \omega$, thus complete by Problem 497.
- The theory of $\langle \omega, 0, S \rangle$ discussed in Problem 552 has also countably many countable models.
- Among the complete theories discussed in Problem 557 the following have countable many countable models: divisible torsion-free Abelian groups (determined by their dimension); algebraically (and real) closed fields of characteristic zero; algebraically closed fields of characteristic p (determined by their transcendence degree).

(c) The theory of discrete orderings without endpoints has continuum many countable models (Problem 368), and it is complete (Problem 553).

Remark. The Baldwin–Lachlan theorem says that if a theory is κ-categorical for some $\kappa > \omega_1$ then it has either 1 or $\aleph_0$ countable models.

582. (a) Let $c_0, c_1, \ldots$ be constant symbols and let Γ be the theory of dense linear orders with formulas added asserting that c_n is strictly increasing. By Problem 555 Γ is complete and has exactly three countable models according to whether $\lim c_n$ is infinite or rational, or irrational.

(b) Consider Γ as in (a) and add a unary relation symbol P to the language. Adjoin new formulas to Γ which express that both P and $\neg P$ is dense, and all c_n are in P. The four models are distinguished whether $\lim c_n$, if rational, is in P or is not in P.

(c) Same as (b), but rather than using two colors (P or not P), use $n - 2$ colors such that between any two points all colors occur, and all c_n are colored with the first color. See also Problem 366.

> **Remark.** By a surprising theorem of Vaught, no complete theory can have exactly two non-isomorphic countable models.

583. Such a theory must have a type of cardinality 2^ω, thus, according to the hint, we can try a structure with ground set ω and a relation symbol R_X for all elements X in an almost disjoint family $\mathcal{F}$. As for different $X, Y \in \mathcal{F}$ there are only finitely many elements satisfying both R_X and R_Y, this condition can be expressed by a formula, thus in any elementary extension R_X and R_Y have no more elements in common. Thus if we could force that in any extension each R_X has additional elements, this would mean $|\mathcal{F}|$ many different elements. Let's see the details.

Let $\mathcal{F}$ be an almost disjoint system of infinite subsets of ω of cardinality continuum. Such a family exists by Problem 1. The similarity type τ will contain unary relation symbols R_X for every $X \in \mathcal{F}$, the binary relation symbol $<$ (for ordering) and constant symbols c_i for $i \in \omega$. The τ-type structure $\mathfrak{A}$ has universe ω, the interpretation of c_i is $i \in \omega$, $<^{\mathfrak{A}}$ is the usual ordering, and for $X \in \mathcal{F}$, $j \in R_X^{\mathfrak{A}}$ iff $j \in X$. Let $\Gamma = \mathrm{Th}(\mathfrak{A})$.

We claim that if $\mathfrak{B} \vDash \Gamma$, then either $\mathfrak{B}$ is isomorphic to $\mathfrak{A}$, or $|\mathfrak{B}| \geq |\mathcal{F}|$. Let $b \in B$ be an element which differs from the interpretation of all constant symbols c_i. (Such a b exists as $\mathfrak{A}$ and $\mathfrak{B}$ are not isomorphic.) The binary relation $<^{\mathfrak{B}}$ is a total discrete order in $\mathfrak{B}$ (the corresponding formulas are true in $\mathfrak{A}$), and for each $i \in \omega$, $c_i^{\mathfrak{B}} < b$. Indeed, the following formula holds in $\mathfrak{A}$, and consequently, in $\mathfrak{B}$: $\forall x (x < c_i \rightarrow x = c_0 \vee \cdots \vee x = c_{i-1})$. It means that elements smaller than $c_i^{\mathfrak{B}}$ are of the form $c_j^{\mathfrak{B}}$.

Each $X \in \mathcal{F}$ is infinite, thus $\mathfrak{A} \vDash \forall x \exists y (y > x \wedge R_X(y))$, consequently the same formula holds in $\mathfrak{B}$. It means that for all $X \in \mathcal{F}$ there is an element $b_X \in B$ such that $b < b_X$ and $b_X \in R_X^{\mathfrak{B}}$. We claim that for different $X, Y \in \mathcal{F}$, b_X and b_Y are also different. Indeed, $X \cap Y$ is finite, say each element in the intersection is smaller than $s = s_{X,Y} \in \omega$. Then

$$\mathfrak{A} \vDash \forall x (R_X(x) \wedge R_Y(x) \rightarrow x < c_s),$$

consequently the same formula holds in $\mathfrak{B}$. It means that if $b' \in R_X^{\mathfrak{B}} \cap R_y^{\mathfrak{B}}$, then b' equals the interpretation of some c_i, and then cannot be bigger than b.

As B contains the distinct elements b_X for $X \in \mathcal{F}$, the cardinality of $\mathfrak{B}$ must be at least $|\mathcal{F}|$, as was claimed.

584. Similarly to Problem 539, for each formula $\psi(\vec{x})$ of the form $\psi(\vec{x}) \equiv \exists y \varphi(y, \vec{x})$ add a new function symbol $f_\psi(\vec{x})$ and define

$$\Gamma' = \Gamma \cup \{(\exists y \varphi(y, \vec{x}) \to \varphi(f_\psi(\vec{x}), \vec{c}) : \varphi \in F(\tau)\}.$$

It is easy to check that Γ' is a conservative extension, and has built-in Skolem functions.

585. Consider the ground set A as nodes of a complete graph, and color the edge ab blue if a and b are ordered the same way by $\leq_1$ and $\leq_2$, and red otherwise. By the infinite Ramsey theorem (Problem 54) there is an infinite homogeneous set. If it is blue, then $\leq_1$ and $\leq_2$ coincide, if red, then they are the reverse of each other.

586. Add the constant symbols c_i to τ, and let Δ be the set of formulas expressing that their interpretation satisfies the requirements. By compactness it suffices to show that $\Gamma \cup \Delta'$ has a model where Δ' is a finite subset of Δ. Let $\mathfrak{A}$ be an arbitrary infinite model of Γ.

(a) Given finitely many formulas $\varphi_1(x), \ldots, \varphi_n(x)$, each element of $\mathfrak{A}$ falls into one of the 2^n many classes depending on whether $\varphi_i(x)$ is true for that element or not. As $\mathfrak{A}$ is infinite, one of the classes is infinite, and interpret the finitely many c_i in Δ' from that class.

(b) Pick different elements $\{a_i : i \in \omega\}$ from $\mathfrak{A}$. Given finitely many formulas $\varphi_1(x, y), \ldots, \varphi_n(x, y)$, each pair (a_i, a_j) with $i < j$ falls into one of the 2^n many classes depending on which φ_i is true for that pair. We need a large homogeneous subset (to be assigned to the finitely many constant symbols in Δ') where every pair is in the same class. But Ramsey's theorem (Problem 54) guarantees an infinite homogeneous subset, so we are done.

587. Apply Definition 8.20 to the formula $\varphi(x, y) \equiv (x < y)$.

588. By Problem 570 the truth of a formula $\varphi(\vec{x})$ in $\mathfrak{A}$ depends only on the $<^{\mathfrak{A}}$-order of the free variables, which is the same for two $\langle a_{i_j} \rangle$ sequences where the indices i_j are increasing.

589. Any permutation of X induces an automorphism of $\mathfrak{A}$. Take $\varphi(\vec{x}) \in F(\tau)$ and $\vec{a}, \vec{b} \in X$. Let π be the automorphism that maps $\vec{a}$ to $\vec{b}$ (they have the same number of elements, so there is a permutation on X which maps $\vec{a}$ to $\vec{b}$). As automorphisms preserve formulas we have $\mathfrak{A} \vDash \varphi[\vec{a}]$ iff $\mathfrak{A} \vDash \varphi[\pi(\vec{a})]$, as required.

590. By Problem 554 the conservative extension of type $\langle \leq, 0, S \rangle$ has quantifier elimination, thus only quantifier-free formulas (on the extended language) should be considered when checking indiscernibility. Points of H are infinitely far apart, thus applications of S (and zero) can be discarded. The ordering of the tuples, however, are the same by assumption.

591. Each element of the generated substructure $\mathfrak{B}$ is of the form $t^{\mathfrak{A}}[\vec{b}]$ where $\vec{b} \in H$. Consequently any type realized in $\mathfrak{B}$ is realized by the tuple $\langle t_i^{\mathfrak{A}}[\vec{b}] : i < n \rangle$. Since $\ll$ is well-ordering, the tuple $\vec{b}$ is order-isomorphic to some $\langle h_0, h_1, \ldots h_m \rangle \in H$ where these elements are among the first m elements of H. As H is indiscernible, the type realized by the tuple $\langle t_i^{\mathfrak{A}}[\vec{b}] \rangle$ is the same which is realized by $\langle t_i^{\mathfrak{A}}[\vec{h}] \rangle$. As there are at most $|F(\tau)|$ many finite sequences of τ-terms, it gives an upper bound on the number of realized types.

592. Add new constants denoting elements of H. It is enough to show that

$$\Delta_{\mathfrak{A}} \cup \{\varphi(\vec{c}) \leftrightarrow \varphi(\vec{d}) : \vec{c}, \vec{d} \in H \text{ are } \ll\text{-increasing tuples}\}$$

has a model $\mathfrak{B}$. Here $\Delta_{\mathfrak{A}}$ is the diagram of $\mathfrak{A}$, see Definition 8.8. Recall that any $\mathfrak{B} \vDash \Delta_{\mathfrak{A}}$ is a proper elementary extension of $\mathfrak{A}$, see Problem 507(b).

By compactness it is enough to show that

$$\Sigma = \Delta_{\mathfrak{A}} \cup \{\varphi(\vec{c}) \leftrightarrow \varphi(\vec{d}) : \vec{c}, \vec{d} \in H_0, \varphi \in \Delta\},$$

is consistent, where H_0 and Δ are finite sets. We may assume that $|\vec{c}| = n$. Define an equivalence $\sim$ on elements of $[A]^n$ by $\vec{a} \sim \vec{b}$ iff

$$\mathfrak{A} \vDash \varphi[\vec{a}] \leftrightarrow \varphi[\vec{b}] \quad \text{for all } \varphi \in \Delta$$

where $\vec{a}$ and $\vec{b}$ are tuples in $<$-increasing order.

$\sim$ has at most $2^{|\Delta|}$ (finitely many) equivalence classes, therefore Ramsey's theorem implies the existence of an infinite $C \subseteq A$ such that all n-element subsets of C belong to the same $\sim$-equivalence class. Interpret $c \in H_0$ by elements $b_c \in C$ ordered in the same way as c. Then $\langle \mathfrak{A}, b_c \rangle_{c \in H_0}$ is a model for Σ.

593. The lexicographic ordering on $\kappa \times \mathbb{Q}$ has 2^κ automorphisms as for each $A \subseteq \kappa$,

$$\pi_A(\langle \alpha, q \rangle) = \begin{cases} \langle \alpha, q \rangle & \text{if } \alpha \in A \\ \langle \alpha, q + 17 \rangle & \text{otherwise} \end{cases}$$

is an automorphism and $\pi_A \neq \pi_B$ for $A \neq B \subset \kappa$.

594. Elements of the substructure generated by B are the terms $t^{\mathfrak{A}}[\vec{b}]$ where $\vec{b} \in B$. The map π extends to these values by taking $\pi(t^{\mathfrak{A}}[\vec{b}]) = t^{\mathfrak{A}}[\pi(\vec{b})]$. This is a sound definition as $t_1^{\mathfrak{A}}[\vec{b}] = t_2^{\mathfrak{A}}[\vec{b}]$ iff $t_1^{\mathfrak{A}}[\pi(\vec{b})] = t_2^{\mathfrak{A}}[\pi(\vec{b})]$ by the property of indiscernibles. For the same reason π preserves the interpretation of function and relation symbols, thus it is an isomorphism between the two substructures.

595. Let Γ' be the conservative extension of Γ which has built-in Skolem functions (see Problem 584) and let $\mathfrak{A} \vDash \Gamma'$. Take the ordering $(H, \ll)$ from Problem 593. By Problem 592 $\mathfrak{A}$ has an elementary extension $\mathfrak{B}$ such that $H \subseteq B$ and $(H, \ll)$ is indiscernible in $\mathfrak{B}$. Let $\mathfrak{C}$ be the substructure of $\mathfrak{B}$ generated by H. As Γ' has built-in Skolem functions, $\mathfrak{C}$ is an elementary substructure (see Problem 491). If we prove that automorphisms of the ordering $(H, \ll)$ extend to automorphisms of $\mathfrak{C}$, then taking the reduct of $\mathfrak{C}$ to the original language completes the solution. This latter claim, however, follows from Problem 594.

596. Similarly to Problem 595, we can assume that Γ has built-in Skolem functions (Problem 584), thus every substructure is elementary (Problem 491). Let $(H, \ll)$ be a well-ordering of size κ, and $\mathfrak{A}$ be a model of Γ in which $(H, \ll)$ is indiscernible (Theorem 8.21). If $\mathfrak{B}$ is the substructure generated by H, then (i) it is an elementary substructure of $\mathfrak{A}$, thus a model of Γ, (ii) as τ is countable, it has cardinality κ, and (iii) by Problem 591 only countably many different types are realized in $\mathfrak{B}$.

12.9   ULTRAPRODUCTS

597. Straightforward induction of the complexity of the term t.

598. Let $\mathfrak{A} = \langle \omega, < \rangle$ and let φ be the formula $\forall x \forall y ((x < y) \vee (x = y) \vee (y < x))$. In $\mathfrak{A} \times \mathfrak{A}$ the elements $\langle 1, 0 \rangle$ and $\langle 0, 1 \rangle$ are not comparable w.r.t $<^{\mathfrak{A} \times \mathfrak{A}}$.

By Problem 597 and by definition variable-free atomic formulas are preserved under taking a product.

599. Let φ be $\exists x (R(x) \vee Q(x))$, where R and Q are relation symbols. Then $\Pi_i \mathfrak{A}_i \vDash \exists x (R(x) \vee Q(x))$ if and only if there exists $a \in \Pi_i A_i$ such that $\Pi_i \mathfrak{A}_i \vDash (R[a] \vee Q[a])$ which holds if and only if $\Pi_i \mathfrak{A}_i \vDash R[a]$ or $\Pi_i \mathfrak{A}_i \vDash Q[a]$. By definition of the product, this holds iff we have either either $\mathfrak{A}_i \vDash R[a(i)]$ for all i, or $\mathfrak{A}_i \vDash Q[a(i)]$ for all $i \in I$. This is not the same thing as requiring $\mathfrak{A}_i \vDash (R[a(i)] \vee Q[a(i)])$ for all $i \in I$.

An explicit counterexample is the following. Let $\mathfrak{A}_n = \langle \omega, R^{\mathfrak{A}_n}, Q^{\mathfrak{A}_n} \rangle$ be such that

$$R^{\mathfrak{A}_n} = \begin{cases} \omega & \text{if } n \text{ is odd,} \\ \varnothing & \text{otherwise,} \end{cases} \qquad Q^{\mathfrak{A}_n} = \begin{cases} \varnothing & \text{if } n \text{ is odd,} \\ \omega & \text{otherwise.} \end{cases}$$

Then for all $n \in \omega$ we have $\mathfrak{A}_n \vDash \exists x (R(x) \vee Q(x))$. But $x \in R^{\Pi \mathfrak{A}_n}$ if and only if $x(n) \in R^{\mathfrak{A}_n}$ for all n, thus $R^{\Pi \mathfrak{A}_n} = \varnothing$ (and similarly $Q^{\Pi \mathfrak{A}_n} = \varnothing$), consequently $\Pi_n \mathfrak{A}_n \nvDash \exists x (R(x) \vee Q(x))$.

600. Let ϑ_n be the formula $(R_1(\bar{x}) \to R_2(\bar{x}))$. For each evaluation e over $\Pi_i \mathfrak{A}_i$ the following implication holds: if $\mathfrak{A}_i \vDash \vartheta_n[e(i)]$ for all $i \in I$, then $\Pi_i \mathfrak{A}_i \vDash \vartheta_n[e]$. Indeed, if $\Pi_i \mathfrak{A}_i \vDash R_1[e]$, then $\mathfrak{A}_i \vDash R_1[e(i)]$ for all $i \in I$ (by the definition of $R_1^{\Pi \mathfrak{A}}$), thus the assumption gives $\mathfrak{A}_i \vDash R_2[e(i)]$ for all i, and then $\Pi_i \mathfrak{A}_i \vDash R_2[e]$ again by the definition of $R_2^{\Pi \mathfrak{A}}$.

Now proceed by induction on the number of quantifiers at the beginning of the formula. Assume the following statement is true: "for all evaluation e over $\Pi_i \mathfrak{A}_i$, if $\mathfrak{A}_i \vDash \vartheta(y, \bar{x})[e(i)]$ for all $i \in I$, then $\Pi_i \mathfrak{A}_i \vDash \vartheta(y, \bar{x})[e]$." Then the same statement is true for $\exists y \vartheta(y, \vec{x})$ and $\forall y \vartheta(y, \vec{x})$. Fix the evaluation e, and assume $\mathfrak{A}_i \vDash \exists y \vartheta(y, \vec{x})[e(i)]$ for all $i \in I$. There is a vector $a \in \Pi_i A_i$ such that $a(i) \in A_i$ is such an element: $\mathfrak{A}_i \vDash \vartheta[e(i)(y/a(i))]$. Apply the inductive hypothesis to ϑ and the evaluation $e(y/a)$ to get $\Pi_i \mathfrak{A}_i \vDash \vartheta[e(y/a)]$, thus $\Pi_i \mathfrak{A}_i \vDash \exists y \vartheta[e]$, as required. Similar reasoning works for the universal quantifier.

601. (a) $a \in a/\mathcal{F}$ as $I \in \mathcal{F}$. If $b \in a/\mathcal{F}$, then $b/\mathcal{F} = a/\mathcal{F}$. Indeed, the indices of a and b agree on a set $I_1 \in \mathcal{F}$. If b' and b (b' and a) agree on the set $I_2 \in \mathcal{F}$, then b' and a (b' and b) agree on a set which is a superset of $I_1 \cap I_2$, thus an element of $\mathcal{F}$, meaning $b' \in a/\mathcal{F}$ ($b' \in b/\mathcal{F}$).

(b) Let $f \in \tau$ be an n-place function symbol, and suppose $\vec{a}/\mathcal{F} = \vec{b}/\mathcal{F}$. Then there is a $J \in \mathcal{F}$ such that for all $i \in J$, $\vec{a}(i) = \vec{b}(i)$. But then $f^{\Pi \mathfrak{A}}(\vec{a})$ and $f^{\Pi \mathfrak{A}}(\vec{b})$ take the same value for all $i \in J$, thus they are in the same equivalence class as well.

602. By Problem 601 the value of a function is independent of the choice of the representatives in $\vec{a}/\mathcal{U}$. Thus it suffices to check that the definition of a relation does not depend on the choice of the representatives. Namely, if $\vec{a}/\mathcal{U} = \vec{b}/\mathcal{U}$, then $I_a = \{i \in I : \vec{a}(i) \in r^{\mathfrak{A}_i}\}$ and $I_b = \{i \in I; \vec{b}(i) \in r^{\mathfrak{A}_i}\}$ are in $\mathcal{U}$ at the same time. This follows immediately from the observation that there is a $J \in \mathcal{U}$ such that for all $i \in J$, $\vec{a}(i) = \vec{b}(i)$ (the intersection of the sets from $\mathcal{U}$ where $a_1(i) = b_1(i), \ldots, a_n(i) = b_n(i)$). As $I_a \cap J = I_b \cap J$ we have $I_a \in \mathcal{U}$ iff $I_a \cap J \in \mathcal{U}$ iff $I_b \cap J \in \mathcal{U}$ iff $I_b \in \mathcal{U}$, as required.

603. We have to prove that $\Pi_{i<n} \mathfrak{A}_i / \mathcal{U}$ is isomorphic to $\mathfrak{A}_j$ for some $j < n$. As every ultrafilter on a finite set is trivial (Problem 30), we have $\mathcal{U} = \{A \subseteq n : j \in A\}$ for some $j < n$. We claim that $\Pi_{i<n} \mathfrak{A}_i / \mathcal{U} \cong \mathfrak{A}_j$. Let $\Phi : \Pi_{i<n} A_i / \mathcal{U} \to A_j$ be the function that maps $a/\mathcal{U}$ into $a(j)$. This map is well defined and injective:

$$a/\mathcal{U} = b/\mathcal{U} \Leftrightarrow \{i < n : a(i) = b(i)\} \in \mathcal{U}$$
$$\Leftrightarrow a(j) = b(j)$$
$$\Leftrightarrow \Phi(a/\mathcal{U}) = \Phi(b/\mathcal{U}).$$

It also is surjective: for each $x \in A_j$ there is an $a \in \Pi_{i<n} A_i$ such that $a(j) = x$. Finally, for a relation symbol R and a function symbol f we have

$$\vec{a}/\mathcal{U} \in R^{\Pi_i \mathfrak{A}_i / \mathcal{U}} \Leftrightarrow \{i < n : \vec{a}(i) \in R^{\mathfrak{A}_i}\} \in \mathcal{U}$$
$$\Leftrightarrow \vec{a}(j) \in R^{\mathfrak{A}_j}$$
$$\Leftrightarrow \Phi(\vec{a}/\mathcal{U}) \in R^{\mathfrak{A}_j},$$

and

$$\Phi\big(f^{\Pi_i \mathfrak{A}_i / \mathcal{U}}(\vec{a}/\mathcal{U})\big) = \Phi\big(\langle f^{\mathfrak{A}_i}(\vec{a}(i)) : i < n\rangle / \mathcal{U}\big)$$
$$= f^{\mathfrak{A}_j}(\vec{a}(j))$$
$$= f^{\mathfrak{A}_j}(\Phi(\vec{a})).$$

Consequently, Φ is an isomorphism.

604. For $i \in I$ let J_i be disjoint sets, and let $\mathcal{V}_i$ be an ultrafilter on J_i. Elements of the ultraproduct $\Pi_i(\Pi_j \mathfrak{A}_{i,j}/\mathcal{V}_j)/\mathcal{U}$ are equivalence classes of vectors of elements of $\Pi_j \mathfrak{A}_{i,j}/\mathcal{V}_j$, which, in turn, are equivalence classes of vectors of elements of $\mathfrak{A}_{i,j}$. Thus two vectors $\langle a(i,j) : i \in I; j \in J_i \rangle$ and $\langle b(i,j) \rangle$ are equivalent if the set of indices $i \in I$ for which $\{j \in J_i : a(i,j) = b(i,j)\} \in \mathcal{V}_i$ is in $\mathcal{U}$. Let $\mathcal{W}$ be the collection of those subsets X of indices (i,j) for which the following holds:

$$\{i \in I : \{j \in J_i : (i,j) \in X\} \in V_i\} \in \mathcal{U}.$$

It is easy to check that $\mathcal{W}$ is an ultrafilter, and $a/\mathcal{W} = b/\mathcal{W}$ just in case a and b represent the same element in $\Pi_i(\Pi_j \mathfrak{A}_{i,j}/\mathcal{V}_j)/\mathcal{U}$. Thus the ultraproduct $\Pi_{i,j} \mathfrak{A}_{i,j}/\mathcal{W}$ is isomorphic to the ultraproduct of the ultraproducts.

605. Suppose $\{\mathfrak{A}_i : i \in I\} = \{\mathfrak{B}_1, \ldots, \mathfrak{B}_k\}$ and for $j < k$ let $X_j = \{i \in I : \mathfrak{A}_i = \mathfrak{B}_j\}$. Then $X_1, \ldots, X_k$ is a finite partition of I and hence exactly one of them, say X_k, belongs to $\mathcal{U}$ (Problem 31). Write $B_k = \{b_0, \ldots, b_\ell\}$ (the ground set of $\mathfrak{B}_k$) and choose $a_0, \ldots, a_\ell \in \Pi_{i \in I} A_i$ such that for all $i \in X_k$

$$a_0(i) = b_0, \quad a_1(i) = b_1, \quad \ldots, \quad a_\ell(i) = b_\ell.$$

Then for all $a \in \Pi_{i \in I} A_i$ there is a unique a_j such that $a/\mathcal{U} = a_j/\mathcal{U}$ since

$$\{i \in I : a(i) = a_0(i)\} \cup \{i \in I : a(i) = a_1(i)\} \cup \ldots \cup \{i \in I : a(i) = a_k(i)\} \supseteq X_k,$$

thus exactly one of the above sets belongs to $\mathcal{U}$. It is not hard then to check that the map

$$\Phi : \Pi_i A_i / \mathcal{U} \to B_k,$$
$$\Phi(a_j/\mathcal{U}) = b_j$$

is an isomorphism.

606. Let $\bar{\varphi}$ be the universal closure of φ. Then for a structure $\mathfrak{A}$ and evaluation over $\mathfrak{A}$, $\mathfrak{A} \vDash \varphi$ iff $\mathfrak{A} \vDash \varphi[e]$, see Problem 311. So let e be an arbitrary evaluation over $\Pi_i \mathfrak{A}_i$, then

$$\Pi_i \mathfrak{A}_i / \mathcal{U} \vDash \varphi \Leftrightarrow \Pi_i \mathfrak{A}_i \vDash \bar{\varphi}[e/\mathcal{U}] \Leftrightarrow \{i \in I : \mathfrak{A}_i \vDash \bar{\varphi}[e(i)]\} \in \mathcal{U}$$
$$\Leftrightarrow \{i \in I : \mathfrak{A}_i \vDash \bar{\varphi}\} \in \mathcal{U} \Leftrightarrow \{i \in i : \mathfrak{A}_i \vDash \varphi\} \in \mathcal{U}.$$

607. As the whole index set is always in $\mathcal{U}$ for any formula φ we have

$$^\kappa \mathfrak{A} \vDash \varphi \Leftrightarrow \{i \in \kappa : \mathfrak{A} \vDash \varphi\} \in \mathcal{U} \Leftrightarrow \mathfrak{A} \vDash \varphi$$

according to Łoś lemma. Here either use closed formulas only, or refer to Problem 606.

608. Solution 1. It is enough to prove part (c). Write $X = \{n_i : i < |X|\}$ and for each $i < |X|$ let

$$X_i = \{\alpha \in I : |A_\alpha| = n_i\} \in \mathcal{U}, \qquad Y = \{\alpha \in I : |A_\alpha| \notin X\} \notin \mathcal{U}.$$

Exactly one block of the finite partition $I = Y \cup \bigcup_i^* X_i$ belongs to $\mathcal{U}$, say X_0 (Problem 31). For simplicity we may assume that for each $\alpha \in X_0$ the underlying set of $\mathfrak{A}_\alpha$ is $\{b_1, \ldots, b_{n_0}\}$. Choose elements $a_1, \ldots, a_{n_0} \in \Pi_\alpha A_\alpha$ such that

$$a_1(\alpha) = b_1, \quad a_1(\alpha) = b_1, \quad \ldots, \quad a_{n_0}(\alpha) = b_{n_0} \quad \text{for all } \alpha \in X_0.$$

Then for all $a \in \Pi_\alpha A_\alpha$ there is an a_j such that $a/\mathcal{U} = a_j/\mathcal{U}$ because

$$\bigcup_{j < n_0} \{\alpha \in I : a(\alpha) = a_j(\alpha)\} \supseteq X_0.$$

This means that A has exactly n_0 elements, hence $|\mathfrak{A}| \in X$.

Solution 2 (Using Łoś lemma). (a) The first-order formula which expresses that there are at most n elements is true in all $\mathfrak{A}_\alpha$, therefore by Łoś lemma it is true in $\mathfrak{A}$, as well.

(b), (c) Let $X = \{n_1, \ldots, n_k\}$ and let φ be the formula that expresses "the size of the structure is one of the n_k's". Then for an index set in $\mathcal{U}$ we have $\mathfrak{A}_i \vDash \varphi$, and by Łoś lemma, $\mathfrak{A} \vDash \varphi$.

609. The ultraproduct has cardinality at most continuum as the product $\Pi_{i<\omega} A_i$ of countable sets has continuum. We show that there are continuum many different elements in the ultraproduct.

Let $\mathcal{F}$ be an almost disjoint family of cardinality continuum on ω having infinite members only (see Problem 1). Let $j : \omega \to A$ be an injection. For $X \in \mathcal{F}$ with $X = \{s_0, s_1, \ldots\}$ enumerated increasingly define $a_X \in {}^\omega A$ by $a_X(i) = j(s_i)$. For different $X, Y \in \mathcal{F}$ $a_X/\mathcal{U}$ and $a_Y/\mathcal{U}$ are different elements of the ultraproduct as $\{i \in \omega : a_X(i) = a_Y(i)\}$ is a subset of $X \cap Y$, which is finite.

610. We use the idea from Problem 13. Assume first that $2^n \leq n_i$. For a sequence $s : \omega \to 2$ let $s{\upharpoonright}n$ be the natural number whose binary expansion is given by the first n digits of s, and let $a_s \in \Pi_i A_i$ be the sequence whose n-th element is the $s{\upharpoonright}n$-th element of A_n. Clearly $a_s/\mathcal{U}$ gives continuum many distinct elements, as for different s_1, s_2 the sequences a_{s_1} and a_{s_2} agree at finitely many places only.

This construction can be modified for the case when $\lim n_i = \infty$. In this case there is an increasing sequence $m_0 < m_1 < \cdots$ such that $n_i \geq 2^n$ when $i \geq m_n$. For a sequence $s : \omega \to 2$ let $a_s(i)$ be the $s{\upharpoonright}\ell$-th element of A_i where $m_\ell \leq i < m_{\ell+1}$. Again, different sequences a_{s_1} and a_{s_2} agree at finitely many indices only, thus they give continuum many different elements in the ultraproduct.

If $\limsup n_i = \infty$, then it may happen that the sets $X_n = \{i : n_i > n\}$ do not belong to $\mathcal{U}$. Consider for example the following sequence

$$n_i = \begin{cases} 2 & \text{if } i \text{ is even,} \\ i & \text{otherwise.} \end{cases}$$

$\limsup n_i = \infty$ but if $\{2n : n \in \omega\} \in \mathcal{U}$, then we have $|\Pi_{i<\omega} n_i/\mathcal{U}| = 2$.

611. That δ is a bijection between $\mathfrak{A}$ and $\delta[\mathfrak{A}]$ is straightforward. For $a \in A$ write $\epsilon(a) = \langle a : i \in \kappa \rangle$, then $\delta(a) = \epsilon(a)/\mathcal{U}$. As for the relation and function symbols R and f we have

$$\langle \delta(a_1), \ldots, \delta(a_n) \rangle \in R^{\kappa \mathfrak{A}/\mathcal{U}}$$
$$\Leftrightarrow \{ i \in I : \langle \epsilon(a_1)(i), \ldots, \epsilon(a_n)(i) \rangle \in R^{\mathfrak{A}} \} \in \mathcal{U}$$
$$\Leftrightarrow \langle a_1, \ldots, a_n \rangle \in R^{\mathfrak{A}},$$

and

$$\begin{aligned} f^{\kappa \mathfrak{A}/\mathcal{U}}\big(\delta(a_1), \ldots, \delta(a_n)\big) &= \langle f^{\mathfrak{A}}\big(\epsilon(a_1)(i), \ldots, \epsilon(a_n)(i)\big) : i \in I \rangle /\mathcal{U} \\ &= \langle f^{\mathfrak{A}}\big(a_1, \ldots, a_n\big) : i \in I \rangle /\mathcal{U} \\ &= \delta\big(f^{\mathfrak{A}}(a_1, \ldots, a_n)\big). \end{aligned}$$

Consequently $\delta[\mathfrak{A}]$ is isomorphic to $\mathfrak{A}$. To show that $\delta[\mathfrak{A}]$ is an elementary substructure of $^{\kappa}\mathfrak{A}/\mathcal{U}$ we use the Tarski–Vaught test (Theorem 8.3). Let $\vec{a} \in A$ and $\varphi(\vec{x}, y)$ be a formula. By Łoś lemma we have

$$^{\kappa}\mathfrak{A}/\mathcal{U} \vDash \exists y \varphi[\delta(\vec{a})] \Leftrightarrow \{ i \in \kappa : \mathfrak{A} \vDash \exists y \varphi[\epsilon(\vec{a})(i)] \} \in \mathcal{U}$$
$$\Leftrightarrow \{ i \in \kappa : \mathfrak{A} \vDash \exists y \varphi[\vec{a}] \} \in \mathcal{U}$$
$$\Leftrightarrow \mathfrak{A} \vDash \exists y \varphi[\vec{a}].$$

Therefore, if for some $b \in A$ we have $\mathfrak{A} \vDash \varphi[\vec{a}, b]$ and then again by Łoś lemma we get $^{\kappa}\mathfrak{A}/\mathcal{U} \vDash \varphi[\delta(\vec{a}), \delta(b)]$. Since $\delta(b) \in \delta[A]$ conditions of the Tarski–Vaught test are satisfied, consequently $\delta[\mathfrak{A}]$ is an elementary substructure of $^{\kappa}\mathfrak{A}/\mathcal{U}$.

612. The embedding $\delta : \mathfrak{A} \to {}^{\omega}\mathfrak{A}/\mathcal{U}$ is an elementary embedding by Problem 611. Let $\epsilon(a) = \langle a : i \in \omega \rangle \in {}^{\omega}A$, then $\delta(a) = \epsilon(a)/\mathcal{U}$. We only need to show that there is an element in the ultrapower which differs from all $\delta(a)$. By assumption $\mathfrak{A}$ is infinite, and take $b = \langle a_i : i \in \omega \rangle$ where all $a_i \in A$ are different. We claim that $b/\mathcal{U}$ is not in the range of δ. Indeed, $\delta(a)$ and $b/\mathcal{U}$ are different as they agree on at most one place, while $\mathcal{U}$ is non-trivial.

613. By Problem 609, for a non-trivial ultrafilter $\mathcal{U}$, $^{\omega}\mathfrak{A}/\mathcal{U}$ has cardinality 2^{ω}. By Problem 611, it is an elementary extension of $\mathfrak{A}$.

614. The construction from Problem 19 gives 2^{κ} many elements a_{ξ} from the product $\Pi_i A_i$ such that choosing finitely many $a_{\xi_1}, \ldots, a_{\xi_n}$, there is an index $i \in I$ where all $a_{\xi_j}(i)$ differ. When factoring the product by the ultrafilter $\mathcal{U}$, these elements will be different if all the sets

$$I_{\xi, \eta} = \{ i \in I : a_{\xi}(i) \neq a_{\eta}(i) \}$$

are in $\mathcal{U}$. Such an ultrafilter $\mathcal{U}$ exists if and only if this collection has the FIP property, namely given finitely many pairs, there is an $i \in I$ which is an element of each $I_{\xi, \eta}$. But clearly this is the case, so we are done.

615. The ultrapower $^\kappa\mathfrak{A}/\mathcal{U}$ has cardinality at least 2^κ for the ultrafilter from Problem 613. From $|A| \leq 2^\kappa$ it follows that the product space $^\kappa A$ has cardinality 2^κ, thus the ultrapower is of cardinality 2^κ, and is an elementary extension of $\mathfrak{A}$ by Problem 611.

616. The ω-ultrapower of a set X of cardinality continuum has cardinality continuum: $|X| \leq |^\omega X/\mathcal{U}| \leq |^\omega X| = |X|$. Therefore there is a bijection between X and $^\omega X/\mathcal{U}$, and any bijection is an isomorphism in the empty language (i.e. considering sets without further structure). Each infinite set can be properly embedded into itself.

617. The field of complex numbers is algebraically closed – a property that can be expressed by first-order formulas, see Problems 339, 504, 557. It follows that $^\omega\mathbb{C}/\mathcal{U}$ is algebraically closed, has characteristic zero, and has cardinality continuum. Thus both the characteristic and the transcendental degree of the algebraically closed fields $\mathbb{C}$ and $^\omega\mathbb{C}/\mathcal{U}$ are the same (zero and continuum respectively), consequently they are isomorphic.

618. Let $\mathfrak{A}_i$ and $\mathfrak{B}_j$ be structures in the empty language (pure sets) such that $|\mathfrak{A}_i| = 2^{i+1}$ and $|\mathfrak{B}_j| = 3^{j+1}$. $\mathfrak{A}_i \not\equiv \mathfrak{B}_j$ is straightforward and both $\Pi_i\mathfrak{A}_i/\mathcal{U}$ and $\Pi_i\mathfrak{B}_i/\mathcal{U}$ have cardinality continuum (Problem 610), consequently they are isomorphic.

> **Remark.** There is nothing in particular choosing pure sets. If C_n denotes the cyclic graph on n vertices, then $\Pi_n C_{2n}/\mathcal{U}$ and $\Pi_n C_{2n+1}/\mathcal{U}$ are isomorphic as both consist of continuum many infinite paths (see Problems 629).

619. By assumption $\{\mathrm{Th}(\mathfrak{A}_i) : i \in I\}$ is finite, say it is equal to $\{T_0, \ldots, T_{n-1}\}$. For $j < n$ let $X_j = \{i \in I : \mathrm{Th}(\mathfrak{A}_i) = T_j\}$. As $I = \bigcup_j^* X_j$ is a finite partition, there is a unique $j < n$ such that $X_j \in \mathcal{U}$ (see Problem 31). Then $\mathrm{Th}\left(\Pi_i\mathfrak{A}_i/\mathcal{U}\right) = T_j$, which can be seen as follows: $\varphi \in T_j \Leftrightarrow (\forall i \in X_j)\ \mathfrak{A}_i \vDash \varphi \Leftrightarrow X_j \subseteq \{i \in I : \mathfrak{A}_i \vDash \varphi\} \in \mathcal{U} \Leftrightarrow \Pi_i\mathfrak{A}_i/\mathcal{U} \vDash \varphi$.

620. Our similarity type will consist of countably many constant symbols $\langle c_n : n < \omega\rangle$. The ultraproduct of two element structures has two elements, thus for simplicity suppose that the ground set of both structures is $\{0, 1\}$. Let the interpretations of the constants as follows

$$c_n^{\mathfrak{A}_i} = \begin{cases} 1 & \text{if } i = n, \\ 0 & \text{otherwise.} \end{cases}$$

Then for all $n < \omega$ we have $\{i < \omega : c_n^{\mathfrak{A}_i} = 0\} \in \mathcal{U}$, therefore $c_n^{\Pi\mathfrak{A}_i/\mathcal{U}} = 0$. But in each $\mathfrak{A}_i$ the interpretation of the constant c_i is 1, thus $\mathfrak{A}_i$ is not elementarily equivalent to the ultraproduct.

621. Recall that $\mathcal{V} = \{Z \cap J : Z \in \mathcal{U}\}$ is an ultrafilter on J (see Problem 37). For $a/\mathcal{V} \in \Pi_{j \in J}\mathfrak{A}_j/\mathcal{V}$ write

$$a'(i) = \begin{cases} a(i) & \text{if } i \in J, \\ \text{arbitrary} \in A_i & \text{otherwise.} \end{cases}$$

The mapping $\Phi(a/\mathcal{V}) = a'/\mathcal{U}$ is well defined (does not depend on the choice of representative of $a/\mathcal{V}$) and is an isomorphism.

622. One direction is obvious. For the other direction, write

$$I = [\Gamma]^{<\omega} = \{\Sigma \subseteq \Gamma : \Sigma \text{ is finite}\}.$$

By assumption, for all $i \in I$ there is a model $\mathfrak{A}_i \vDash i$. Let $\mathfrak{A} = \prod_{i \in I} \mathfrak{A}_i / \mathcal{U}$ for an ultrafilter $\mathcal{U}$ over I. We want to choose $\mathcal{U}$ so that for all $\gamma \in \Gamma$ we have $\mathfrak{A} \vDash \gamma$. By Łoś lemma this amounts to

$$\{i \in I : \mathfrak{A}_i \vDash \gamma\} \in \mathcal{U}.$$

But if $\gamma \in i$, then $\mathfrak{A}_i \vDash \gamma$, therefore it is enough to guarantee that

$$I_\gamma = \{i \in I : \gamma \in i\} \in \mathcal{U}.$$

Note that the family $\{I_\gamma : \gamma \in \Gamma\}$ has the finite intersection property, therefore it can be extended to an ultrafilter $\mathcal{U}$.

623. For each n let $\mathfrak{A}_n$ be a finite model of Γ of cardinality at least n. The ultraproduct $\prod_n \mathfrak{A}_n / \mathcal{U}$ for a non-trivial ultrafilter $\mathcal{U}$ has cardinality continuum (Problem 610) and is a model of Γ.

624. By Łoś lemma. For a fix $e \in E$ the set

$$\{\xi \in I : \mathfrak{A}_\xi \vDash \varphi_e[f_\xi]\}$$

contains $e \subseteq I$ by assumption, and $e \in \mathcal{U}$.

625. As $\mathcal{U}$ is regular, for every $\xi \in I$ there are only finitely many $e \in E$ such that $\xi \in e$. Choose $\mathfrak{A}_\xi \in \mathcal{K}$ to be a model of the finite set $\{\varphi_e : \xi \in e\}$. Problem 624 claims that $\prod_\xi \mathfrak{A}_\xi \vDash \varphi_e$ for all $e \in E$.

626. Solution 1. Let $\mathcal{U}$ be a regular filter over I with the corresponding set $E \subset \mathcal{U}$ such that $|E| = |A|$. Fix a bijection $f : A \to E$. For every $\xi \in I$ there are finitely many $a \in A$ such that $\xi \in f(a)$; let $\mathfrak{A}_\xi$ be the substructure generated by these elements. In particular, $a \in A_\xi$ whenever $\xi \in f(a)$. As $f(a) \in \mathcal{U}$, it means that given finitely many $a_1, \ldots, a_n \in A$, each a_i is in A_ξ for every $\xi \in f(a_1) \cap \cdots \cap f(a_n) \in \mathcal{U}$.

We claim that $\mathfrak{A}$ can be embedded as substructure to the ultraproduct $\prod_\xi \mathfrak{A}_\xi / \mathcal{U}$. For $a \in A$ define $x_a \in \prod_\xi A_\xi$ as

$$x_a(\xi) = \begin{cases} a & \text{if } a \in A_\xi, \\ \text{arbitrary} \in A_\xi & \text{otherwise.} \end{cases}$$

The function $\Phi : a \mapsto x_a/\mathcal{U}$ is such an embedding. For different $a_1, a_1 \in A$ we have $x_{a_1}(\xi) \neq x_{a_2}(\xi)$ for all $\xi \in f(a_1) \cap f(a_2) \in \mathcal{U}$, hence Φ is injective.

Using the observation above, for any $\vec{a} \in A$ the substructure generated by $\vec{a}$ is a substructure of A_ξ for every $\xi \in \bigcap_i f(a_i) \in \mathcal{U}$. By Łoś lemma it means that

$\Pi_\xi \mathfrak{A}_\xi / \mathcal{U} \models \varphi[\Phi(\bar{a})]$ and $\mathfrak{A} \models \varphi[\bar{a}]$ are equivalent for atomic formulas. Consequently Φ preserves the truth of atomic formulas, thus it is an embedding indeed.

Solution 2. Let $\mathfrak{A}$ be the structure, and $\Gamma = \Delta^0_{\mathfrak{A}}$ be its atomic diagram, see Definition 8.8. By Problem 507 $\mathfrak{A}$ can be embedded into any model of Γ, and by Problem 625 there is a model of Γ which is an ultraproduct of structures from the collection $\mathcal{K}$ assuming that each finite subset $\Gamma' \subseteq \Gamma$ has a model in $\mathcal{K}$. In our case $\mathcal{K}$ is the collection of all finitely generated substructures of $\mathfrak{A}$. As each finite subset of $\Delta^0_{\mathfrak{A}}$ mentions finitely many elements of $\mathfrak{A}$ only, that subset holds in that finitely generated substructure. Consequently the required ultraproduct exists.

627. As all finite subsets of $T(x)$ can be realized in $\Pi_i \mathfrak{A}_i / \mathcal{U}$, for each $k \in \omega$ the set

$$X_k = \{i \geq k : \mathfrak{A}_i \models \exists x (\varphi_0(x) \wedge \cdots \wedge \varphi_{k-1}(x))\}$$

belongs to $\mathcal{U}$. For $i \in X_k - X_{k+1}$ choose any realization $a(i)$ of $\varphi_0 \wedge \cdots \wedge \varphi_{k-1}$ in $\mathfrak{A}_i$. Then $\langle a(i) : i \in \omega \rangle / \mathcal{U}$ realizes $T(x)$ as for any $k \in \omega$ fixed we have

$$X_{k+1} \subseteq \{i \in \omega : \mathfrak{A}_i \models \varphi_k[a(i)]\} \in \mathcal{U}.$$

Remark. See also Problem 677.

628. It is clear that ${}^\kappa \mathfrak{A}_k / \mathcal{U}$ is a substructure of ${}^\omega \mathfrak{A} / \mathcal{U}$, and also that ${}^\kappa \mathfrak{A}_1 / \mathcal{U}$ and ${}^\kappa \mathfrak{A}_2 / \mathcal{U}$ have disjoint ground sets (as the product structures are disjoint). Thus it suffices to show that every element in ${}^\kappa \mathfrak{A} / \mathcal{U}$ is also an element either of ${}^\kappa \mathfrak{A}_1 / \mathcal{U}$ or ${}^\kappa \mathfrak{A}_2 / \mathcal{U}$. Let $a \in {}^\kappa A$ and define

$$X_k = \{i \in \kappa : a(i) \in A_k\}.$$

As X_1 and X_2 are complements, exactly one of them is in $\mathcal{U}$, say X_1. Pick $s \in A_1$, and define $a' \in \kappa A$ as

$$a'(i) = \begin{cases} a(i) & \text{if } i \in X_1, \\ s & \text{if } i \in X_2. \end{cases}$$

Then $a' / \mathcal{U} = a / \mathcal{U}$, thus $a / \mathcal{U} \in {}^\kappa \mathfrak{A}_1 / \mathcal{U}$, as required.

629. Let φ be the formula that expresses "each vertex has degree two" and let ψ_k express "no k vertices form a circle." Łoś lemma and $G_i \models \varphi$ implies $\Pi_{i \in \omega} G_i / \mathcal{U} \models \varphi$, further $\Pi_{i \in \omega} G_i / \mathcal{U} \models \psi_k$ because $\{i \in \omega : G_i \models \psi_k\}$ is co-finite. By Problem 610, $|\Pi_{i < \omega} G_i / \mathcal{U}| = 2^\omega$.

It follows that $\Pi_{i \in \omega} G_i / \mathcal{U}$ is a cycle-free graph of cardinality continuum, each of its vertices has exactly two neighbors. Such a graph is isomorphic to a union of continuum many infinite paths without endpoints.

630. Let φ be the formula expressing "all vertices have exactly two neighbors, except for two which have one", and let ψ_k express "there are no k vertices that form a circle." Clearly $G_n \models \varphi, \psi_k$, hence by Łoś lemma we have

$\Pi_{n\in\omega}G_n/\mathcal{U} \vDash \varphi, \psi_k$. By Problem 610 the cardinality of $\Pi_{n\in\omega}G_n/\mathcal{U}$ is continuum. It follows that $\Pi_{n\in\omega}G_n/\mathcal{U}$ is a cycle-free graph of cardinality continuum, each of its vertices has exactly two neighbors, except for two vertices which have one neighbor. Such a graph consists of infinitely many infinite paths without endpoints together either with two infinite paths that have endpoints or with one finite path (that has two endpoints). In our situation this latter cannot be the case, which can be seen as follows. Let ϑ_k express "the distance of the two vertices that has one neighbor is k". Then for all n, by Łoś lemma, we have $\Pi_{n\in\omega}G_n/\mathcal{U} \nvDash \vartheta_k$, hence the ultraproduct does not contain any finite path.

631. We prove that if G is such that for any $n \in \omega$ there exist two vertices such that the shortest path between them has length at least n, then $^\omega G/\mathcal{U}$ is not connected. For $n \in \omega$ let v_n and w_n be vertices witnessing that the diameter of G is at least n. Then for all k we have that the set

$$\{n \in \omega : G_n \vDash \text{"the shortest path between } v_n \text{ and } w_n \text{ has length at most } k\text{"}\}$$

does not belong to $\mathcal{U}$. It follows that there is no finite path between $v = \langle v_n : n \in \omega\rangle/\mathcal{U}$ and $w = \langle w_n : n\in\omega\rangle/\mathcal{U}$, consequently $\Pi_{n\in\omega}G_n/\mathcal{U}$ is not connected.

632. Let G_1 be the countable 3-regular, cycle free, connected graph (an infinite tree with degrees three: such a graph is unique), and let G_2 be the disjunct union of two copies of G_1. As G_2 is not connected, G_1 and G_2 are not isomorphic. The ultrapower is again 3-regular and cycle-free, therefore all of its connected components are isomorphic to G_1. The cardinality of $^\omega G_i/\mathcal{U}$ is continuum (Problem 609), hence it is isomorphic to the union of continuum many copies of G_1. Thus $^\omega G_1/\mathcal{U} \cong {}^\omega G_2/\mathcal{U}$. See also Problem 360.

The case of 4-regular cycle-free graphs is completely analogous. Up to isomorphism there is a unique connected graph of this type, and the ultrapower consists of continuum many disjoint instances.

633. In $^\omega H/\mathcal{U}$ there is exactly one vertex that has two neighbors and all the other vertices have three neighbors. Also $^\omega H/\mathcal{U}$ must be cycle-free as H does not contain cycles. Since H is embeddable into $^\omega H/\mathcal{U}$ (Problem 611), the connected component which contains the vertex that has two neighbors is isomorphic to H. All the other components are connected 3-regular cycle-free graphs, which are unique, up to isomorphism. By Problem 609 the cardinality of $^\omega H/\mathcal{U}$ is continuum, therefore

$$^\omega H/\mathcal{U} \cong H \cup \bigcup_{\alpha < \mathfrak{c}} \text{(the 3-regular connected cycle-free graph)}.$$

634. $^\omega G/\mathcal{U}$ has cardinality continuum. The following sentences are valid in G thus they are valid in $^\omega G/\mathcal{U}$:

  (i) each vertex has either one, two or more than three neighbors,

  (ii) no n points form a cycle,

 (iii) there is exactly one vertex that has more than three neighbors,

(iv) the degree-one vertices are connected to the vertex that has more than three neighbors.

In what follows v denotes the vertex of G that has infinitely many neighbors, the x_i's are the degree-one vertices, and the y_i's refer to degree-two vertices, as pictured below. $\delta : G \to {}^\omega G/\mathcal{U}$ is the diagonal embedding (see Problem 611).

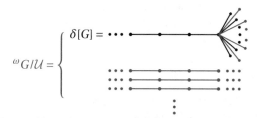

The product ${}^\omega G/\mathcal{U}$ has cardinality continuum, contains $\delta[G]$, thus there must be additional vertices, which must be of degree one or degree two (sentences (i) and (iii)). As there are no cycles in G, the additional degree-two vertices form infinite paths, and the degree one vertices are connected to $\delta(v)$.

A new degree one vertex which is connected to $\delta(v)$ is $\langle x_i : i \in \omega \rangle / \mathcal{U}$. If X denotes $\{x_i : i \in \omega\}$, then it is not hard to check that degree-one neighbors of $\delta(v)$ are exactly elements of ${}^\omega X/\mathcal{U}$. As the cardinality of ${}^\omega X/\mathcal{U}$ is 2^ω, we get continuum many new neighbors.

As for degree-two vertices, notice that $\langle y_i : i \in \omega \rangle / \mathcal{U}$ and $\langle y_{i+1} : i \in \omega \rangle / \mathcal{U}$ are connected, thus there must appear new infinite paths in ${}^\omega G/\mathcal{U}$. Similarly as above, if $Y = \{y_i : i \in \omega\}$, then elements of ${}^\omega Y/U$ are the degree-two vertices, which form continuum many infinite paths. In summary, the ultraproduct is depicted below with $\delta[G]$ in black and the new items in red.

635. By the Erdős–deBruijn theorem (Problem 265) there must exist a finite subgraph $H \subset \Pi_{i \in \omega} G_i/\mathcal{U}$ that has chromatic number k. The graph H, being finite, can be described, up to isomorphism, by a single formula $\varphi_H(\vec{x})$ having $|H|$-many free variables. (For example, up to equivalence the diagram Δ_H contains finitely many formulas, take their conjunction, and replace the constants by variables.) By Łoś lemma it follows that

$$J = \{n \in \omega : G_n \vDash \exists \vec{x} \varphi_H(\vec{x})\} \in \mathcal{U}.$$

For each $i \in J$, the graph H can be embedded into G_i.

636. Solution 1. Each element $\hat{a} = \langle a(i) : i \in \omega \rangle / \mathcal{U}$ has an immediate successor $\hat{a} + 1 = \langle a(i) + 1, : i \in \omega \rangle / \mathcal{U}$ and each non-zero element has an immediate predecessor $\hat{a} - 1 = \langle a(i) \dotdiv 1 : i \in \omega \rangle / \mathcal{U}$. The cardinality of $^\omega\mathfrak{A}/\mathcal{U}$ is continuum (Problem 609), hence if we denote by $\mathfrak{B}$ the integers with the successor function, then we get

$$^\omega\mathfrak{A}/\mathcal{U} \cong \mathfrak{A} \cup \bigcup_{\alpha < \mathfrak{c}} \mathfrak{B}.$$

Solution 2. Problem 501 describes all models of a theory Γ and proves that Γ is κ-categorical for each $\kappa \geq \omega_1$. As $\mathfrak{A} \vDash \Gamma$, Łoś lemma gives that the ultrapowers are also models Γ. The cardinality of any non-trivial ultrapower is $2^\omega \geq \omega_1$, thus all these ultrapowers are isomorphic to the structure described above.

637. Γ can be the set

$$0 \neq S(x), \quad x \neq 0 \to \exists y (x = S(y)), \quad x \neq y \to S(x) \neq S(y).$$

Let $\mathfrak{A} \vDash \Gamma$. Following the method used in Problem 501, draw a directed edge from x to $S(x)$. From the formulas it follows that every node, except for $0^\mathfrak{A}$ has in-degree exactly one. Consequently $\mathfrak{A}$ consists of an infinite path starting from $0^\mathfrak{A}$ (which is isomorphic to ω), paths infinite in both directions (isomorphic to $\mathbb{Z}$), and loops of size one (when $f(x) = x$), two, etc. Therefore a countable model of Γ is determined uniquely by the number of loops and infinite paths: each such number is either a non-negative integer, or countably infinite.

Let $\mathfrak{A} \vDash \Gamma$ be countable. Then $\mathfrak{A}_1 = \langle \omega, S \rangle$ is a substructure of $\mathfrak{A}$, thus by Problem 628 $^\omega\mathfrak{A}/\mathcal{U}$ contains $^\omega\mathfrak{A}_1/\mathcal{U}$ as a substructure. According to Problem 636, this ultrapower has a single instance of $\langle \omega, S \rangle$, and continuum many instances of $\langle \mathbb{Z}, S \rangle$. Consequently the ultrapower $^\omega\mathfrak{A}/\mathcal{U}$ also contains continuum many instances of $\langle \mathbb{Z}, S \rangle$, and not more, as its cardinality is continuum.

It remained to determine the number of loops of size $k \geq 1$ in $^\omega\mathfrak{A}/\mathcal{U}$. Suppose there are $s \in \omega$ such loops in $\mathfrak{A}$. There is a formula $\varphi_{s,k}$ expressing that there are exactly s loops of size k (including the case $s = 0$. As the same formula holds in the ultrapower by Łoś lemma, there will be exactly s loops of size k there as well.

In the case when there are countably many loops of size k in $\mathfrak{A}$, then $\mathfrak{A}$ is the disjoint union of structures $\mathfrak{A}_1^k$ and $\mathfrak{A}_2^k$ where $\mathfrak{A}_1^k$ contains the k-loops only. Now the ultrapower $^\omega\mathfrak{A}_1^k/\mathcal{U}$ has cardinality continuum, consists of k-loops only (as this is a property expressible by a formula), thus has continuum many k-loops. It means that $^\omega\mathfrak{A}/\mathcal{U}$ has also continuum many k-loops.

In summary: in the ultrapower $^\omega\mathfrak{A}/\mathcal{U}$ there is a single copy of $\langle \omega, S \rangle$, continuum many copies of $\langle \mathbb{Z}, S \rangle$, and for each $k \geq 1$, either there are finitely many k-loops (exactly as many as in $\mathfrak{A}$), or there are continuum many k-loops (if the number of k-loops in $\mathfrak{A}$ is infinite).

638. $^\omega\mathfrak{A}/\mathcal{U}$ is a linear order of cardinality continuum (Problem 609) with a smallest element $\langle 0 : i \in \omega \rangle / \mathcal{U}$ and without largest element. For $a \in {}^\omega A$ the immediate successor of $a/\mathcal{U}$ is $\langle a(i) + 1 : i \in \omega \rangle / \mathcal{U}$, and the immediate predecessor is $\langle a(i) \dot{-} 1 : i \in \omega \rangle / \mathcal{U}$. The distance from $a_1 / \mathcal{U}$ to $a_2 / \mathcal{U}$ is infinite if for all $n \in \omega$ we have

$$X_n = \{i \in \omega : a_2(i) - a_1(i) > n\} \in \mathcal{U}.$$

Set $c(i)$ such that for all $n \in \omega$

$$\{i \in \omega : a_2(i) - c(i) > n/3 \text{ and } c(i) - a_1(i) > n/3\} \in \mathcal{U}$$

(e.g., set $c(i) = [a(i) + b(i)/2]$). Then $c/\mathcal{U}$ is as desired.

639. Similarly to Problem 638, for $a \in {}^\omega \mathbb{Z}$, the immediate predecessor of $a/\mathcal{U}$ is $\langle a(i) - 1 : i \in \omega \rangle / \mathcal{U}$, and its immediate successor if $\langle a(i) + 1 : i \in \omega \rangle / \mathcal{U}$. The distance between $a_1 / \mathcal{U}$ and $a_2 / \mathcal{U}$ is finite if they are on the same copy of $\mathbb{Z}$, which happens if and only if there is an $n \in \omega$ such that

$$\{i \in \omega : |a_1(i) - a_2(i)| < n\} \in \mathcal{U}.$$

Let $a' = \langle a(i) - i : i \in \omega \rangle$, then $a'/\mathcal{U} < a/\mathcal{U}$, and the distance between them is infinite – thus M has no smallest element. Similarly, $\langle a(i) + i : i \in \omega \rangle / \mathcal{U}$ shows that M has no largest element either.

Finally if $a_1 / \mathcal{U} < a_2 / \mathcal{U}$ are not on the same copy of $\mathbb{Z}$, then their distance is infinite. Taking $c(i) = [(a_1(i) + a_2(i))/2]$, $c/\mathcal{U}$ is between $a_1 / \mathcal{U}$ and $a_2 / \mathcal{U}$, and the distance between $a_i / \mathcal{U}$ and $c/\mathcal{U}$ is infinite – showing that M is dense.

> **Remark.** If the continuum hypothesis holds, then ultrapowers ${}^\omega \mathfrak{A}/\mathcal{U}$ for different non-trivial ultrafilters $\mathcal{U}$ are actually isomorphic (Keisler's theorem, see Problem 679). If the continuum hypothesis fails, then there are 2^{2^ω} pairwise non-isomorphic ultrapowers ${}^\omega \mathfrak{A}/\mathcal{U}$. This result of S. Shelah holds for any countable structure $\mathfrak{A}$ on a countable similarity type in which an infinite total order is definable.

640. Let $B = \{b_i / \mathcal{U} : i < \omega\}$ be a countable set of elements from the ultraproduct, and write $x(n) = \max\{b_i(n) : i \leq n\} + 1$. Then for each $i \in \omega$ the set $\{n \in \omega : x(n) > b_i(n)\}$ is co-finite, hence belongs to $\mathcal{U}$, therefore $x/\mathcal{U}$ is an upper bound for B.

> **Remark.** This is a standard diagonal argument, see also Problem 677.

As for the second part, we define a strictly increasing sequence $\langle x_\alpha : \alpha < \omega_1 \rangle$ as follows. Let $x_0 \in {}^\omega A/\mathcal{U}$ be arbitrary and suppose for $\beta < \omega_1$ the sequence $\langle x_\alpha : \alpha < \beta \rangle$ has already been defined. Then the set $B_\beta = \{x_\alpha : \alpha < \beta\}$ is countable, thus by the first part it is bounded. Choose x_β to be any upper bound for B_β.

641. Let $\mathfrak{A} = \langle \omega, < \rangle$, we claim that the ultrapower ${}^\omega \mathfrak{A}/\mathcal{U}$ has this property. Let $(a_n / \mathcal{U}, b_n / \mathcal{U})$ be the endpoints of the strictly decreasing interval sequence. First note that there are infinitely many distinct elements between $a_n / \mathcal{U}$ and $b_n / \mathcal{U}$, thus $\{i \in \omega : a_n(i) + 10 < b_n(i)\} \in \mathcal{U}$, see Solution 638. Also, $a_{n-1} / \mathcal{U} < a_n / \mathcal{U}$ and $b_n / \mathcal{U} < b_{n-1} / \mathcal{U}$, therefore

$$X_n = \{i \in \omega : a_{n-1}(i) < a_n(i) < a_n(i) + 10 < b_n(i) < b_{n-1}(i)\} \in \mathcal{U}.$$

Let $P_k = X_1 \cap \cdots \cap X_k \in \mathcal{U}$. Observe that if $i \in P_k$ then

$$a_0(i) < \cdots < a_k(i) < a_k(i) + 10 < b_k(i) < \cdots < b_0(i).$$

For each i let k be the maximal such that $i \in P_k$ (there can only be finitely many such k), and define $x(i) = a_k(i) + 1$, $y(i) = b_k(i) - 1$. If no such a k exists, let $x(i) = y(i) = 0$. Now $a_k(i) < x(i) < y(i) < b_k(i)$ whenever $i \in P_k \in \mathcal{U}$, thus $a_k/\mathcal{U} < x/\mathcal{U} < y/\mathcal{U} < b_k/\mathcal{U}$. It means that the non-empty interval $(x/\mathcal{U}, y/\mathcal{U})$ is in the intersection of all $(a_k/\mathcal{U}, b_k/\mathcal{U})$.

642. It is straightforward that $\mathbb{Q}$ does not have the desired property, therefore, using Problem 365, such an ordering must be uncountable.

The ultrapower $\mathfrak{M} = {}^\omega\mathbb{Q}/\mathcal{U}$, where $\mathcal{U}$ is a non-trivial ultrafilter on ω, is a dense ordering of cardinality $2^{\aleph_0}$. Suppose $A = \{a_i/\mathcal{U} : i \in \omega\}$ and $B = \{b_i/\mathcal{U} : i \in \omega\}$ are countable subsets of M such that $A < B$. It is enough to define two elements $x/\mathcal{U}$ and $y/\mathcal{U}$ in the ultrapower so that $A < x/\mathcal{U} < y/\mathcal{U} < B$. Since $a_i/\mathcal{U} < b_j/\mathcal{U}$ for all i, j, we have that

$$X_{k,\ell} = \{i \geq k, \ell : a_k(i) < b_\ell(i)\} \in \mathcal{U},$$
$$X_k = \bigcap_{\ell \leq k} X_{k,\ell} = \{i \geq k : a_k(i) < b_0(i), \ldots, b_k(i)\} \in \mathcal{U},$$
$$P_k = \bigcap_{j \leq k} X_j = \{i \geq k : a_0(i), \ldots, a_k(i) < b_0(i), \ldots, b_k(i)\} \in \mathcal{U}.$$

Clearly $P_0 \supseteq P_1 \supseteq P_2 \supseteq \cdots$. For each $i \in \omega$ let k be the largest number such that $i \in P_k$, and define $x(i)$ and $y(i)$ such that

$$\max\{a_j(i) : j \leq k\} < x(i) < y(i) < \min\{b_j(i) : j \leq k\}.$$

This can be done as $\mathbb{Q}$ is dense. Finally, note that for all k we have $\{i \geq k : a_k(i) < x(i) < y(i) < b_i(i)\} \supseteq P_i \in \mathcal{U}$, consequently $A < x/\mathcal{U} < y/\mathcal{U} < B$.

> **Remark.** Problem 681 provides an alternative construction as a non-trivial ultrapower of $\mathbb{Q}$ is σ-saturated by Problem 677 or Problem 682.

643. (a) Let the decreasing sequence be $\langle a_n/\mathcal{U} : n \in \omega \rangle$ where $a_n \in {}^\omega\mathbb{R}$. For each $k \in \omega$ the set

$$X_k = \{i \geq k : 0 < a_k(i) < a_{k-1}(i) < \cdots < a_0(i)\}$$

belongs to $\mathcal{U}$. Choose $b(i)$ so that

$$0 < b(i) < a_k(i) < a_{k-1}(i) < \cdots < a_0(i) \quad \text{if} \quad i \in X_k - X_{k+1}.$$

Then for all $k \in \omega$ we get

$$X_k \subseteq \{i \in \omega : 0 < b(i) < a_k(i)\} \in \mathcal{U},$$

in particular $0^{\mathfrak{A}} < b/\mathcal{U} < a_k/\mathcal{U}$ for all $k \in \omega$.

(b) Let $B' = \{\langle b : i \in \omega \rangle / \mathcal{U} : b \in B\}$ be the image of B w.r.t the diagonal embedding, see Problem 611. B' has an upper bound (e.g. $\langle n : n \in \omega \rangle / \mathcal{U}$) but has no least upper bound. For $B' < x/\mathcal{U}$ if and only if the sequence $\langle x(i) : i \in \omega \rangle$ is unbounded on some element of $\mathcal{U}$, and then $B' < x'/\mathcal{U}$, where $x'(i) = x(i) - 1$ (and, of course, $x'/\mathcal{U} < x/\mathcal{U}$).

644. There must exist a universal bound for the length of E-chains. If there is some $n \in \omega$ such that all E-chains are of length $\leq n$ in $\mathfrak{A}$, then there can be no E-chain in $^\omega\mathfrak{A}/\mathcal{U}$ longer than n, thus each non-empty subset of $^\omega\mathfrak{A}/\mathcal{U}$ has an E-minimal element. Conversely, suppose for each n there is an E-sequence $a_n^n \mathrel{E} \cdots \mathrel{E} a_2^n \mathrel{E} a_1^n$ of length n. Choose an arbitrary $a \in A$ and for $j > n$ write $a_j^n = a$. Let $b_i = \langle a_i^n : n < \omega \rangle/\mathcal{U}$. Then $\cdots \mathrel{E} b_n \mathrel{E} \cdots \mathrel{E} b_1$ is an infinite decreasing E-chain in $^\omega\mathfrak{A}/\mathcal{U}$ which does not have a minimal element.

645. (a) Let $X = \{i_1, i_2, \ldots\}$. By assumption, for all $n \in \omega$ there exists $a(n) \in A$ such that $a(n)$ satisfies $R_{i_1}, \ldots, R_{i_n}$ but does not satisfy R_j for $j \notin X$. It follows that if $n > k$, then $a(n)$ satisfies R_{i_k} and thus the set

$$\{n \in \omega : a(n) \in R_{i_k}^{\mathfrak{A}}\}$$

is co-finite and belongs to $\mathcal{U}$. By Łoś lemma we get $a/\mathcal{U} \in R_{i_k}^{^\omega\mathfrak{A}/\mathcal{U}}$. Similarly, $a(n) \notin R_j^{\mathfrak{A}}$ for $j \notin X$, and hence $a/\mathcal{U} \notin R_j^{^\omega\mathfrak{A}/\mathcal{U}}$.

(b), (c) We prove first that for $\{i_1, \ldots, i_n\} \cap \{j_1, \ldots, j_n\} = \emptyset$ the intersection

$$A_1 = R_{i_1}^{\mathfrak{A}} \cap \cdots \cap R_{i_n}^{\mathfrak{A}} \cap (A - R_{j_1}^{\mathfrak{A}}) \cap \cdots \cap (A - R_{j_m}^{\mathfrak{A}})$$

is always infinite. Indeed, elements which satisfy $R_{i_1}, \ldots, R_{i_n}, R_{k_1}, \ldots, R_{k_\ell}$ where the latter relation symbols differ from each R_i and R_j, are all different and are in the intersection.

Let $X = \{i_1, i_2, \ldots\}$ and $\omega - X = \{j_1, j_2, \ldots\}$ and consider the ultraproduct

$$\mathfrak{B} = \Big(\prod_{n \in \omega} R_{i_1}^{\mathfrak{A}} \cap \cdots \cap R_{i_n}^{\mathfrak{A}} \cap (A - R_{j_1}^{\mathfrak{A}}) \cap \cdots \cap (A - R_{j_n}^{\mathfrak{A}}) \Big)/\mathcal{U} \subseteq {}^\omega\mathfrak{A}/\mathcal{U}.$$

A moment of thought shows that each element $b \in B$ satisfies $R_k^{\mathfrak{B}}$ if and only if $k \in X$. By Problem 609, $\mathfrak{B}$ has cardinality continuum.

646. Let p_n be the n-th prime and denote by G_n the cyclic group of order p_n. If $a/\mathcal{U} \in \Pi_n G_n/\mathcal{U}$ is not the unit element and $n \in \omega$, then

$$\begin{aligned} (a/\mathcal{U})^n = 1/\mathcal{U} \quad &\Leftrightarrow \quad \langle a(i)^n : i \in \omega \rangle/\mathcal{U} = 1/\mathcal{U} \\ &\Leftrightarrow \quad \{i \in \omega : a(i)^n = 1\} \in \mathcal{U} \\ &\Leftrightarrow \quad \{i \in \omega : p_i | n\} \in \mathcal{U} \end{aligned}$$

Obviously, no natural number is divisible by infinitely many primes, hence $a/\mathcal{U}$ has infinite order, thus $\Pi_n G_n/\mathcal{U}$ is torsion free.

647. The element $\langle n! : n \in \omega \rangle/\mathcal{U}$ is infinitely divisible.

As for the free group G on generators X, note that all groups with generators X are homomorphic images of G, and the homomorphic image of a divisible group is a divisible group. As there are non-divisible groups generated by X, G cannot be divisible. However, G might still contain infinitely divisible elements. To see why it is not the case observe that cyclically reduced words, i.e. words of the form $w = x_1 x_2 \ldots x_n$, where $x_i \neq x_j^{-1}$ (the x_i's are the generators), are torsion free. As every word is the conjugate of cyclically reduced words (cyclically reducing a word amounts to conjugating by the indices which cancel), every word must be torsion free.

If g is any of the generators, then $\langle g^{i!} : i \in \omega \rangle / \mathcal{U}$ is infinitely divisible.

648. (a) Correct, see Problem 610.

(b) Incorrect. Each cyclic group of even order contains elements of order two. The equivalence class of the element defined below has order two in $\Pi_i C_i / \mathcal{U}$, provided $\{2i : i \in \omega\} \in \mathcal{U}$.

$$g(i) = \begin{cases} \text{arbitrary} \in C_i & \text{if } i \text{ is odd,} \\ \text{element of order two} \in C_i & \text{if } i \text{ is even.} \end{cases}$$

(c) Correct. Suppose $P = \{p : p \text{ is prime}\} \in \mathcal{U}$ and let $a/\mathcal{U} \in \Pi_i C_i / \mathcal{U}$ be a non-zero element. We wish to show that for all n we have $n \cdot a/\mathcal{U} \neq 0/\mathcal{U}$. By assumption $N = \{i \in \omega : a(i) \neq 0\} \in \mathcal{U}$, thus $n \cdot a(i) \neq 0$ for all $i \in N \cap \{p \in P : n < p\}$. As $N \cap \{p \in P : n < p\} \in \mathcal{U}$, we get

$$N \cap \{p \in P : n < p\} \subseteq \{i \in \omega : n \cdot a(i) \neq 0\} \in \mathcal{U}.$$

(d) Incorrect. $\Pi_i C_i / \mathcal{U}$ always contains an element of infinite order: If $g_i \in C_i$ is an element of order i, then $\langle g_i : i \in \omega \rangle / \mathcal{U}$ has infinite order.

649. If $\{n : \mathfrak{F}_n \text{ is a prime order field}\} \in \mathcal{U}$, then $\Pi_n \mathfrak{F}_n / \mathcal{U}$ has characteristics 0. Similarly, if $\{n : \mathfrak{F}_n \text{ has order } p^k\} \in \mathcal{U}$ then $\Pi_n \mathfrak{F}_n / \mathcal{U}$ has characteristics p.

650. We prove that $\Pi_{p \text{ prime}} \mathbb{F}_p / \mathcal{U}$ is not algebraically closed. In fact, there is an element which has no square root. For each prime $p \geq 3$ there is an element $x_p \in \mathbb{F}_p$ which is not a square in $\mathbb{F}_p$ (not a quadratic residue mod p). Then $\langle x_p : p \text{ is prime} \rangle / \mathcal{U}$ does not have a square root in $\Pi_{p \text{ prime}} \mathbb{F}_p / \mathcal{U}$.

651. Take a polynomial $f(x) = a_n x^d + \cdots + a_1 x + a_0 \in \mathbb{F}_p[x]$ with non-zero leading coefficient. The sentence "f has a root" can be expressed as

$$\varphi_f \equiv \exists x (t(a_d) x^d + \cdots + t(a_1) x + t(a_0) = 0),$$

where $t(a_i)$ is the term $(0 + 1 + \cdots + 1)$ which equals to a_i in any field of characteristic p. Now f has all its roots in $\mathbb{F}_{p^n}$ for some n, and then in $\mathbb{F}_{p^{nk}}$ for all $k \geq 1$. If $\mathcal{U}$ contains the set $\{k! : k \in \omega\}$, then we have $\{i \in \omega : \mathbb{F}_{p^i} \models \varphi_f\} \in \mathcal{U}$. Therefore each $f \in \mathbb{F}_p[x]$ has all its roots in $\Pi_{i \in \omega} \mathbb{F}_{p^i} / \mathcal{U}$.

652. There is a quadratic equation which has no root. Let $p \geq 3$, then $x \mapsto x^2$ is not an injection in $\mathbb{F}_p^n$ (as x and $-x$ has the same square), thus $\mathbb{F}_p^n \models \exists a \forall x (x^2 - a \neq 0)$. Consequently the same formula is true in the ultraproduct, showing that it is not algebraically closed.

When $p = 2$ then one has to take the polynomial $x^2 + x$. As $x^2 + x = (x + 1)^2 + (x + 1)$, in every finite field of characteristic two there is an element which is not of this form, thus the ultraproduct satisfies $\exists a \forall x (x^2 + x + a \neq 0)$.

653. By way of contradiction suppose that there are fields of arbitrarily large characteristic p such that $\mathfrak{F}_p \nvDash \varphi$. As the characteristics of the fields $\mathfrak{F}_p$ are different, the ultraproduct $\Pi_p \mathfrak{F}_p / \mathcal{U}$ has characteristics 0 for any non-trivial ultrafilter $\mathcal{U}$. By Łoś lemma $\Pi_p \mathfrak{F}_p / \mathcal{U} \nvDash \varphi$, contradicting the assumption.

654. Suppose each polynomial $f \in \mathfrak{F}_n[x]$ of degree at most n has a root in $\mathfrak{F}_n$ (which is of characteristic p), while $\mathfrak{F}_n$ is not algebraically closed (i.e. higher degree polynomials can be irreducible). This property can be expressed by the first-order formula $\varphi_1 \wedge \varphi_2 \wedge \cdots \wedge \varphi_n$ where

$$\varphi_j \equiv \forall a_0 \ldots \forall a_j (a_j \neq 0 \rightarrow \exists x (a_j x^j + \cdots + a_1 x + a_0 = 0)).$$

Let $\mathfrak{F} = \Pi_n \mathfrak{F}_n / \mathcal{U}$ for any non-trivial ultrafilter $\mathcal{U}$ on ω. Then $\mathfrak{F} \vDash \varphi_n$ for all $n \in \omega$, which means that $\mathfrak{F}$ is algebraically closed. Note that $\mathfrak{F}$ also has characteristics p. So it remained to construct the fields $\mathfrak{F}_n$.

Let $\mathbb{F}_p$ be the p-element field (in case $p = 0$ start from $\mathbb{Q}$), let n be fixed and consider the field extensions

$$\mathbb{F}_p = F_0^n \subset F_1^n \subset F_2^n \subseteq \cdots$$

where F_{i+1}^n is the algebraic extension by the roots of all polynomials of degree at most n with coefficients from F_i^n. The degree of every element in F_{i+1}^n over F_i^n is at most n, consequently the degree of $a \in F_i^n$ over F_0^n is the product of i numbers each of which is at most n.

Write $\mathfrak{F}_n$ for the union $\bigcup_i F_i^n$. All coefficients of $f(x) \in \mathfrak{F}_n[x]$ are in F_i^n for some i, thus if f has degree at most n, then f has roots in $F_{i+1}^n \subset \mathfrak{F}_n$. To prove that $\mathfrak{F}_n$ is not algebraically closed, observe that the degree of each element in $\mathfrak{F}_n$ over $F_0^n = \mathbb{F}_p$ is a product of numbers $\leq n$. Thus any irreducible polynomial over $\mathbb{F}_p$ of degree a prime larger than n has no root in $\mathfrak{F}_n$.

655. Remark first that there are regular ultrafilters with $|E| = \kappa$ for every infinite cardinal κ (Problem 57), thus the infinite collection of polynomials can always be indexed by elements of such an E.

If $\mathcal{U}$ is regular, then $\{p_e = 0 : \xi \in e\}$ is finite for every $\xi \in I$, thus there is indeed a common solution f_ξ which satisfies $p_e[f_\xi] = 0$ when $\xi \in e$. By Problem 624 we have $p_e[\langle f_\xi \rangle / \mathcal{U}] = 0$ for all $e \in E$. As each f_ξ maps the variables to the finite set $\mathbb{F}$ (elements of the finite field), $\langle f_\xi \rangle / \mathcal{U}$ also maps the variables to the same finite set, providing the required common solution.

   Remark. Compare this solution to that of Problem 626.

656. The argument of Solution 655 also works here. Since E is countable, the regular ultrafilter $\mathcal{U}$ can be chosen over the countable set ω. Each partial solution f_ξ maps the variables to the structure $\mathbb{C}$, and, according to Problem 624, $\langle f_\xi \rangle / \mathcal{U}$ maps the variables to the ultrapower $^\omega \mathbb{C} / \mathcal{U}$. As this ultrapower is isomorphic to $\mathbb{C}$, the system has a solution in $\mathbb{C}$ as well.

For a different solution see Problem 680.

657. Fix any polynomial $f \in F[x]$ and denote by δ the diagonal embedding $\delta : \mathfrak{F} \to \mathfrak{B}$ (see Problem 611). We prove that no element in $B \smallsetminus F$ is a root of $\delta(f)$. For if $\mathfrak{B} \vdash \delta(f)[g/\mathcal{U}] = 0/\mathcal{U}$ for some $g/\mathcal{U} \notin F$, then

$$\{i \in \kappa : f[g(i)] = 0\} \in \mathcal{U}.$$

This set must be infinite (as $\mathcal{U}$ is non-trivial), and since g is not a diagonal element, it follows that f must have infinitely many different roots.

658. Note that in an ordered field squares are non-negative, and $0 < 1$ and $-1 < 0$ always hold. The set of non-negative elements determines the ordering uniquely as $x \leq y$ iff $0 \leq y - x$.

(a) Each square is non-negative, and for each real x either x or $-x$ is a square.

(b), (d) Finite fields have characteristics p for some prime number p, so it is enough to prove that fields of non-zero characteristics cannot be turned into ordered fields. This follows easily from the equation $0 < (p-1) \cdot 1 = -1 < 0$.

(c) The property that a field is orderable can be expressed by a first-order formula (in the language extended by a relation for $\leq$). If $\Pi_{n \in \omega} \mathbb{F}_n / \mathcal{U}$ were orderable, then the set of $\mathbb{F}_n$'s that are orderable would belong to $\mathcal{U}$, in particular, it would be non-empty, which contradicts (b).

(e) $\mathbb{Q}(\sqrt{2})$ denotes the extension of $\mathbb{Q}$ with $\sqrt{2}$. Each element of $\mathbb{Q}(\sqrt{2})$ can be uniquely written in the form $a + b\sqrt{2}$, where $a, b \in \mathbb{Q}$. Let $<_1$ be the ordering inherited from the reals, and let $<_2$ be defined such that $\sqrt{2} <_2 0$. Then $<_2$ is a field-ordering which can be seen using that the automorphism $f(a + b\sqrt{2}) = a - b\sqrt{2}$ of $\mathbb{Q}(\sqrt{2})$ turns $<_1$ to $<_2$.

(f) That the condition is necessary is straightforward. Let P the set of sums of squares

$$P = \left\{ \sum_{i \in n} a_i^2 : a_i \in F, n \in \omega \right\}.$$

It can be checked that P is closed under addition and multiplication, and by assumption $-1 \notin P$ (such sets are called *cones*). In fact, $P - \{0\}$ is a subgroup of the multiplicative group of $\mathfrak{F}$. Using Zorn's lemma we can find a maximal cone P' extending P. Then $F = P' \cup (-P')$ and thus an ordering can be defined as

$$0 < a \quad \Leftrightarrow \quad a \in P'.$$

659. Let $\mathcal{U}$ be a regular ultrafilter on κ (see Problem 57) with the corresponding set $E \subseteq \mathcal{U}$, $|E| = \kappa$ for which $\gamma(\alpha) = \{e \in E : \alpha \in e\}$ is finite for all $\alpha \in \kappa$. It is straightforward to see that

$$\left| {}^\kappa A / \mathcal{U} \right| \leq \left| {}^\kappa A \right| = \lambda^\kappa.$$

Since $|A| = \lambda$ is infinite, it follows that $|A| = |{}^{\gamma(\alpha)} A|$, therefore we may replace A with ${}^{\gamma(\alpha)} A$ and it is enough to prove

$$\left| \left(\Pi_{\alpha < \kappa} {}^{\gamma(\alpha)} A \right) / \mathcal{U} \right| \geq \lambda^\kappa.$$

Let $\Phi : {}^E A \to \left(\Pi_{\alpha < \kappa} {}^{\gamma(\alpha)} A \right) / \mathcal{U}$ be the mapping $\Phi(s) = \langle s \restriction \gamma(\alpha) : \alpha < \kappa \rangle / \mathcal{U}$. For $s \neq t \in {}^E A$ we have

$$\{ \alpha \in \kappa : s \restriction \gamma(\alpha) = t \restriction \gamma(\alpha) \} \notin \mathcal{U},$$

thus Φ is injective, meaning that

$$\lambda^\kappa = \left| {}^E A \right| \leq \left| \left(\Pi_{\alpha < \kappa} {}^{\gamma(\alpha)} A \right) / \mathcal{U} \right|.$$

Remark. See also Solution 609.

660. If $\omega \leq \lambda < \kappa$, then there exists a non-uniform ultrafilter $\mathcal{U}$ on κ which has an element $X \in \mathcal{U}$ of cardinality λ (Problem 59), and there also exists a regular ultrafilter $\mathcal{V}$ on κ (Problem 57; and note that regular ultrafilters are uniform: Problem 58). We know by Problem 659 that $|{}^\kappa \mathfrak{A} / \mathcal{V}| = 2^\kappa$ and it is not hard to see that $|{}^\kappa \mathfrak{A} / \mathcal{U}| \leq |{}^X A| = \kappa^\lambda$. It follows that ${}^\kappa \mathfrak{A} / \mathcal{U}$ and ${}^\kappa \mathfrak{A} / \mathcal{V}$ have different cardinalities provided $\kappa^\lambda < 2^\kappa$. This can be achieved e.g. with the choice $\lambda = \omega$ and $\kappa = 2^\omega$.

661. Take the two discrete orderings $\mathfrak{A} = \omega_2 \times \mathbb{Z}$ and $\mathfrak{B} = (\omega_2 + \omega_1) \times \mathbb{Z}$. The theory of discrete linear orderings without endpoints is complete (see Problem 553, thus $\mathfrak{A} \equiv \mathfrak{B}$. Clearly $\mathrm{cf}(\mathfrak{A}) = \omega_2$ and $\mathrm{cf}(\mathfrak{B}) = \omega_1$, thus $\mathfrak{A}$ and $\mathfrak{B}$ are not isomorphic. Let $c_\alpha : \omega \to \{\alpha\}$ denote the constant function with value α. No countable sequence is cofinal in ω_2, therefore for every $f : \omega \to \omega_2$ there is an $\alpha < \omega_2$ such that $\forall n \in \omega \; f(n) < c_\alpha(n)$. It follows that the sequence $(c_\alpha, 0)/\mathcal{U}$ $(\alpha < \omega_2)$ is a cofinal sequence in ${}^\omega \mathfrak{A} / \mathcal{U}$ (and thus $\mathrm{cf}({}^\omega \mathfrak{A} / \mathcal{U}) = \omega_2$). Similarly, $(c_\alpha, 0)/\mathcal{V}$ $(\alpha < \omega_1)$ is a cofinal sequence in ${}^\omega \mathfrak{B} / \mathcal{V}$, thus the ultrapowers cannot be isomorphic.

662. Let R be a unary relation symbol and put $\mathfrak{A} = \langle \omega_2, R^{\mathfrak{A}} \rangle$ and $\mathfrak{B} = \langle \omega_2, R^{\mathfrak{B}} \rangle$ with $|R^{\mathfrak{A}}| = \aleph_0$ and $|R^{\mathfrak{B}}| = \aleph_1$. Each non-trivial ultrafilter on ω is regular (Problem 58(b)), hence the ω-ultrapower of a set of cardinality λ has cardinality λ^ω (see Problems 659 and 609). It follows that $|{}^\omega \mathfrak{A} / \mathcal{U}| = |{}^\omega \mathfrak{B} / \mathcal{V}| = \omega_2^\omega$ and

$$|R^{{}^\omega \mathfrak{A} / \mathcal{U}}| = |{}^\omega R^{\mathfrak{A}} / \mathcal{U}| = \omega^\omega, \quad \text{and} \quad |R^{{}^\omega \mathfrak{B} / \mathcal{V}}| = |{}^\omega R^{\mathfrak{B}} / \mathcal{V}| = \omega_1^\omega.$$

Note that $\omega^\omega = \omega_1^\omega$ which ensures that ${}^\omega \mathfrak{A} / \mathcal{U}$ and ${}^\omega \mathfrak{B} / \mathcal{V}$ are isomorphic.

663. Let R_X be a unary relation symbol for each $X \subseteq \omega$ and let $\mathfrak{A}$ be the structure with ground set ω in which each $R_X^{\mathfrak{A}}$ is equal to X. Let $\mathcal{U}$ be an ultrafilter on ω and $\mathcal{F} \subseteq \wp(\omega)$ a system of sets. We say that $\mathcal{F}$ is *realized in* ${}^\omega \mathfrak{A} / \mathcal{U}$ if there is an $a/\mathcal{U} \in {}^\omega \mathfrak{A} / \mathcal{U}$ such that

$$\mathcal{F} = \{ X \subset \omega : {}^\omega \mathfrak{A} / \mathcal{U} \vDash R_X[a/\mathcal{U}] \}.$$

Observe that $\mathcal{U}$ can be realized in ${}^\omega \mathfrak{A} / \mathcal{U}$: if $a : \omega \to \omega$ is the identity function $a(n) = n$, then

$$\mathcal{U} = \{ X \subset \omega : {}^\omega \mathfrak{A} / \mathcal{U} \vDash R_X[a/\mathcal{U}] \}.$$

Notice that only 2^ω many set systems can be realized in each non-trivial ultrapower $^\omega\mathfrak{A}/\mathcal{U}$, because $|^\omega\mathfrak{A}/\mathcal{U}| = 2^\omega$ (see Problem 609). As there are 2^{2^ω} many non-trivial ultrafilters on ω (Problem 33) and each ultrafilter $\mathcal{U}$ can be realized in $^\omega\mathfrak{A}/\mathcal{U}$, there must exist 2^{2^ω} non-isomorphic ultrapowers of $\mathfrak{A}$ (Note: isomorphic ultrapowers realize the same systems of sets).

664. Let $E \subseteq \mathcal{U}$, $|E| = \kappa$ witness the regularity of $\mathcal{U}$ where $\{e \in E : \xi \in e\}$ is finite for all $\xi < \kappa$ (see Definition 1.17). We define the increasing sequence $\langle x_\alpha/\mathcal{U} : \alpha < \kappa^+\rangle$ by transfinite recursion. Suppose it has been defined up to $\alpha < \kappa^+$. As $|\alpha| \le \kappa$, there is an injection $f_\alpha : \alpha \to E$. Put

$$x_\alpha(\xi) = 1 + \max\{x_\beta(\xi) : \beta < \alpha \text{ and } \xi \in f_\alpha(\beta)\}.$$

As for each ξ there are finitely many β for which $\xi \in f_\alpha(\beta)$, this is a sound definition. Now for all $\beta < \alpha$ we have

$$\{\xi < \kappa : x_\beta(\xi) < x_\alpha(\xi)\} \supseteq \{\xi < \kappa : \xi \in f_\alpha(\beta)\} = f_\alpha(\beta) \in \mathcal{U}.$$

Therefore $x_\beta/\mathcal{U} < x_\alpha/\mathcal{U}$, as required.

> **Remark.** Each non-trivial ultrafilter $\mathcal{U}$ on ω is regular (Problem 58), thus Problem 640 is a special case.

665. Let $\text{Fin}(\mathfrak{A}) = \{\mathfrak{A}_i : i < \kappa\}$. By Problem 626 $\mathfrak{A}$ can be embedded to an appropriate ultraproduct $\Pi_{i<\kappa}\mathfrak{A}_i/\mathcal{U}$. As each $\mathfrak{A}_i$ is a subgraph of $\mathfrak{B}$, it follows that $\Pi_{i<\kappa}\mathfrak{A}_i/\mathcal{U}$ can be embedded into $^\kappa\mathfrak{B}/\mathcal{U}$, and therefore $\mathfrak{A}$ can be embedded into $^\kappa\mathfrak{B}/\mathcal{U}$.

666. $\mathfrak{B}$ can be embedded into some ultraproduct $\Pi_{i<\kappa}\mathfrak{B}_i/\mathcal{U}$ of its finitely generated substructures $\mathfrak{B}_i$. Each $\mathfrak{B}_i$ embeds into $\mathfrak{A}$, thus $\mathfrak{B}$ can be embedded into $^\kappa\mathfrak{A}/\mathcal{U}$. As $\mathfrak{B}$ is countable, by the downward Löwenheim–Skolem theorem 8.5 $^\kappa\mathfrak{A}/\mathcal{U}$ has a countable elementary substructure $\mathfrak{A}'$ that contains B. But $\mathfrak{A}$ is $\aleph_0$-categorical, thus $\mathfrak{A}$ and $\mathfrak{A}'$ are isomorphic, and so $\mathfrak{B}$ can be embedded into $\mathfrak{A}$.

667. It suffices to check that every finitely generated substructure has such an order. Indeed, if it is so, then their ultraproduct also has a shift-invariant total order (as the corresponding formulas are true), and the inherited ordering is shift-invariant in any substructure. By Problem 626 every structure can be embedded into some ultraproduct of its finitely generated substructures.

According to the hint, it is enough to check the claim for $\mathbb{Z}^n$, in which the lexicographic ordering works.

668. As G can be embedded to an appropriate ultraproduct of its finitely generated subgroups (Problem 626), and all such subgroups are assumed to be n-representable, it suffices to show that any ultraproduct of the groups $G_\xi = \text{GL}(n, \mathfrak{F}_\xi)$ is n-representable.

Let $\mathfrak{F} = \Pi_\xi \mathfrak{F}_\xi$. Elements of the group $H = \text{GL}(n, \mathfrak{F})$ can be identified with $g/\mathcal{U}$, where $g(\xi) \in \text{GL}(n, \mathfrak{F}_\xi)$. This is so, as elements of $\text{GL}(n, \mathfrak{F})$ are actually

n^2-tuples of elements of $\mathfrak{F}$, and similarly for $g(\xi)$. (If $g/\mathcal{U} \in H$, then $g(\xi)$ is non-singular at $\mathcal{U}$-many places, thus we may as assume that it is non-singular everywhere.) Thus $\Pi_\xi G_\xi$ is isomorphic to H, and then n-representable.

669. Problem 536 says that the dense linear order without endpoints has quantifier elimination. Thus any type over a finite subset X describes the ordering between elements of X and the variables. Realization of such a type requires the existence of other elements which are in a given order w.r.t. elements of X. If the type is finitely satisfiable in $\mathbb{Q}$, then such elements exist in $\mathbb{Q}$.

670. For each natural number $n \in \omega$ let $\pi(n)$ be the term which has the value n. The type

$$T(x) = \left\{ \pi(p) < \pi(q) \cdot x : \frac{p}{q} < \sqrt{2} \right\} \cup \left\{ \pi(p) > \pi(q) \cdot x : \frac{p}{q} > \sqrt{2} \right\}$$

is clearly finitely satisfiable, but not satisfiable in $\mathbb{Q}$.

671. $\langle \mathbb{R}, < \rangle$ is a dense linear ordering, thus $\aleph_0$-saturatedness can be proved exactly the same way as in 669. To see that $\mathbb{R}$ is not $\aleph_1$-saturated, consider the following set of formulas with $X = \{0, 1/n : n \geq q\} \subset \mathbb{R}$:

$$T(x) = \{x < 1/n : n \in \omega\} \cup \{0 < x\}.$$

It is finitely satisfiable, but not satisfiable in $\mathbb{R}$.

672. Suppose $\mathfrak{A}$ is $|A|^+$-saturated. Then each type $T(x) \subset \tau_A$ can be realized in $\mathfrak{A}$. Since $\mathfrak{A}$ is infinite, the set $T(x) = \{x \neq a : a \in A\}$ is finitely satisfiable in $\mathfrak{A}$. But it is impossible to realize $T(x)$ in $\mathfrak{A}$.

673. An infinite structure $\mathfrak{A}$ cannot be $|A|^+$-saturated by Problem 672.

674. If $X \in [A]^{<\kappa}$ and $T(x) \subset F(\tau_X)$, then $|T(x)| < \kappa \cdot |F(\tau)|^+$.

675. First let $T(x, y) \subset F(\tau_X)$ be a 2-type, and let $T'(x)$ be the set of formulas $\exists y (\varphi_1(x, y) \wedge \ldots \varphi_k(x, y))$ where $\{\varphi_1, \ldots, \varphi_k\}$ runs over the finite subsets of $T(x, y)$. As every finite subset of $T(x, y)$ is satisfiable in $\mathfrak{A}$, the same is true for $T'(x)$. Consequently $T'(x)$ is a 1-type, and suppose $a \in A$ realizes it. Now consider $T''(y) = \{\varphi(c_a, y) : \varphi(x, y) \in T\}$. This $T''(y)$ is a formula set over $F(\tau_{X \cup \{a\}})$, and it is finitely realizable as a realizes every element of $T'(x)$. Thus $T''(y)$ is realized in $\mathfrak{A}$, which means that $T(x, y)$ is also realized in $\mathfrak{A}$.

Similar ideas work in general. One can fix the value of variables x_1, x_2, etc., in this order so that after changing the variable to the realizing constant in all formulas of T, T remains finitely satisfiable. As every formula contains finitely many variables only, every element of T will be realized eventually.

676. For any $\epsilon \in {}^\omega 2$ the type $T_\epsilon(x) = \{R_i^{\epsilon(i)}(x)\}$ (negate / don't negate R_i) is finitely satisfiable in Σ, thus a saturated model must have such an element. It means that a saturated model has continuum many elements at least.

Similarly, for any subset $X \subseteq A$ the type $T'(x) = T_{\epsilon}(x) \cup \{x \neq c_a : a \in X\}$ is finitely satisfiable. Therefore each type $T_{\epsilon}(x)$ must be satisfied by $|A|$ many different elements. This motivates the definition $\mathfrak{A} = \langle {}^{\omega}\mathbb{R}, R_0^{\mathfrak{A}}, R_1^{\mathfrak{A}}, \ldots \rangle$ with the interpretation

$$R_i^{\mathfrak{A}} = \{x \in {}^{\omega}\mathbb{R} : x(i) \geq 0\}.$$

Then $\mathfrak{A} \models \Sigma$ as for a given $\{i_0, \ldots, i_n\} \cap \{j_0, \ldots, j_m\} = \emptyset$ the element

$$x = \langle \ldots, \overset{i_0}{a}, \ldots, \overset{j_k}{-a}, \ldots \overset{i_n}{a}, \overset{j_m}{-a}, \ldots \rangle$$

realizes $R_{i_0}(v) \wedge \ldots \wedge R_{i_n}(v) \wedge \neg R_{j_0}(v) \wedge \ldots \wedge \neg R_{j_m}(v)$ for every positive real a. To show that $\mathfrak{A}$ is saturated let $X \subset A$, $|X| < |A| = |\mathbb{R}|$, and $T(\vec{x}) \subset F(\tau_X)$ be a type. As there are only unary relation symbols and no function symbols, the type can only express that what unary relations the variables in $\vec{x}$ satisfy and what they do not, and whether they are equal or different from elements in X. As $|X|$ has cardinality smaller than $|\mathbb{R}|$, such an element always exists in $\mathfrak{A}$.

677. (a) Let X be a subset of ${}^{\omega}A/\mathcal{U}$ and $T(\vec{x}) = \{\varphi_i(\vec{x}) : i < \omega\}$ be a countable type over X. As finite subsets of $T(\vec{x})$ are realizable in ${}^{\omega}\mathfrak{A}/\mathcal{U}$, for each $k \in \omega$ the set

$$X_k = \{i \geq k : \mathfrak{A} \models \exists \vec{x}(\varphi_0 \wedge \cdots \wedge \varphi_{k-1})\}$$

belongs fo $\mathcal{U}$. For $i \in X_k - X_{k+1}$ choose any realization $\vec{a}(i)$ of $\varphi_0 \wedge \cdots \wedge \varphi_{k-1}$ in $\mathfrak{A}$. Then $\vec{a} = \langle \vec{a}(i) : i \in \omega \rangle/\mathcal{U}$ realizes T in ${}^{\omega}\mathfrak{A}/\mathcal{U}$ as for $k \in \omega$ fixed we have

$$X_{k+1} \subseteq \{i \in \omega : \mathfrak{A} \models \varphi_k[\vec{a}(i)]\} \in \mathcal{U}.$$

(b) If the similarity type of $\mathfrak{A}$ is countable then each type over a countable set is countable. It follows that in this case ${}^{\omega}\mathfrak{A}/\mathcal{U}$ is σ-saturated and by Problem 611 it is an elementary extension of $\mathfrak{A}$.

678. If $\mathfrak{A}$ or $\mathfrak{B}$ is finite and elementarily equivalent, then they are isomorphic by Problem 465. So suppose $|\mathfrak{A}| = |\mathfrak{B}| = \kappa$ is infinite, and let $\{c_{\xi} : \xi < \kappa\}$ be new constant symbols which will be used to mark corresponding elements of $\mathfrak{A}$ and $\mathfrak{B}$. The interpretation will be assigned using the standard back and forth method, see Problems 365, 528, 531, 555, 570. Thus assume $\alpha < \kappa$ and the constant symbols $\{c_{\xi} : \xi < \alpha\}$ has been added to the type τ such that $\mathfrak{A}$ and $\mathfrak{B}$ are still elementarily equivalent in the extended τ_{α} type. Let $a \in A$ be the first element in $\mathfrak{A}$ not assigned a constant symbol yet (forth), and take the type $T(x) = \{\varphi(x) \in F(\tau_{\alpha}) : \mathfrak{A} \models \varphi[a]\}$. As $\mathfrak{A}$ and $\mathfrak{B}$ are elementarily equivalent, $T(x)$ is a type in $\mathfrak{B}$, and as it is saturated, there is a $b \in B$ such that $\mathfrak{B} \models T[b]$. Assign the constant symbol c_a to a in $\mathfrak{A}$, and to b in $\mathfrak{B}$. Then choose the first unassigned element $b \in B$ (back), and repeat the above process.

679. If any of the structures is finite, then both are finite (Problem 465), and the ultrapowers of finite structures are isomorphic to the original structures (Problem 605). So suppose $\mathfrak{A}$ and $\mathfrak{B}$ are countably infinite. Both ${}^{\omega}\mathfrak{A}/\mathcal{U}$ and ${}^{\omega}\mathfrak{B}/\mathcal{U}$ are $\aleph_1$-saturated (Problems 677 and 674), elementarily equivalent (Problem 607) and have cardinality $2^{\aleph_0}$ (Problem 609). As $2^{\aleph_0} = \aleph_1$, Theorem 9.8 applies.

680. (a) Problem 617 says that $\mathbb{C}$ and $^\omega\mathbb{C}/\mathcal{U}$ are isomorphic. Problem 677 says that this ultrapower is σ-saturated. Consequently $\mathbb{C}$, as a field, is σ-saturated.

(b) Immediate from the definition of saturated models.

681. Let $\mathfrak{M}$ be a σ-saturated dense ordering, A be the initial points of the intervals, and B be the endpoints of the intervals. Write

$$T(x, y) = \{a < x < y < b : a \in A, b \in B\}.$$

Then $T(x, y)$ is a countable set of formulas with parameters from A and B and free variables x, y. Since $\mathfrak{M}$ is dense, $T(x, y)$ is finitely satisfiable, hence, by σ-saturatedness, $T(x, y)$ is satisfiable in $\mathfrak{M}$. This means that the interval $[x, y]$ is contained in each I_i.

682. Let T be a type over A with $|T| \le \kappa$. Let $E \subset \mathcal{U}$, $|E| = \kappa$ witness regularity of $\mathcal{U}$, i.e. the sets $\{e \in E : \xi \in e\}$ are finite for all $\xi < \kappa$. As $|T| \le \kappa$, it can be enumerated as $\{\varphi_e(\vec{x}) : e \in E\}$. We want to apply Problem 624. Let $\xi \in I$, and consider the finite subset $T_\xi = \{\varphi_e(\vec{x}) : \xi \in e\}$ of T. This is realizable in $^\kappa\mathfrak{A}/\mathcal{U}$ by assumption. Note that $\mathfrak{A}$ is an elementary substructure of $^\kappa\mathfrak{A}/\mathcal{U}$ (Problem 611), and since formulas in T_ξ contain parameters only from $\mathfrak{A}$, T_ξ is realizable in $\mathfrak{A}$ as well, say by $f_\xi \colon \mathfrak{A} \vDash \varphi_e[f_\xi]$ for every $\xi \in e$. Then $\langle f_\xi \rangle / \mathcal{U}$ realizes T as $^\kappa\mathfrak{A}/\mathcal{U} \vDash \varphi_e[\langle f_\xi \rangle / \mathcal{U}]$ for all $e \in E$ by Problem 624.

683. Straightforward application of the Downward Löwenheim–Skolem theorem 8.5.

684. (a) If Γ has arbitrarily large finite models, then it has an infinite model by Problems 623 and 419.

(b) The set of formulas $\{\varphi_n : n \in \omega\}$, where

$$\varphi_n \equiv \exists x_1 \ldots \exists x_n \bigwedge_{i \ne j} x_i \ne x_j$$

axiomatizes infinite structures.

685. Every infinite $\mathfrak{A} \in \mathcal{K}$ has an elementary extension of arbitrary large cardinality (Problems 420, 615), which cannot be isomorphic to $\mathfrak{A}$ for cardinality reasons. For the other direction see Problem 465.

686. (a), (b) No, $\mathcal{K}_1$ is never axiomatizable. An axiomatizable class contains all isomorphic images of its members. Elementary substructures of $\mathfrak{B}$ are not closed for isomorphism, thus cannot form an axiomatizable class.

(c) Yes, let $\mathfrak{A}$ be a countable $\emptyset$-type structure (a set). Then $\mathcal{K}_2$ is the class of all infinite sets which can be axiomatized.

(d) Yes, let $\mathfrak{A}$ be an $\emptyset$-type structure of cardinality continuum. Then $\mathcal{K}_2$ is the class of all sets having cardinality $\ge$ continuum but any axiomatizable $\emptyset$-type class should contain a countable structure (cf. Problem 683).

687. Assume that the set of closed formulas $\Delta = \{\varphi_i : i < \omega\}$ axiomatizes $\mathcal{K}$, and contains no formulas valid on every structure. If Δ would be empty or finite, then it is equivalent to a single formula, which is independent.

Define by induction on n the formulas ψ_n such that $\psi_0 = \varphi_0$ and ψ_{n+1} is the least φ_k such that $\{\psi_0, \ldots, \psi_n\} \not\vdash \varphi_k$. It is easy to see that $\{\varphi_i : i < \omega\}$ and $\{\psi_i : i < \omega\}$ are equivalent. Let $\vartheta_0 = \psi_0$ and $\vartheta_{n+1} = \bigwedge_{m \leq n} \psi_m \to \psi_{n+1}$ and put $\Gamma = \{\vartheta_i : i < \omega\}$. Since $\{\psi_0, \ldots, \psi_n\} \not\vdash \psi_{n+1}$, there is some model $\mathfrak{A} \vDash \psi_0, \ldots, \psi_n, \neg\psi_{n+1}$. Then $\mathfrak{A} \vDash \vartheta_0, \ldots, \vartheta_n$ but $\mathfrak{A} \not\vDash \vartheta_{n+1}$. For $m > n+1$ we have that the antecedent of ϑ_m is not true in $\mathfrak{A}$, therefore

$$\mathfrak{A} \vDash \bigwedge_{m \neq n+1} \vartheta_m \wedge \neg\vartheta_{n+1}.$$

Consequently $\{\vartheta_i : i < \omega\}$ is independent. Checking $\mathcal{K}$ is axiomatized by Γ is straightforward.

688. Every ultraproduct is a model of Γ by Łoś lemma. For the other direction let $\mathfrak{A} \vDash \Gamma$. Problem 625 claims that if every finite subset of $\mathrm{Th}(\mathfrak{A})$ has a model in $\mathcal{K}$, then there is an ultraproduct $\Pi_\xi \mathfrak{A}_\xi / \mathcal{U}$ of elements of $\mathcal{K}$ which is a model of $\mathrm{Th}(\mathfrak{A})$, and this model is clearly elementarily equivalent to $\mathfrak{A}$. Thus suppose, by way of contradiction, that no $\mathfrak{A}_\xi \in \mathcal{K}$ is a model of $\{\varphi_1, \ldots, \varphi_n\} \subseteq \mathrm{Th}(\mathfrak{A})$. Then

$$\neg\bar\varphi_1 \vee \cdots \vee \neg\bar\varphi_n \in \Gamma$$

as this formula is true in every element of $\mathcal{K}$, contradicting that $\mathfrak{A} \vDash \Gamma$.

689. (a) We will use Problem 625. Suppose Γ axiomatizes $\mathcal{K}$. To show that there is *some* structure in $\mathcal{K}$ which is an ultraproduct of structures not in $\mathcal{K}$, we need to find, for each finite subset of Γ, a model not in $\mathcal{K}$. But such a model is provided by the assumption that $\mathcal{K}$ is not axiomatized by any finite subset of Γ.

(b) Not true. Let R_i be unary relation symbols, $\varphi \equiv \forall x R_0(x)$ and $\psi_i \equiv \varphi \vee \forall x R_i(x)$. Let $\mathcal{K}$ be axiomatized by $\{\varphi, \psi_i : i \geq 1\}$. Clearly $\mathcal{K}$ cannot be axiomatized finitely, and no structure outside of $\mathcal{K}$ is a model of φ. Thus no structure with $\mathfrak{A} \vDash \varphi$ (clearly an element of $\mathcal{K}$) can be elementarily equivalent to an ultraproduct of structures not in $\mathcal{K}$.

690. (a) If $\mathcal{K}$ is axiomatizable, then it is clearly closed under elementary equivalence and ultraproduct. The converse follows from Problem 688: $\mathcal{K}$ is axiomatized by the theory $\Gamma = \{\varphi : A \vDash \varphi \text{ for each } \mathfrak{A} \in \mathcal{K}\}$.

(b) If $\mathcal{K}$ is finitely axiomatizable, then its complement is finitely axiomatizable as well. The converse is covered in Problem 689: if $\mathcal{K}$ is axiomatizable but not finitely axiomatizable, then its complement is not closed for ultraproducts.

691. (a) Let $\mathcal{K}$ contain all sets of cardinality $\geq$ continuum.

(b) Let $\mathcal{K}$ consist of all finite sets. $\mathcal{K}$ is closed under elementary equivalence but not closed under ultraproducts (see Problem 609).

(c) No such a $\mathcal{K}$ exists. For, if $\mathfrak{A} \in \mathcal{K}$, $\mathfrak{B} \notin \mathcal{K}$ and $\mathfrak{A} \equiv \mathfrak{B}$ then for some ultrafilter $\mathcal{U}$ we get $\mathcal{K} \ni {}^{\kappa}\mathfrak{A}/\mathcal{U} \cong {}^{\kappa}\mathfrak{B}/\mathcal{U} \notin \mathcal{K}$.

692. The class $\mathcal{K}$ of models of Γ_1 is closed under elementary equivalence and ultraproducts, thus it is enough to check whether the complement of $\mathcal{K}$ is closed under ultraproduct (Theorem 9.10). Take any $\mathfrak{A}_i \notin \mathrm{Mod}(\Gamma_1)$. Then $\mathfrak{A}_i \vDash \Gamma_2$ hence $\Pi\mathfrak{A}_i/\mathcal{U} \vDash \Gamma_2$ therefore $\Pi\mathfrak{A}_i/\mathcal{U} \notin \mathcal{K}$.

693. If no finite subset of Γ axiomatizes $\mathcal{K}$, then every such subset has a model in the complement of $\mathcal{K}$. In this case Problem 625 shows that some of their ultraproducts will be a model of Γ, contradicting the assumption (and Theorem 9.10) that $\mathcal{K}$ is finitely axiomatizable.

694. (a) It is easy to see that universal formulas are preserved under substructures. For the other direction suppose that $\mathcal{K}$ is preserved under substructures. Let $\Sigma = \{\varphi : \Gamma \vDash \varphi$ and φ is closed universal$\}$, then every element of $\mathcal{K}$ clearly models Σ. Now let $\mathfrak{A} \vDash \Sigma$. By Problem 514 $\mathfrak{A}$ is a substructure of some model of Γ, and as $\mathcal{K}$ is closed under substructures, it is an element of $\mathcal{K}$ as well.

(b) This is the dual statement of (a) and follows from the observations that (i) a formula is equivalent to an existential formula if and only if its negation is equivalent to a universal formula; and (ii) a formula is preserved under extensions if and only if its negation is preserved under substructures.

695. It is enough to prove that a closed $\forall\exists$-formula is preserved under unions of chains. For, consider the $\forall\exists$-formula $\varphi = \forall\vec{x}\exists\vec{y}\delta(\vec{c},\vec{y})$, where δ is quantifier-free. Let $\langle \mathfrak{A}_\xi : \xi < \kappa \rangle$ be an increasing chain of models of φ and set $\mathfrak{A} = \bigcup_\xi \mathfrak{A}_\xi$. For each $\vec{a} \in A$ we show $\mathfrak{A} \vDash \exists\vec{y}\delta[\vec{a},\vec{y}]$. Let ξ be the minimal ordinal such that $\vec{a} \in A_\xi$. Since $\mathfrak{A}_\xi \vDash \varphi$ it follows that $\mathfrak{A}_\xi \vDash \exists\vec{y}\delta[\vec{a},\vec{y}]$, and thus for some $\vec{b} \in A_\xi$ we have $\mathfrak{A}_\xi \vDash \delta[\vec{a},\vec{b}]$. As δ is quantifier-free and $\mathfrak{A}$ is an extension of $\mathfrak{A}_\xi$, $\mathfrak{A} \vDash \delta[\vec{a},\vec{b}]$ follows.

> **Remark.** Models of theory are closed under unions of chains if and only if the theory is equivalent to a set of $\forall\exists$ formulas.

696. Using Problem 694(a). The $\langle 0, S, > \rangle$-type model $\omega + \mathbb{Z}$ has a substructure in which an element, different from 0, has no immediate predecessor. To prevent this construction, add another unary function symbol S^{-1} which assigns the predecessor to each element (and $S^{-1}(0) = 0$). Now every substructure is a discrete ordering with 0 as the initial element.

697. (a), (b), (c), (d), (g), (j) $\mathcal{K}$ is not closed under ultraproduct, see Problems 629, 631, 610, 646, 649, 644,

(e), (h), (k) $\mathcal{K}$ is not closed under elementary equivalence, see Problems 563, 561, 568.

(f) Cyclic groups of prime order are simple but a non-trivial ultraproduct of such groups contains elements of infinite order, thus it cannot be simple. See Problem 648.

(i) $\mathcal{K}$ is not closed under ultraproduct: Pick any non-trivial ultrafilter and observe that the element

$$\langle x, x + x^2, x + x^2 + x^3, \ldots \rangle / \mathcal{U} \in \prod F[x]/\mathcal{U}$$

cannot be written as a *finite* sum $a_0 x + \ldots a_n x^n$ for some $a_i \in F$. (Note that $\Pi F_i[x]/\mathcal{U}$ is usually not isomorphic to $(\Pi F_i/\mathcal{U})[x]$).

698. The set of three universal formulas

$$\forall x \forall y ((x \leq y \wedge y \leq x) \rightarrow x = y),$$
$$\forall x \forall y \forall z ((x \leq y \wedge y \leq z) \rightarrow x \leq z),$$
$$\forall x \forall y (x \leq y \vee y \leq x)$$

axiomatizes linear orderings. Since each dense ordering has a non-dense subordering, by Problem 694(a) the class of dense linear orderings without endpoints cannot be axiomatized by universal formulas.

699. The class of totally ordered Dedekind-complete fields is not closed under elementary equivalence. By Problem 683 it suffices to show that no countably infinite ordered field is Dedekind-complete. Firstly, an ordered field $\mathfrak{F}$ has characteristics zero (Problem 658). It follows that $\mathbb{Q}$ is a subfield of $\mathfrak{F}$. For a real number α let $C_\alpha = \{q \in \mathbb{Q} : q < \alpha\}$. Then each $C_\alpha \subseteq \mathbb{Q} \subseteq \mathfrak{F}$ must have a supremum, and this implies $|\mathfrak{F}| \geq |\mathbb{R}|$.

700. Let $k \geq 3$ be fixed and pick, for $i \in \omega$, graphs G_i such that $\chi(G_i) = k$ (or $\chi(G_i) > k$ in the case of $\mathcal{K}_{>k}$) and the girth of G_i is at least i. If $\mathcal{U}$ is a non-principal ultrafilter on ω then the ultraproduct $\Pi_{i \in \omega} G_i / \mathcal{U}$ does not contain cycles of any length because the set

$$\{i \in \omega : G_i \text{ contains a cycle of length } \ell\}$$

is finite, thus by the fundamental theorem of ultraproducts we get, for each ℓ,

$$\Pi_{i \in \omega} G_i / \mathcal{U} \models \text{there is no cycle of length } \ell$$

(see also Problem 629). It follows that the chromatic number of $\Pi_{i \in \omega} G_i / \mathcal{U}$ is 2. Consequently $\mathcal{K}_{=k}$ and $\mathcal{K}_{>k}$ are not closed under ultraproducts, hence they are not axiomatizable.

As for $\mathcal{K}_{\neq k}$ let H_i be the disjoint union of the complete graph K_k on k points and a graph G_i with girth at least i and chromatic number $\chi(G_i) > k$. Then $\chi(H_i) \neq k$. The ultraproduct $\Pi_{i \in \omega} H_i / \mathcal{U}$ contains K_k, and an argument similar to the previous one shows that apart from K_k there are no cycles in the ultraproduct. Therefore $\chi(\Pi_{i \in \omega} H_i / \mathcal{U}) = k$, hence $\mathcal{K}_{\neg k}$ is not axiomatizable.

A graph has chromatic number 2 iff it contains no cycles of odd length. So $\Gamma = \{\varphi_{2k+1} : k \in \omega\}$ axiomatizes $\mathcal{K}_2$, where φ_n expresses that there is no cycle of length n, see Problem 330(a).

701. There are infinite planar graphs, and no graph of cardinality larger than continuum can be planar.

702. By the Kuratowski theorem a countable graph is planar if and only if it contains no (finite) subdivision of K_5 or $K_{3,3}$ as a subgraph. For each such finite graph G there is a formula φ_G expressing "G is not a subgraph". Thus the set

$$\Gamma = \{\varphi_G : G \text{ is a finite subdivision of } K_5 \text{ or } K_{3,3}\}$$

axiomatizes the theory of planar graphs.

> **Remark.** The class of planar graphs is not axiomatizable, see Problem 701.

As for finite axiomatizability, denote by G_i the graph obtained from K_5 by dividing some edge with i new points. Clearly, G_i is a subdivision of K_5 and is not planar. But $\Pi_{i\in\omega} G_i / \mathcal{U} \models \varphi_G$ for every finite subdivision G of K_5 or $K_{3,3}$. Therefore the complement of the class of graphs which are elementarily equivalent to a planar graph is not closed under ultraproducts, hence this theory is not finitely axiomatizable.

703. The complement of $\mathcal{K}$ is not closed under ultraproducts: $\Pi_{n\in\omega} K_n / \mathcal{U}$ is a complete graph of cardinality continuum (cf. Problem 610), hence it belongs to $\mathcal{K}$. Notice that $\mathcal{K}$ is axiomatizable by the infinite set of formulas "each vertex has at least n neighbors".

704. For better visualization edges and non-edges are flipped, thus the requirement is that the graph has infinitely many vertices and each vertex is *not connected* to infinitely many other vertices.

As indicated in the hint, Problem 449 describes a method which can be used to embed a linear ordering into the graph. Let Φ be the graph formula which forces the graph to have the structure as described there, namely there is a unique vertex a with a single one-degree neighbor, another single vertex b with two one-degree neighbors. Further neighbors of a are A-vertices, and further neighbors of b are B-vertices; every A-vertex is connected to a unique C-vertex of degree two, and, similarly, every B-vertex is connected to a unique C-vertex. Further edges are between A and B-vertices.

The ordering is defined on A-vertices: u and v are in the relation R (in this order), if there is an edge from u to v', where v' is the unique B=vertex connected to v via a C vertex. Stipulating that this relation is a strict order and adding that $\forall x \exists y R(x, y)$ yields the required finitely axiomatizable class.

705. In each case Γ will be an infinite set of axioms for the class in question and we show that the complement of the class is not closed under ultraproducts, therefore is not finitely axiomatizable.

(a) The ultraproduct $\Pi_{n\in\omega} C_n / \mathcal{U}$ of cyclic groups of order n is infinite (Problem 610).

$$\Gamma = \{\text{group axioms}\} \cup \{\exists x_1 \cdots \exists x_n (\bigwedge_{i\neq j} x_i \neq x_j) : n \in \omega\}.$$

(b) Non-trivial ultraproducts of cyclic groups of prime order are torsion-free, see Problem 646.

$\Gamma = \{\text{group axioms}\} \cup \{\forall g (g \neq 0 \rightarrow n \cdot g \neq 0) : n \in \omega\}$.

(c) Let p_j be the j-th prime and $G_i = \langle \mathbb{Q}_i, + \rangle$ where $\mathbb{Q}_i = \{n/m : n, m \in \mathbb{Z},$ $p_j \nmid m$ for $j > i\}$. For each $i \in \omega$ the group G_i is not divisible while $\Pi_i G_i / \mathcal{U}$ is divisible.

$\Gamma = \{\text{group axioms}\} \cup \{\forall g \exists y (n \cdot y = g), \ \forall g (g \neq 0 \rightarrow n \cdot g \neq 0) : n \in \omega\}$.

(d) If $\mathfrak{F}_n$ is the n-th finite field then $\Pi_n \mathfrak{F}_n / \mathcal{U}$ can be of characteristics 0 (Problem 649).

$\Gamma = \{\text{field axioms}\} \cup \{n \cdot 1 \neq 0 : n \in \omega\}$.

(e) See Problem 654.

$\Gamma = \{\text{field axioms}\} \cup$

$\{\forall a_0 \cdots \forall a_n \exists x (a_n \neq 0 \rightarrow a_n x^n + \ldots + a_1 x + a_0 = 0) : n \in \omega\}$.

(f) Similarly to Solution 654 start with $F_0 = \mathbb{Q}$, and for each fixed n take the field extensions $F_0 \subseteq F_1^n \subseteq F_2^n \subseteq \ldots$ where all polynomials of odd degree at most n with coefficients in F_i^n have a root in F_{i+1}^n. Let $\mathfrak{F}_n = \bigcup_i F_i^n$. It is not real closed, but the ultraproduct $\Pi_n \mathfrak{F}_n / \mathcal{U}$ is.

$\Gamma = \{\text{field axioms}\} \cup \{\forall a_0 \ldots \forall a_{n-1} (a_0^2 + \cdots + a_{n-1}^2 \neq -1) : n \in \omega\} \cup$

$\{\forall a_0 \cdots \forall a_{2n+1} \exists x (a_{2n+1} x^{2n+1} + \ldots + a_1 x + a_0 = 0) : n \in \omega\}$.

(g) With C_n the cyclic graph on n vertices and $\mathcal{U}$ non-trivial, the graph $\Pi_{n \in \omega} C_n / \mathcal{U}$ is cycle-free.

$\Gamma = \{\text{graph axioms}\} \cup \{\forall x_1 \cdots \forall x_n (\text{``}x_1, \ldots, x_n \text{ is not a cycle''}) : n \in \omega\}$.

(h) Let G_n be the following graph: take a circle on n vertices and to each of these vertices join an infinite binary tree. The resulting graph is 3-regular and contains exactly one circle. Then the ultraproduct $\Pi_n G_n / \mathcal{U}$ is cycle-free and 3-regular.

$\Gamma = \{\text{graph axioms}\} \cup \{\text{``all vertices have degree 3''}\} \cup$

$\{\forall x_1 \cdots \forall x_n (\text{``}x_1, \ldots, x_n \text{ is not a cycle''}) : n \in \omega\}$.

706. Let $\mathfrak{A}_n$ be the model that consists of ω with the usual 0 and S and n new elements $a_0, \ldots, a_{n-1}$ for which S is defined by

$$S(a_k) = a_{k+1 \bmod n}.$$

Clearly $\mathfrak{A}_n$ is a model of Γ except for the last set of formulas. For each k we have that the set

$n \in \omega : \mathfrak{A}_n$ contains an element which is the k-th successor of itself$\}$

is finite, therefore $\Pi_{n \in \omega} \mathfrak{A}_n / \mathcal{U} \vDash \Gamma$ for all non-trivial ultrafilter $\mathcal{U}$ on ω. It follows that the complement of $\mathcal{K}$ is not closed under ultraproducts, consequently $\mathcal{K}$ is not finitely axiomatizable.

12.10   ARITHMETIC

707. It is clear by inspection.

708. If $0 = 1$, then $0 + 0$ would be equal to $0 + 1$ but by Q1 we have $0 + 1 \neq 0$, and by Q3 we have $0 + 0 = 0$.

709. (a) If $x \neq 0$ then by Q7 $x = y + 1$, and then $0 + x = 0 + (y + 1) = (0 + y) + 1$ by Q4, but this cannot be 0 by Q1.

(b) By Problem 708 $0 \neq 1$, thus Q7 gives $1 = x + 1$ for some x. Then $0 + 1 = 0 + (x + 1) = (0 + x) + 1$ by Q4, then $0 = 0 + x$ by Q2. Therefore $x = 0$ by (a).

710. (a) By induction on p. For $p = 0$ we have $\pi_p = 0$ and by Q3, $x + 0 = x$. Thus $\pi_n + 0 = \pi_n$. In the inductive step we use Q4 to get

$$\pi_n + \pi_{p+1} = \pi_n + (\pi_p + 1) = (\pi_n + \pi_p) + 1 = \pi_{n+p} + 1.$$

(b) Similar as in (a) but use Q5 and Q6 in place of Q3 and Q4.

711. We proceed by induction on n. The $n = 0$ case follows from Q3. Suppose we know that $Q \vdash x \leq \pi_n \vee \pi_n \leq x$, we need to show that $x \leq \pi_{n+1} \vee \pi_{n+1} \leq x$ for all x. If $x = 0$, then $x \leq \pi_{n+1}$ by Q3. If $x \neq 0$, then by Q7 there is an y so that $x = y + 1$. Apply the inductive hypothesis to y to get $y \leq \pi_n \vee \pi_n \leq y$, that is, $z + y = \pi_n$ or $z + \pi_n = y$ for some z. In the first case $z + x = z + (y + 1) = (z + y) + 1 = \pi_{n+1}$ by Q4, thus $x \leq \pi_{n+1}$. In the second case $z + \pi_{n+1} = (z + \pi_n) + 1 = y + 1 = x$, therefore $\pi_{n+1} \leq x$.

712. By Q7 and Q4 if $z + x = 0$ and $x \neq 0$, then there is a y so that $x = y + 1 \wedge (z + y) + 1 = 0$. Q1 implies $(z + y) + 1 \neq 0$, thus in the case $z + x = 0$ we get $x = 0$. That is, $Q \vdash x \leq \pi_0 \rightarrow x = \pi_0$. In general, by Q7 and Q4 we obtain

$$Q \vdash z + x = \pi_{n+1} \wedge x \neq 0 \rightarrow \exists y (x = y + 1 \wedge (z + y) + 1 = \pi_n + 1),$$

Q2 then ensures

$$Q \vDash z + x = \pi_{n+1} \rightarrow x = 0 \vee \exists y (x = y + 1 \wedge y \leq \pi_n).$$

From here we get the result by induction.

713. Suppose, by contradiction, that there is a model $\mathfrak{A} \vDash Q$ and $\mathfrak{A} \vDash \pi_n = \pi_p$ for different n and p. Let n be minimal such that $\mathfrak{A} \vDash \pi_n = \pi_p$ for some $n \neq p$. Then $n < p$, and $n \neq 0$ by Q1. Thus $\pi_n = \pi_{n-1} + 1 = \pi_p = \pi_{p-1} + 1$, and then $\mathfrak{A} \vDash \pi_{n-1} = \pi_{p-1}$ by Q2, contradicting the minimality of n.

By Problem 710 for $n \leq p$ we have $Q \vdash \pi_{p-n} + \pi_n = \pi_p$, thus $Q \vDash \pi_n \leq \pi_p$. On the other hand if $p < n$ and $\mathfrak{A} \vDash \pi_n \leq \pi_p$ in some model of Q, then by 712 we would get

$$\mathfrak{A} \vDash \pi_n = \pi_0 \vee \cdots \vee \pi_n = \pi_p,$$

though, by the first part, in models of Q we have $\neq$ everywhere above.

714. The condition $a + b = \pi_n$ means $b \leq \pi_n$, thus by Problem 712 $\mathfrak{A} \models b = \pi_m$ for some $m \leq n$. Applying Q4 and Q2 m times we get $\mathfrak{A} \models a + \pi_0 = \pi_{n-m}$, thus $a = \pi_{n-m}$ by Q3.

715. Take ω with the usual operations and an additional element ∞. The result of addition and multiplication involving ∞ is ∞ except for $\infty \cdot 0 = 0$.

716. The universe of the model is $\omega \cup \{a, b\}$. The interpretations of the operations on ω are the usual ones, while we let $a + n = a \cdot n = a$ and $b + n = b \cdot n = b$ for all natural numbers n except for $a \cdot 0 = b \cdot 0 = 0$. Further, for any x we put $x + a = x \cdot a = b$ and $x + b = x \cdot b = a$. Then $b + a = b$ while $a + b = a$.

717. No. The model in Solution 716 has $0 + a \neq a$.

718. Keep 0, 1, addition and multiplication in ω. The result of both operations on $n \in \omega$ and $a \in D$ could be a with the exception of multiplying by zero, which should be zero.

719. Consider the model created in Problem 718. In that model $\mathfrak{A}$ for two infinite elements $a, b \in D$ we have $a \leq^{\mathfrak{A}} b$ iff $a \leq^D b$.

(a) Any such model works as $a + a = a$ for all infinite elements.

(b) $\leq^{\mathfrak{A}}$ is not an ordering if D is not totally ordered.

(c) Any D works which has two incomparable elements.

(d) In models considered above $a \leq b$ and $b \leq a$ always implies $a = b$. However, in the model of Problem 716 there are $a + b = a$ and $b + a = b$, thus both $b \leq^{\mathfrak{A}} a$ and $a \leq^{\mathfrak{A}} b$ holds, while a and b are different.

720. The model consists of one copy of ω, and three copies of $\mathbb{Z}$ with elements $a(i)$, $b(i)$, $c(i)$ for $i \in \mathbb{Z}$. In each copy addition and multiplication is done locally, for example $z(i) \cdot z(j) = z(i \cdot j)$, where z is any of a, b, or c. Similarly, addition and multiplication by elements of ω is $n + x(i) = x(i) + n = x(n + i)$, and $n \cdot x(i) = x(i) \cdot n = x(n \cdot i)$, except for $x(i) \cdot 0 = 0$. Finally, addition and multiplication across the copies of $\mathbb{Z}$ is defined as

$$
\begin{array}{lll}
a(i) + b(j) = b(i + j) & b(i) + a(j) = a(i + j) & c(i) + a(j) = b(j) \\
a(i) + c(j) = a(i + j) & b(i) + c(j) = b(i + j) & c(i) + b(j) = a(j)
\end{array}
$$

$$
\begin{array}{lll}
a(i) \cdot b(j) = a(i \cdot j) & b(i) \cdot a(j) = b(i \cdot j) & c(i) \cdot a(j) = a(i \cdot j) \\
a(i) \cdot c(j) = a(i \cdot j) & b(i) \cdot c(j) = b(i \cdot j) & c(i) \cdot b(j) = b(i \cdot j).
\end{array}
$$

It is tedious but otherwise routine to check that this is a model of Q. Now we have $a(i) \leq b(i)$ (and $b(i) \leq a(i)$), and then $a(i) < b(i)$ and $a(i)$ as $b(i)$ are different. However $a(i) + 1 = a(i + 1)$ and $b(i)$ are incomparable.

721. The set $B = \{\pi_n^{\mathfrak{A}} : n \in \omega\}$ is a substructure isomorphic to $\mathfrak{N}$ by Problems 709 and 710. Problems 711 and 712 show that $B \leq^{\mathfrak{A}} A \smallsetminus B$, thus $\mathfrak{A}$ is an end-extension of $\mathfrak{B}$.

722. By Problem 721 the standard model $\mathfrak{N}$ is an initial segment of every model of Q. Quantified variables bounded by elements of $\mathfrak{N}$ run over elements in $\mathfrak{N}$, thus if φ is true in $\mathfrak{N}$, then it is true in every model of Q.

723. By Problems 713 and 715 the formula $x + 1 \neq x$ works. Another example is $0 + x = x$ by Problem 717.

724. Suppose first that Γ is recursive, and consider the (code of the) sequences $\langle \varphi_1, \varphi_2, \ldots, \varphi_n \rangle$ which are valid Hilbert-type derivations from Γ, see Definition 5.19. It is tedious, but otherwise routine to check that they form a recursive set (see also Problem 381). Whether a formula is in Γ is recursive by assumption: each axiom scheme and the validity of the derivation rules can be recognized by recursive relations. To enumerate all consequences of Γ, check each sequence whether it is a valid derivation from Γ. If yes, return the last element of the sequence, otherwise return $\top$.

If Γ is recursively enumerable, then for each pair (u, n) check that u is a valid derivation from the first n elements of Γ as provided by the enumeration.

725. Extend the set with $\bot$. Then $\{\varphi \in F(\tau) : \Gamma, \bot \vdash \varphi\}$ is the set of all τ-type formulas, which is a recursive set by Problem 381(a).

726. If Γ has no infinite models, then, by the compactness theorem, there is a natural number N (depending on Γ) such that every $\mathfrak{A} \models \Gamma$ has size $|A| \leq N$. As τ is finite, there are finitely many possible models of size at most N, consequently there are only finitely many models of Γ. $\Gamma \models \varphi$ if φ is true in that finitely many models. By Problem 385 whether a formula is true in a finite model is decidable, thus $\Gamma \models \varphi$ is decidable as well.

727. (a) If the similarity type is empty, then atomic formulas are of the form $x = y$ for variable symbols x and y (not necessarily distinct). Thus, $\vdash \varphi$ can be decided by checking whether $\mathfrak{A} \models \varphi$ for every set $\mathfrak{A}$ of size at most the number of variable symbols in φ, see Problems 385 and 726.

(b) Let $\tau = \langle R_1, \ldots, R_k \rangle$. By Problem 362 $\emptyset \vdash \varphi$ iff $\mathfrak{A} \models \varphi$ for every τ-structure $\mathfrak{A}$ of cardinality $|A| \leq n2^k$ where n is the length of φ. Given n one can recursively enumerate all such structures (observe: k is a fixed number, only n varies), and then check each of them whether φ holds or not, see Problem 385.

(c) By Problem 536 this theory admits elimination of quantifiers. The method outlined in the solution can be turned into a recursive procedure: maintain, for each subformula of φ, the equivalent quantifier-free formula, and then check that logical operations and quantifications are performed correctly.

Another approach could be defining a function which assigns to each subformula of φ the collection of those orders of its free variables which makes the subformula true. Clearly this function can be defined by a course-of-value recursion using recursive functions and relations.

728. Similarly to Solution 307 let g be a recursive function whose range is not recursive. Let φ_k be the formula which says that it has no k-element models. The i-th formula in Γ is $\varphi_k \vee \cdots \vee \varphi_k$ ($i+1$ times). Then Γ is recursive, and $\Gamma \vdash \varphi_k$ iff $k \in \text{dom}(g)$. As the function $k \mapsto \alpha(\varphi_k)$ is clearly recursive, if $\{\alpha(\varphi) : \Gamma \vdash \varphi\}$ is recursive, then $\text{dom}(g)$ were recursive, a contradiction.

729. The same idea works as in the Solution 728. Let $\{\varphi_i : i \in \omega\}$ be a recursive enumeration of Γ. Σ will be the set

$$\{\underbrace{\top \wedge \top \wedge \cdots \wedge \top}_{i+1 \text{ times}} \wedge (\bot \vee \varphi_i) : i \in \omega\}.$$

Clearly this set has the same consequences as Γ. Given any formula one can decide whether it has the form above, recover i by a recursive function, and check that φ_i is indeed the i-the element of the enumeration.

730. This is a variant of Problem 157 which says that if a set and its complement are both recursively enumerable, then the set is recursive.

Let $\tilde{\varphi}$ be the universal closure of φ. By Problem 381(d) the function $\alpha(\varphi) \mapsto \alpha(\tilde{\varphi})$ is recursive, thus clearly so the function $g(\alpha(\varphi)) \mapsto \alpha(\neg\tilde{\varphi})$. As Γ is complete, for every $\varphi \in F(\tau)$ either $\Gamma \vdash \varphi$ or $\Gamma \vdash \neg\tilde{\varphi}$. By Problem 724, the set $A = \{\alpha(\varphi) : \Gamma \vdash \varphi\}$ is recursively enumerable, say it is the range of the recursive function h. Now let

$$f(i) = \mu\{u : h(u) = i \text{ or } h(u) = g(i) \text{ or } i \text{ is not a formula code}\}.$$

This is a recursive function (for each $i \in \omega$ there is a u satisfying the recursive condition) and $\Gamma \vdash \varphi$ iff $h(f(\alpha(\varphi))) = \alpha(\varphi)$, clearly a recursive relation.

731. (a) By the Łoś–Vaught test 8.7, Γ is complete. Apply Theorem 10.3 to complete the proof.

(b) Let τ be the empty type, and φ_k be the formula which says that the universe has exactly k elements. For every $A \subset \omega$ the theory $\Gamma = \{\neg\varphi_k : k \in A\}$ is ω-categorical. If A is not recursive, then Γ is undecidable. By Problem 729 Γ can be replaced by a recursive set with the same consequences if it is recursively enumerable.

732. It is clear that each theory is recursively axiomatizable, and has infinite models only. By Problems 557, 553, 531 these theories are complete. Recursive and complete theories with infinite models only are decidable by Theorem 10.3.

733. (a) As Δ is finite, it can be assumed to be a single closed formula Φ (the conjunct of the closure of the formulas in Δ). Now $\Gamma \cup \Delta \vdash \varphi$ iff $\Gamma \vdash \Phi \to \varphi$ by the deduction theorem 7.1.

By way of contradiction assume that the set $\{\Gamma \vDash \varphi\}$ is recursive. As the function $\alpha(\varphi) \mapsto \alpha(\Phi \to \varphi)$ is clearly recursive, we get that the set $\{\varphi : \Gamma \vDash \Phi \to \varphi\}$ is also recursive. This set, however, is just the consequences of $\Gamma \cup \Delta$ which was assumed not to be recursive, a contradiction.

(b) Immediate from (a) using $\Gamma = \emptyset$.

734. The main problem is that Γ has many more consequences in type $\tau' = \tau \cup \{c\}$ than it has in type τ. So assume Γ is decidable, and let $\psi \in F(\tau')$ be arbitrary. Now ψ can be written as $\varphi[x/c]$ for some $\varphi \in F(\tau)$ where x is a variable symbol not present in ψ. The code of φ is clearly a recursive function of the code of ψ. By Problem 400, $\Gamma \vdash \psi$ iff $\Gamma \vdash \forall x \varphi$, and this is recursive as Γ is decidable.

735. (a) Suppose $\mathfrak{B} \models \Delta$ can be defined semantically in $\mathfrak{A} \models \Gamma$. The ground set of $\mathfrak{B}$ is defined by some τ'-formula $\vartheta(x)$, and all symbols in τ have similar definitions. As τ is finite, there is a single closed formula $\Phi \in \mathcal{F}(\tau')$ which says that the elements satisfying $\vartheta(x)$ form a τ-type structure (it is closed for the interpretation of function symbols in τ). Clearly, for every $\varphi \in F(\tau)$ there is a "translation" $\varphi^* \in F(\tau')$ such that in the semantical substructure determined by Φ the formula φ holds iff φ^* holds in $\mathfrak{A}$.

Let $\Delta^* = \{\delta^* : \delta \in \Delta\} \cup \{\Phi\}$, and consider the formula set $\Sigma = \{\varphi \in F(\tau) : \Gamma \cup \Delta^* \vdash \varphi^*\}$. This set contains Δ, and is consistent as every element holds in the embedded structure $\mathfrak{B}$. As Δ is essentially undecidable, Σ is undecidable. If Γ were decidable, then so would be $\Gamma \cup \Delta^*$ by Problem 733(a), and as the function $\alpha(\varphi) \mapsto \alpha(\varphi^*)$ is clearly recursive, the same would be true for Σ, a contradiction.

(b) Δ^* is essentially undecidable. For any $\Gamma \subseteq F(\tau')$, any model of $\Delta^* \cup \Gamma$ has a semantical substructure described by $\Phi \in \Delta^*$ which will be a model of Δ. Then (a) applies.

736. The undecidable theory Δ created in Problem 728 works. It is defined in the empty type, and excludes certain finite models only. Thus the structure with infinite ground set is a model of Δ, and it is also a model of the decidable theory of the empty set, Problem 727(a).

737. (a) and (b) are clear. For (c) remark that a formula can represent at most one function or relation, and the type τ is finite.

(d) Let R be a triadic relation symbol, and Γ be the set of formulas

$$\{\pi_k \neq \pi_\ell : k \neq \ell\} \cup$$
$$\{\forall y (R(\pi_i, \pi_j, y) \leftrightarrow y = \pi_k) : f_i(j) = k\}.$$

(e) Suppose φ represents f. By Problem 724 the consequences of Γ is enumerable. For each $\vec{a} \in \omega^n$ consider the first formula in the enumeration which has the form $\varphi(\pi_{\vec{a}}, y) \leftrightarrow y = \pi_j$ for some $j \in \omega$. This is a recursive condition, and by the assumption of the representability, there is such a consequence of Γ. Thus an application of the μ operator shows that this j is a recursive function of $\vec{a}$. Now $j \in \omega$ is the value of $f(\vec{a})$, as otherwise, using that $\Gamma \vdash \pi_k \neq \pi_\ell$ for different k and ℓ, Γ would be contradictory.

738. For addition and multiplication the formulas $x_1 + x_2 = y$ and $x_1 \cdot x_2 = y$ work by Problems 710 and 713. For the function $K_<(x_1, x_2)$ Problem 713 indicates that the formula $(x_2 \leq x_1 \rightarrow y = \pi_1) \wedge (\neg(x_2 \leq x_1) \rightarrow y = \pi_0)$ works. The projection functions are immediate. Checking that

$$\exists z_1 \dots \exists z_\ell (\psi_1(\vec{x}, z_1) \wedge \cdots \wedge \psi_\ell(\vec{x}, z_\ell) \wedge \varphi(\vec{z}, y))$$

works for the composition operator $\mathrm{Comp}(g, h_1 \dots, h_\ell)$ is routine. Finally $f = \mu(g)$ can be represented by

$$\varphi(\vec{x}, y) \equiv \psi(\vec{x}, y, 0) \wedge (\forall z < y)\exists u(\psi(\vec{x}, z, u) \wedge u \neq 0)$$

where $\psi(\vec{x}, y, u)$ represents g. Suppose $f(\vec{a}) = b$, then $g(\vec{a}, b) = 0$ and $g(\vec{a}, d) \neq 0$ for all $d < b$. First we show that $Q \vdash \varphi(\pi_{\vec{a}}, \pi_b)$. By induction hypothesis, $Q \vdash \psi(\pi_{\vec{a}}, \pi_b, 0)$ and for all $d < b$, $Q \vdash \varphi(\pi_{\vec{a}}, \pi_d, u) \rightarrow u \neq 0$. This fact combined with $Q \vdash x < \pi_b \rightarrow \bigwedge_{d<b} x = \pi_d$ (Problem 712) shows that $Q \vdash \varphi(\pi_{\vec{a}}, \pi_b)$ indeed. To check that $Q \vdash \varphi(\pi_{\vec{a}}, y) \rightarrow y = \pi_b$, observe that Problem 711 says that $Q \vdash (z \leq \pi_b) \vee (\pi_b \leq z)$. For $z < \pi_b$ $Q \vdash \neg\psi(\pi_{\vec{a}}, z)$ by induction (as in this case z is one of π_d for $d < b$). For $\pi_b < z$ again by induction, $Q \vdash \neg\exists u(\psi(\vec{a}, \pi_b, u) \wedge u \neq 0)$, thus Q proves that the second part of $\varphi(\pi_{\vec{a}}, y)$ fails.

739. ($\Rightarrow$) Suppose $\varphi(\vec{x})$ represents the relation R, then

$$\psi(\vec{x}, y) \equiv (\varphi(\vec{x}) \rightarrow y = \pi_0) \wedge (\neg\varphi(\vec{x}) \rightarrow y = \pi_1)$$

clearly represents χ_R as $\Gamma \vdash \pi_0 \neq \pi_1$.

($\Leftarrow$) Suppose $\psi(\vec{x}, y)$ represents the function χ_R. Then a possible representation of R is

$$\varphi(\vec{x}) \equiv \forall y(\psi(\vec{x}, y) \leftrightarrow y = \pi_1).$$

740. Let Γ be the theory of $\langle \omega, 0, S \rangle$ where S is the successor function (see Problem 552), and, for the sake of concreteness, define $x + y = x \cdot y = S(x)$. Then $\Gamma \vdash \pi_k \neq \pi_\ell$ for $k \neq \ell$. Let R be a new unary relation symbol, and add the set $\{R(\pi_n) : n \text{ is even}\}$ to Γ. We claim that the set of even numbers is not representable. For if $\varphi(x)$ represents this set, then $\Gamma \vdash \neg\varphi(\pi_k)$ for all odd integers k. Now take the model $\mathfrak{A} = \langle \omega, 0, S \rangle$ where $R^{\mathfrak{A}}$ is identically true; this is a model of Γ as well. Then in $\mathfrak{A}$ the set of even integers is defined by the formula $\varphi(x)$. As in this model the relation R is identically true, φ is equivalent in $\mathfrak{A}$ to a formula of type $\langle 0, S \rangle$ which defines the set of even numbers. But by Problem 552(c) such a formula does not exist.

741. Let Γ extend Q. Enumerate all infinite recursive sets. As Q represents all recursive functions (Problem 738), pick the formula $\varphi_i(x)$ which represents the i-th infinite recursive set. Let R be a new binary relation symbol, and N be a new unary relation symbol. The new representations will be the formulas $R(\pi_i, x)$. Thus add these formulas to Γ for different i and j:

$$\{\exists x (\neg R(\pi_i, x) \wedge R(\pi_j, x)) : i \neq j\}. \tag{$\star$}$$

We need $R(\pi_i, x)$ and $\varphi_i(x)$ be equivalent when x is one of the terms π_n, but not in other cases. Thus add to Γ that all terms π_i satisfy N, and for elements in N these formulas are equivalent:

$$\{N(\pi_i) : i \in \omega\} \cup \{\forall x\big(N(x) \to (R(\pi_i, x) \leftrightarrow \varphi_i(x))\big) : i \in \omega\}. \qquad (\star\star)$$

Then Γ clearly satisfies the requirements. It is consistent. Take a model of Q which has infinitely many infinite elements. Interpret N as the set of natural numbers. For each i, j pick $c_{i,j}$ as different infinite elements. Finally, interpret R on natural numbers so that it satisfies $(\star\star)$, and otherwise it holds for the pairs $(j, c_{i,j})$ only, giving $(\star)$.

742. Denote the set by $R \subset \omega^2$. By Problem 383 the function $n \mapsto \alpha(\pi_n)$ is recursive, and then the function $f : \langle \alpha(\varphi), n\rangle \mapsto \alpha(\varphi(\pi_n))$ is also recursive using the recursivity of the substitution, Problem 381(e). Now $(i, j) \in R$ if i is a code of a formula φ with x as the only free variable (recursive condition by Problems 381(a) and 381(b)), and $f(i, n) \notin \{\alpha(\psi) : \Gamma \vdash \psi\}$. As Γ is decidable, this latter set is recursive, thus R is recursive as well.

743. Suppose, by contradiction, that Γ is decidable. By Problem 742 the set $R = \{\langle \alpha(\varphi), n\rangle : \Gamma \nvdash \varphi(\pi_n)\} \subset \omega^2$ is a recursive set, thus so is the diagonal $P = \{n : \langle n, n\rangle \in R\}$. As Γ represents all recursive functions, according to Problem 739 there is a formula $\varphi(x) \in F(\tau)$ such that $n \in P$ implies $\Gamma \vdash \varphi(\pi_n)$, and $n \notin P$ implies $\Gamma \vdash \neg\varphi(\pi_n)$. Let $m = \alpha(\varphi)$. The question is whether $m \in P$ or not.

If $m \in P$, then $(\alpha(\varphi), m) \in R$, that is, $\Gamma \nvdash \varphi(\pi_m)$. As φ represents P, we have $\Gamma \vdash \varphi(\pi_m)$, which is impossible.

If $m \notin P$, then $(\alpha(\varphi), m) \notin R$, that is, $\Gamma \vdash \varphi(\pi_m)$. As φ represents P, we have $\Gamma \vdash \neg\varphi(\pi_m)$, thus Γ is not consistent.

744. If Γ is inconsistent, then for every formula φ we have $\Gamma \vdash \varphi$, and then Γ is decidable. On the other hand, there are consistent and decidable theories, e.g. the empty theory in the empty similarity type (see Problem 727).

745. Q represents every recursive functions by Problem 738. As it has a model it is consistent, thus Church's Theorem 10.5 applies. By Problem 737 extensions of Q represent recursive functions as well.

746. $\Gamma \cup Q$ is consistent (as $\mathfrak{A} \vDash \Gamma \cup Q$) and Q represents recursive functions (Problem 738), hence $\Gamma \cup Q$ is undecidable by Church's theorem 10.5. So Γ has a finite consistent extension $\Gamma \cup Q$ which is undecidable. Problem 733 implies that in this case Γ is undecidable.

747. Q is essentially undecidable as it is finite, and every consistent extension of Q is undecidable (Problem 745), and $\mathfrak{N}$ is a model of Q. By Problem 735(a) Γ is undecidable.

748. There is a model of the theory in which $\mathfrak{N}$ can be defined semantically. This is clear for (a), (b). For (c), (d), (e) it was proved in Problems 444, 450, 451.

By Problem 453 the theory of groups is undecidable in the language with an additional constant symbol. But then Problem 734 says that it is undecidable without that symbol as well. (g) follows from J. Robinson's theorem 7.17.

749. By Problem 450 there is a graph in which $\mathfrak{N}$ can be defined semantically. This graph is a model of the empty theory with a binary relation symbol, thus that theory is undecidable by Problem 747. If τ contains a binary function symbol, then use Problem 445 instead.

750. It suffices to show that every graph can be semantically defined in some structure with unary function symbols f and g. Suppose the graph is (V, E) where V is the set of vertices and is E the set of edges. The ground set of our structure will be $V \cup E$. For $v \in V$ $f(v) = g(v) = v$, and for an edge $e \in E$, $f(e) = u$ and $g(e) = v$ where u and v are the endpoints of the edge e.

751. By Church's theorem 10.5, if Γ is consistent and represents recursive functions, then Γ is undecidable. By Theorem 10.3, Γ cannot be complete, because complete and recursive theories are decidable.

752. By the incompleteness theorem there is a closed ϑ which is independent of Γ. Then $\varphi(x) \equiv (\vartheta \to x = 0) \land (\neg\vartheta \to x = 1)$, works.

753. Let ϑ be independent of Γ, and define $\varphi(x, y) \equiv (\vartheta \to y = 0) \land (\neg\vartheta \to y = 1)$. This formula defines either the constant 0 or the constant 1 function, but does not decide between the two values.

754. (a) By Theorem 10.3 any recursive and complete theory suffices. For a list see Problem 732.

(b) No such theory exists by Theorem 10.3.

(c) Algebraically closed fields without the characteristic specified. This theory is clearly recursive, and is incomplete shown by the formula $\forall x (x + x = 0)$. Decidability follows from 557(c).

(d) Robinson's Q, Peano arithmetic or ZFC are the standard examples. Both represent recursive functions (Problem 738), hence they are undecidable by Church's theorem 10.5. Both are recursive, thus they are incomplete by Gödel's first incompleteness theorem 10.6.

(e) Take a finite, complete and decidable Γ (e.g. dense linear ordering without endpoints). There are continuum many subsets Γ' with $\Gamma \subset \Gamma' \subset \text{Cons}_\Gamma$, thus there must exist a non-recursive such Γ'. Clearly $\text{Cons}_\Gamma = \text{Cons}_{\Gamma'}$, and Γ' is complete.

(f) $\text{Th}(\mathfrak{N})$ is complete, but not recursive and undecidable. $\text{Th}(\mathfrak{N})$ represents recursive function, thus it is undecidable (Church's theorem 10.5). The theory of any model is complete. It cannot be recursive by Theorem 10.3.

(g) The similarity type τ consists of uncountably many constant symbols c_α for $\alpha < \kappa$. Let $\Gamma = \{c_\alpha = c_\beta : \alpha, \beta < \kappa\}$. Γ and Cons_Γ being uncountable are non-recursive. However Γ is clearly complete.

(h) The similarity type τ consists of uncountably many constant symbols c_α for $\alpha < \kappa$. Let $\Gamma = \{c_\alpha = c_\alpha : \alpha < \kappa\}$. As uncountable sets cannot be recursive, Γ is not recursive and not decidable. It is clearly incomplete (e.g. $c_0 = c_1$ is independent from Γ).

755. Let $j = \alpha(\varphi(\pi_m))$. As $\alpha(\varphi) = m$, the value of the function at the pair (m, m) is j, thus representability gives

$$\Gamma \vdash \forall y (\chi(\pi_m, \pi_m, y) \leftrightarrow y = \pi_j). \tag{$\star$}$$

By definition $\Psi(\pi_m)$ is the same as $\forall y (\chi(\pi_m, \pi_m, y) \to \Phi(y))$. According to $(\star)$ in every model $\mathfrak{A}$ of Γ there is a unique y satisfying $\chi(\pi_m, \pi_m, y)$, namely π_j. Thus $\Psi(\pi_m)$ holds in $\mathfrak{A}$ if and only if this value satisfies Φ, which means $\mathfrak{A} \vDash \Psi(\pi_m) \leftrightarrow \Phi(\pi_j)$. As this is true in every model of Γ, it can be derived from Γ.

756. According to Problem 755 there is a formula $\Psi(x)$ such that $\Gamma \vdash \Psi(\pi_m) \leftrightarrow \Phi(\ulcorner \varphi(\pi_m) \urcorner)$ for any $\varphi(x)$ with $m = \alpha(\varphi)$. Choose $\varphi \equiv \Psi$ and $\nu \equiv \Psi(\pi_m)$.

757. Suppose otherwise. By the fixed point theorem there is a formula ν such that $\Gamma \vdash \nu \leftrightarrow \neg\Phi(\ulcorner \nu \urcorner)$. If $\Gamma \vdash \nu$ then $\Gamma \vdash \Phi(\ulcorner \nu \urcorner)$ by assumption, and $\Gamma \vdash \neg\Phi(\ulcorner \nu \urcorner)$ as $\Gamma \vdash \nu$. This means that Γ is inconsistent. If $\Gamma \nvdash \nu$, then $\Gamma \vdash \neg\Phi(\ulcorner \nu \urcorner)$ by assumption, and then $\Gamma \vdash \nu$, a contradiction.

758. Suppose by contradiction that $\{\varphi : \Gamma \vdash \varphi\}$ is recursive. The existence of the formula $\Phi(x)$ representing this relation contradicts Problem 757.

759. A fixed point of $\neg\Phi(x)$ is a counterexample.

760. Proceed along the lines of the proof of Theorem 10.7 (see Solution 756), but instead of $\Psi(x) \equiv \forall y (\chi(x, x, y) \to \Phi(y))$ use the formula

$$\Psi^*(x) \equiv \forall y (\chi(x, x, y) \to \underbrace{\top \wedge \top \wedge \cdots \wedge \top}_{n \text{ times}} \wedge \Phi(y)).$$

Each one leads to a different fixed point of $\Phi(x)$.

761. We need to find ν_1 and ν_2 such that

$$\Gamma \vdash \nu_1 \leftrightarrow \Phi_1(\ulcorner \nu_2 \urcorner),$$
$$\Gamma \vdash \nu_2 \leftrightarrow \Phi_2(\ulcorner \nu_1 \urcorner)$$

hold. If we already have ν_1 and ν_2, then by the equivalences we obtain

$$\Gamma \vdash \nu_1 \leftrightarrow \Phi_1(\ulcorner \Phi_2(\ulcorner \nu_1 \urcorner) \urcorner)$$

Indeed, we claim that it is enough to find a formula φ such that

$$\Gamma \vdash \varphi \leftrightarrow \Phi_1(\ulcorner \Phi_2(\ulcorner \varphi \urcorner) \urcorner)$$

as in this case we can take $\nu_1 \equiv \varphi$ and $\nu_2 \equiv \Phi_2(\ulcorner \varphi \urcorner)$.

To prove the existence of such a φ we follow the proof of the fixed point theorem 10.7. The function $f : \langle \alpha(\psi), n \rangle \mapsto \alpha(\Phi_2(\ulcorner \psi(\pi_n) \urcorner))$ is recursive, let the formula $\chi(x_1, x_2, y)$ represent it, and define

$$\Psi(x) \equiv \forall y (\chi(x, x, y) \to \Phi_1(y)).$$

Let $m = \alpha(\Psi)$. Then $f(m, m) = \alpha(\Phi_2(\ulcorner \Psi(\pi_m) \urcorner))$, thus the only y which satisfies $\chi(\pi_m, \pi_m, y)$ is $\ulcorner \Phi_2(\ulcorner \Psi(\pi_m) \urcorner) \urcorner$. Consequently for $\varphi \equiv \Psi(\pi_m)$ we have $\Gamma \vdash \varphi \leftrightarrow \Phi_1(\ulcorner \Phi_2(\ulcorner \varphi \urcorner) \urcorner)$, as required.

762. Let $\chi(x_1, x_2, x_3, z)$ represent the recursive function which maps the triplet $\langle \alpha(\varphi), k, \ell \rangle$ to the code of $\varphi[x/\pi_k, y/\pi_\ell]$. For $i = 1, 2$ let

$$\Psi_i(x, y) \equiv \forall z_1 \forall z_2 (\chi(x, x, y, z_1) \wedge \chi(y, x, y, z_2) \to \Phi_i(z_1, z_2)).$$

Let $k = \alpha(\Psi_1)$, $\ell = \alpha(\Psi_2)$, and replace x by π_k and y by π_ℓ. Then $\chi(x, x, y, z_1)$ holds for $z_1 = \ulcorner \Psi_1(\pi_k, \pi_\ell) \urcorner$ only, and $\chi(y, x, y, z_2)$ holds for $z_2 = \ulcorner \Psi_2(\pi_k, \pi_\ell) \urcorner$ only. Thus the choice $v_i \equiv \Psi_i(\pi_k, \pi_\ell)$ works.

763. Suppose by contradiction that the recursive C separates A and B. Let $\Phi(x)$ represent C in Γ, and v be a fixed point of $\neg \Phi$: $\Gamma \vdash v \leftrightarrow \neg \Phi(\ulcorner v \urcorner)$.

If $\alpha(v) \in C$, then $\Gamma \vdash \Phi(\ulcorner v \urcorner)$, that is, $\Gamma \vdash \neg v$, contradicting that C is disjoint from B. If $\alpha(v) \notin C$, then $\Gamma \vdash \neg \Phi(\ulcorner v \urcorner)$, that is, $\Gamma \vdash v$, contradicting that C contains A.

764. (a) If there is a derivation of φ from Γ, then let $u \in \omega$ be the code of this derivation. As $\langle u, \alpha(\varphi) \rangle \in \mathsf{PP}_\Gamma$, the definition of representability gives $\Gamma \vdash \mathsf{Prov}(\pi_u, \ulcorner \varphi \urcorner)$. From here the claim follows.

(b) If φ cannot be derived from Γ, then there is no $u \in \omega$ with $\langle u, \alpha(\varphi) \rangle \in \mathsf{PP}_\Gamma$. Again, by definition of representability, we have $\Gamma \vdash \neg \mathsf{Prov}(\pi_u, \ulcorner \varphi \urcorner)$.

765. Suppose by contradiction that every formula φ has a derivation with code less than $f(\alpha(\varphi))$. As the relation $\mathsf{PP} \subseteq \omega^2$ is recursive, it means that the set $\{\varphi : \Gamma \vdash \varphi\}$ is recursive, that is, Γ is decidable. But Γ is undecidable by Church theorem 10.5.

766. (a) Let u be the code of the derivation $\Gamma \vdash \forall x \varphi(x)$. By Ax_{10}, for each $n \in \omega$, $\Gamma \vdash \varphi(\pi_n)$, as

1. $\forall x \varphi(x)$ (proof from Γ)
2. $\forall x \varphi(x) \to \varphi(\pi_n)$ (axiom Ax_{11})
3. $\varphi(\pi_n)$ (MP from 1, 2)

It is clear that the code of this proof is a recursive function of u (a fixed number) and n.

(b) No such a formula exists. As Γ is recursive, the provability predicate PP_Γ is recursive. For a fixed formula $\varphi(x)$ the function $n \mapsto \alpha(\varphi(\pi_n))$ is recursive. By assumption $\Gamma \vdash \varphi(\pi_n)$ for all $n \in \omega$, thus for every n there is a $u \in \omega$ such that $\langle u, \alpha(\varphi(\pi_n)) \rangle \in \mathsf{PP}_\Gamma$. The function which picks the smallest such u is recursive.

767. If $\Gamma \vdash v$, then there is a natural number u such that $\Gamma \vdash \mathrm{Prov}(\pi_u, \ulcorner v \urcorner)$. Then $\Gamma \vdash \exists u \, \mathrm{Prov}(u, \ulcorner v \urcorner)$. By the definitions of Pr and v this implies $\Gamma \vdash \neg v$. This, combining with the assumption on v, yields the inconsistency of Γ.

768. Let v be a fixed point of $\neg \mathrm{Pr}(x)$, that is, $\Gamma \vdash v \leftrightarrow \neg \mathrm{Pr}(\ulcorner v \urcorner)$. By assumption, $\Gamma \vdash \mathrm{Pr}(\ulcorner v \urcorner) \to v$, thus $\Gamma \vdash \mathrm{Pr}(\ulcorner v \urcorner) \to \neg \mathrm{Pr}(\ulcorner v \urcorner)$. As it has contradicting consequences, $\Gamma \vdash \neg \mathrm{Pr}(\ulcorner v \urcorner)$. Consequently $\Gamma \vdash v$, while Problem 767 says that it cannot happen for a consistent theory.

769. Let v be a fixed point of $\neg \mathrm{Pr}(x)$, and let $\varphi(u) \equiv \neg \mathrm{Prov}(u, \ulcorner v \urcorner)$. As $\Gamma \nvdash v$ (Problem 767), we have $\Gamma \vdash \neg \mathrm{Prov}(\pi_n, \ulcorner v \urcorner)$ for each $n \in \omega$ (Problem 764(b)). Also,

$$\Gamma \vdash v \leftrightarrow \neg \mathrm{Pr}(\ulcorner v \urcorner) \leftrightarrow \forall u \neg \mathrm{Prov}(u, \ulcorner v \urcorner),$$

as required.

770. If Γ is inconsistent, then there is nothing to prove. So assume Γ is consistent. Suppose first, that $\Gamma \vdash \varphi$ is witnessed by $u \in \omega$, that is, $\langle u, \alpha(\varphi) \rangle \in \mathrm{PP}_\Gamma$. Then $\Gamma \nvdash \neg \tilde\varphi$ by consistency, $\Gamma \vdash \mathrm{Prov}(\pi_u, \ulcorner \varphi \urcorner)$ as Prov represents the provability predicate, and $\Gamma \vdash \neg \mathrm{Prov}(\pi_n, \ulcorner \neg \tilde\varphi \urcorner)$ for every $n \in \omega$ by Problem 764. By the definition of $n(x)$ we have $\Gamma \vdash n(\ulcorner \varphi \urcorner) = \ulcorner \neg \tilde\varphi \urcorner$. As $Q \subseteq \Gamma$ Problem 712 gives

$$\Gamma \vdash v {\le} \pi_u \to (v = \pi_0 \lor v = \pi_1 \lor \cdots \lor v = \pi_u).$$

Combining them together we get that $\Gamma \vdash \mathrm{Prov}^*(\pi_u, \ulcorner \varphi \urcorner)$.

If $\langle u, \alpha(\varphi) \rangle \notin \mathrm{PP}_\Gamma$, then $\Gamma \vdash \neg \mathrm{Prov}(\pi_u, \ulcorner \varphi \urcorner)$, thus Γ also proves the negation of Prov^*.

771. Assume Γ is consistent, otherwise there is nothing to prove. Now $\Gamma \vdash \mathrm{Prov}(\pi_u, \ulcorner \neg \varphi \urcorner)$ for some $u \in \omega$, and $\Gamma \vdash \neg \mathrm{Prov}(\pi_n, \ulcorner \varphi \urcorner)$ for every $n \in \omega$ as Γ is consistent. Assume by contradiction that $\mathfrak{A}$ is a model of Γ where $\mathfrak{A} \vDash \mathrm{Prov}^*(a, \ulcorner \varphi \urcorner)$ for some element $a \in A$. Then $a \neq \pi_n$ for any $n \in \omega$, and then $\mathfrak{A} \vDash \pi_u \le a$ by Problem 711. Also we have $\mathfrak{A} \vDash \mathrm{Prov}(\pi_u, \ulcorner \varphi \urcorner)$, thus

$$\mathfrak{A} \nvDash (\forall v \le u) \neg \mathrm{Prov}(v, \ulcorner \varphi \urcorner),$$

a contradiction.

772. We have $\Gamma \vdash v^* \leftrightarrow \neg \mathrm{Pr}^*(\ulcorner v^* \urcorner)$. The case $\Gamma \nvdash v^*$ goes similarly to Problem 767. If $\Gamma \vdash v^*$, then $\Gamma \vdash \mathrm{Pr}^*(\ulcorner v^* \urcorner)$ by representation, and $\Gamma \vdash \neg \mathrm{Pr}^*(\ulcorner v^* \urcorner)$ by the fixed point, thus Γ is inconsistent.

For the other case use Problem 771. If $\Gamma \vdash \neg v^*$, then $\Gamma \vdash \neg \mathrm{Pr}^*(\ulcorner v^* \urcorner)$. This combined with the fixed point shows that Γ is inconsistent.

773. Clearly, $\Gamma \vdash \neg \bot$, thus $\Gamma \vdash \neg \mathrm{Pr}^*(\ulcorner \bot \urcorner)$ by Problem 771, as required.

774. (a) As $\Gamma \vdash \top$, we also have $\Gamma \vdash \mathrm{Pr}^*(\ulcorner \top \urcorner)$, thus $\top$ is such a fixed point.

(b) By Problem 773 $\Gamma \vdash \neg \mathrm{Pr}^*(\ulcorner \bot \urcorner)$. Thus $\Gamma \vdash \bot \leftrightarrow \mathrm{Pr}^*(\ulcorner \bot \urcorner)$, $\bot$ is also a fixed point, and clearly, $\Gamma \vdash \neg \bot$.

775. (a) Using Rosser's trick. If either A or B is finite (or recursive), then the formula (or its negation) representing that set works. So assume both are infinite, and let $\Phi(x, y)$ and $\Psi(x, y)$ represent the recursive functions enumerating A and B, respectively. We claim that

$$\varphi(y) \equiv \exists x(\Phi(x, y) \wedge (\forall z {\le} x)\neg\Psi(z, y)) \qquad\qquad (\star)$$

is as required. Indeed, if $n \in A$, then there is an $i \in \omega$ such that $\Gamma \vdash \Phi(\pi_i, \pi_n)$, and $\Gamma \vdash \neg\Psi(\pi_j, \pi_n)$ for all $j \le i$ as $n \notin B$. Conversely, if $n \in B$, then $\Gamma \vdash \Psi(\pi_j, \pi_n)$ for some $j \in \omega$, and $\Gamma \vdash (\pi_j \le x) \vee (x \le \pi_j)$. In the first case the second part of $(\star)$ fails. In the first case x is one of π_i with $i \le j$, and $\Gamma \vdash \neg\Phi(\pi_i, \pi_n)$, thus the first part of $(\star)$ fails, showing that $\Gamma \vdash \neg\varphi(\pi_n)$.

(b) For any formula $\vartheta(x)$ the set $\{n : \mathrm{PA} \vdash \vartheta(\pi_n)\}$ is recursively enumerable, and if a set and its complement are recursively enumerable, then the set is recursive. Thus, the additional requirement would guarantee a recursive set C which separates A and B. In Problem 763 two recursively enumerable sets are created which are not separable.

776. We only need to show that $\mathrm{PA} \vDash Q7$. Let $\varphi(y)$ be the formula $y \ne 0 \to \exists x(y = x + 1)$. Clearly $\mathrm{PA} \vDash \varphi(0)$. $\varphi(y + 1)$ always holds, as $y + 1 = y + 1$ (no need to assume $\varphi(y)$). Thus $\mathrm{PA} \vDash \forall y(\varphi(y) \to \varphi(y+1))$, and then the induction axiom (and modus ponens) gives $\mathrm{PA} \vDash \forall y\varphi(y)$ as desired.

777. A structure with a single element (and with $0^{\mathfrak{A}} = 1^{\mathfrak{A}}$) is a model of PA2–PA7.

778. (a) (b) follow by induction on z.

(c) follows from Q By Problem 714, thus it also follows from PA.

(d) To show commutativity, first observe that $0 + 1 = 1 + 0 = 1$ (follows from Q, Problem 708). Prove by induction that $x + 0 = 0 + x$, and $y + 1 = 1 + y$. Then for $x + y = y + x$ the induction step could be

$$x + (y + 1) \overset{(1)}{=} (x + y) + 1 \overset{(2)}{=} (y + x) + 1 \overset{(3)}{=} y + (x + 1) \overset{(4)}{=} y + (1 + x) \overset{(5)}{=} (y + 1) + x,$$

where (1) is PA4, (2) is the induction hypothesis, (3) and (5) are associativity from part (a), and (4) is the special case proved earlier.

(e) Prove first that $(y + 1) \cdot x = y \cdot x + x$ by induction on x, then proceed similarly to (d).

779. $x \le x$ as $x + 0 = x$ by PA3. If $x \le y$ and $y \le z$, then there are a and b such that $x + a = y$ and $y + b = z$. As $x + (a + b) = (x + a) + b$ by 778(a), we have $x \le z$.

For antisymmetry suppose $x \le y$ and $y \le x$. Then $x + a = y$ and $y + b = x$ for suitable a and b. By 778(a) and PA3, $x = y + b = (x + a) + b = x + (a + b) = x + 0$, hence 778(b) implies $a + b = 0$. By 778(c), $a = 0$ or $b = 0$, therefore $x = y + 0$ or $y = x + 0$, which by PA3 leads to $x = y$.

For totality of $\le$ pick arbitrary x and y. We need to show $x \le y$ or $y \le x$. Let $\varphi(x, y)$ be the formula $\exists z(z + x = y \vee z + y = x)$. Then $\varphi(0, y)$ holds and $\varphi(x, y)$ implies $\varphi(x + 1, y)$, thus the induction axioms ensure $\forall x\varphi(x, y)$.

$0 \le x$ for all x as $x + 0 = x$. If $x < y$ then $z + x = y$ for some non-zero z, that is, $z = z' + 1$. But then $z' + (x + 1) = y$, which means $x + 1 \le$. Also, if $x \ne 0$, then $x = x' + 1$ and then x' is the immediate predecessor of x.

780. By induction on b. For $b = 0$ the only good choice is $r = 0$ (you need the distributivity of multiplication to prove that no other element below a is divisible by a). For $b + 1$ use the induction hypothesis, take r which works for b, and then either the predecessor of r is good, or the predecessor of a in case $r = 0$. The unicity also follows from the induction hypothesis.

781. Assume

$$\forall x \big[(\forall y < x) \varphi(y, \vec{p}) \to \varphi(x, \vec{p}) \big], \qquad\qquad (\star\star)$$

and use the induction axiom PA7 for the formula

$$\varphi^*(x, \vec{p}) \equiv (\forall y \le x) \varphi(y, \vec{p}).$$

The conclusion of P7 is $\forall x \varphi^*(x, \vec{p})$ and from here $\forall x \varphi(x, \vec{p})$ follows as $x \le x$ for all x. Thus it suffices to show that both the initial step and the induction step follow from $(\star\star)$. First we check $\varphi^*(0, \vec{p})$. As there is no $y < 0$, $(\star\star)$ gives $\varphi(0, \vec{p})$, and as $x \le 0 \to x = 0$, we are done.

Second we show $\varphi^*(x, \vec{p}) \to \varphi^*(x + 1, \vec{p})$ for all x. So suppose $\varphi^*(x, \vec{p})$ holds for some x. Now $y \le x + 1$ if either $y \le x$, or $y = x + 1$. For the first case $\varphi^*(x, \vec{p})$ implies $\varphi(y.\vec{p})$, and for the second case $(\star\star)$ implies $\varphi^*(x, \vec{p}) \to \varphi(x + 1, \vec{p})$ as all elements strictly below $x + 1$ are $\le x$. No more cases are left, we are done.

782. The base set of $\mathfrak{A}$ is $\omega \times \omega$, and $0^{\mathfrak{A}} = (0, 0)$, $1^{\mathfrak{A}} = (1, 0)$. Define the addition as $(a, i) +^{\mathfrak{A}} (b, j) = (a, b + i + j)$, and the multiplication as $(a, i) \cdot^{\mathfrak{A}} (b, j) = (ab, j(a + i))$. Then $(a, i) +^{\mathfrak{A}} (1, 0) = (a, i + 1)$, and the multiplication has been chosen so that $(a, i) \cdot^{\mathfrak{A}} (b, 0) = (ab, 0)$, and it satisfies PA6:

$$(a, i) \cdot^{\mathfrak{A}} (b, j + 1) = (a, i) \cdot^{\mathfrak{A}} (b, j) +^{\mathfrak{A}} (a, i).$$

It is a routine to check that PA1–PA6 holds. The relation $x \le y$ is defined by $\exists z(z + x = y)$. In this model $(b, j) \le^{\mathfrak{A}} (a, k)$ iff $b + j \le k$. To check the strong induction $(\star)$, assume $\varphi(x, \vec{p})$ fails in $\mathfrak{A}$ for some $x = (a, k)$. Choose this element so that k is minimal, and if there are more elements with the same k, then choose the one with the smallest a. If $(b, j) = y <^{\mathfrak{A}} x = (a, k)$, then $b + j \le k$ (thus $j \le k$), and if $j = k$ then $b = 0 < a$ (as x and y are different). By the choice of x (minimal counterexample) $\mathfrak{A} \vDash \varphi(y, \vec{p})$, contradicting the antecedent of the strong induction.

783. To get the induction axiom for a formula $\varphi(x, \vec{p})$ with parameters, apply the parameter-free induction axiom to the formula

$$\psi(x) \equiv \forall \vec{p} \big[\varphi(0, \vec{p}) \land \forall y (\varphi(y, \vec{p}) \to \varphi(y + 1, \vec{p})) \to \varphi(x, \vec{p}) \big].$$

784. The collection principle is

$$(\forall x < a)(\exists y) \varphi(x, y, \vec{p}) \to (\exists b)(\forall x < a)) \exists y < b) \varphi(x, y, \vec{p}),$$

where b is not free in *phi*. Denote this formula by $\Phi(a)$, we proceed to prove $(\forall a)\Phi(a)$ by induction on a. For $a = 0$ the statement holds vacuously as there is no $x < 0$. So suppose $\Phi(a)$ holds, witnessed by b. For $\Phi(a+1)$ the only element $x < a+1$ not covered by b is $x = a$. Let $y' = 0$ if there is no y with $(\exists y)\varphi(a, y, \vec{p})$, otherwise any such element. Then $b' = 1 + \max(b, y')$ works for $\Phi(a+1)$.

785. ($\Rightarrow$) Let $X = \{a \in A : \mathfrak{A} \vDash \varphi[a, \vec{p}]\}$, and assume that X has no $\leq^{\mathfrak{A}}$-minimal element. Consider the formula

$$\varphi^*(x, \vec{p}) \equiv (\forall y < x)\neg\varphi(x, \vec{p}).$$

Now $\mathfrak{A} \vDash \varphi^*[0, \vec{p}]$, as nothing is smaller than $0^{\mathfrak{A}}$, and $\mathfrak{A} \vDash \varphi^*(x, \vec{p}) \to \varphi^*(x+1, \vec{p})$. This latter holds since the only element not covered in $\varphi^*(x, \vec{p})$ is x, and if $\mathfrak{A} \vDash \neg\varphi(y, \vec{p})$ for all $y < x$ but $\mathfrak{A} \vDash \varphi(x, \vec{p})$, then $x \in X$ would be a minimal element in X. Thus by P7 we have $\mathfrak{A} \vDash \forall x \varphi^*(x, \vec{p})$, which means that X is the empty set.

($\Leftarrow$) Let $\varphi(x, \vec{p})$ be a formula such that $\mathfrak{A} \vDash \varphi(0, \vec{p})$ and $\mathfrak{A} \vDash \varphi(x, \vec{p}) \to \varphi(x+1, \vec{p})$. Assume $X = \{a \in \mathfrak{A} : \mathfrak{A} \vDash \neg\varphi[a, \vec{p}]\}$ is not empty, and let $a \in X$ be $\leq^{\mathfrak{A}}$-minimal. $a \neq 0$ as $\mathfrak{A} \vDash \varphi(0, \vec{p})$, this the predecessor of a is not in X while its successor is in X, which is impossible.

786. (a) Every element of $\mathfrak{N}$ is the value of one of the terms $\{\pi_n : n \in \omega\}$. Replace the parameters in the defining formula with the corresponding terms.

(b) There are countably many formulas, thus there are countably many subsets in any structure which can be defined without parameters. Any one-element set can be trivially defined using parameters: the formula $x = p$ just does it. Thus any uncountable structure has such a subset.

787. Let $n+1$ be the minimal number of parameters $\vec{p}$ which defines a set without a minimal element, and suppose $X = \{a \in A : \mathfrak{A} \vDash \varphi(a, \vec{p})\}$ has no minimal element. This fact is expressed by

$$\Phi(\vec{p}) \equiv (\exists x \varphi(x, \vec{p})) \wedge \forall x(\varphi(x, \vec{p}) \to (\exists y < x)\varphi(y, \vec{p})).$$

Let $\vec{p} = \langle \vec{q}, p \rangle$ then the set $\{a \in A : \mathfrak{A} \vDash \Phi(\vec{q}, a)\}$ is defined by n parameters thus has a minimal element p which is defined by $\Phi(\vec{q}, x) \wedge (\forall y < x)\neg\Phi(\vec{q}, y)$. But then a set without minimal element is defined by the formula $\varphi(x, \vec{q}, p)$ which has only n parameters, a contradiction.

788. If a and b are definable, then so are $a + b$ and $a \cdot b$, thus the definable elements form a substructure. To check that it is an elementary substructure, use the Tarski–Vaught test 8.3. Suppose $\vec{b} \in B$, and $\mathfrak{A} \vDash (\exists x \varphi)[\vec{b}]$. Then the minimal such element c is defined by

$$\psi(x) \equiv \varphi(x, \vec{b}) \wedge \forall(y < x)\neg\varphi(y, \vec{b})$$

(replace $\vec{b}$ by their defining formulas). Thus $c \in B$ and $\mathfrak{A} \vDash \varphi(c, \vec{b})$.

789. As PA models are also models of Q, this claim follows from Problem 721.

790. A non-standard model is a proper extension of $\mathfrak{N}$, thus it has at least one infinite element a. But the $a+1$, $a+2$, … are all different infinite elements.

791. By Problem 785 every definable subset has a minimal element. However, there is no minimal infinite number: if a is infinite, then so is its immediate predecessor.

792. ($\Leftarrow$) If a is infinite, then $\mathfrak{A} \models \pi_n \le a$ for all $n \in \omega$, and then $\mathfrak{A} \models \varphi(\pi_n, \vec{p})$.
($\Rightarrow$) The set $X = \{a \in A : \mathfrak{A} \models \neg\varphi(a, \vec{p})\}$ is definable. Thus either it is empty, in which case any infinite $a \in A$ works, or it has a minimal element $a \in A$ (Problem 785). This a cannot be finite as all finite elements satisfy $\varphi(x, \vec{p})$, and clearly $\mathfrak{A} \models (\forall x < a)\varphi(x, \vec{p})$, as required.

793. Using the omitting type theorem 7.19. By Problem 784 every Peano model satisfies the collection principle, thus Problem 517 yields the result.

794. The set P of primes is infinite. Adjoin a new constant symbol c to the similarity type and for each subset $X \subseteq P$ consider the formulas

$$T_X(c) = \{p \mid c : p \in X\} \cup \{p \nmid c : p \in P \smallsetminus X\}$$

By the compactness theorem each $T_X(c)$ is realized in some Peano model, which has a countable elementary submodel $\mathfrak{A}_X$ by the Löwenheim–Skolem theorem.

Why are $\mathfrak{A}_X$ and $\mathfrak{A}_Y$ not isomorphic for $X \ne Y$? It might happen that for different $X, Y \subseteq P$ they are. But as $\mathfrak{A}_X$ is countable, there can be at most countably many formulas T_Z which are realized in $\mathfrak{A}_X$. Since there are continuum many subsets of P, we can conclude that there are continuum many models among the $\mathfrak{A}_X$'s which are not isomorphic. We used that whenever two models realize different sets of formulas, then they cannot be isomorphic.

795. (i) and (ii) clearly indicate that the multiplication cannot be defined explicitly (see Definition 7.13) by the $\langle 0, 1, + \rangle$-reduct. Use Beth's theorem 7.14.

796. For an infinite element $a \in A$ the *island of a* is the set

$$\ldots, a-2, a-1, a, a+1, a+2, \ldots$$

which is order-isomorphic to $\mathbb{Z}$. We write $a \ll b$ if $a \le b$ and a and b are on different islands, namely, if they correspond to different elements in the linear order $\langle M, \le \rangle$.

(a) Note that for all infinite a we have $a \ll 2a$ since otherwise $a + \pi_n = 2a$ for some $n \in \omega$ and thus $a = \pi_n$ would be finite. Also, a can be assumed to be even, as either a or $a+1$ is even, and then $a/2 \ll a$ for similar reasons. Now, if a and b are infinite, $a \ll b$, then both can be assumed to be even. We have that $a \ll (a+b)/2 \ll b$, otherwise $a + \pi_n = (a+b)/2$ would imply

$2a+2\pi_n = a+b$, therefore $a+2\pi_n = b$ which contradicts $a \ll b$. Consequently we got that between each copy of $\mathbb{Z}$, there is another copy. This proves denseness of M

(b) A countable, dense linear order is isomorphic to the ordering of rational numbers, see Problem 365.

(c) By way of contradiction suppose that the order type of $\mathfrak{A}$ is $\omega + \mathbb{Z} \times \mathbb{R}$. Then each copy of $\mathbb{Z}$ (each island) can be indexed with a corresponding real number r. Let $a \in A$ be infinite and consider the sequence $\pi_n \cdot a$ for $n \in \omega$. For each such a multiple there is an $r_n \in \mathbb{R}$ so that $\pi_n \cdot a \in \mathbb{Z}_{r_n}$. This way a sequence (r_n) of reals is defined. Observe that $\pi_n \cdot a < a^2$ for every $n \in \omega$, therefore if $a^2 \in \mathbb{Z}_p$, then $r_n < p$.

The sequence (r_n) is clearly increasing, and it is also bounded, hence r_n converges to some $r \in \mathbb{R}$. Pick any $b \in \mathbb{Z}_r$ and define the set

$$S = \{x : a \cdot x < b\}.$$

This is a first-order definition of just the standard elements of $\mathfrak{A}$, which is impossible by problem 791.

797. In each PA-model the formula $\varphi_f(\vec{x}, y)$ defines a function, which means that $\text{PA} \vdash \forall \vec{x} \exists! y \varphi_f(\vec{x}, y)$. Similarly, in every model $\varphi_f(\pi_{\vec{a}}, y)$ holds iff $y = \pi_{f(\vec{a})}$, thus $\text{PA} \vdash \varphi_f(\pi_{\vec{a}}, y) \leftrightarrow y = \pi_{f(\vec{a})}$.

798. According to Problem 769, there is a formula $\vartheta(x)$ such that $\text{PA} \vdash \varphi(\pi_n)$ for all $n \in \omega$, while $\text{PA} \nvdash \forall x \vartheta(x)$. Let f be the constant 0 function and consider $\varphi(x, y) \equiv y = 0$, and $\psi(x, y) \equiv \vartheta(x) \rightarrow y = 0$.

799. $x \dot- y = z$ can be defined by the formula

$$(x \le y \wedge z = 0) \vee (y \le x \wedge x + z = y).$$

For the last function let $\text{prime}(y)$ be the formula which says that y is at least one and has no non-trivial divisor. The defining formula can be

$$(x < z \le 2x + 1 \wedge \forall y(\text{prime}(y) \wedge y | z \rightarrow y = 2)).$$

Other functions can be handled similarly. The existence and uniqueness of z in each model can be proved by induction similarly to Problem 780.

800. Use induction on x.

801. (a) Following the solution of Problem 134, the defining formula could be

$$\begin{aligned}
\varphi(x, y) \equiv \forall u \big[(\text{Len}(u) = x + 1 \ \wedge \ \text{Elem}(u, 0) = 1 \ \wedge \\
\wedge \ (\forall i < x) \, \text{Elem}(u, i + 1) = 2 \cdot \text{Elem}(u, i)) \\
\rightarrow y = \text{Elem}(u, x) \big].
\end{aligned}$$

Both the existence of u and the soundness of this definition can be shown by induction on x (using the provable properties of function Elem).

(b) By induction on n, as $\text{PA} \vdash 2 \cdot \pi_{2^n} = \pi_{2^{n+1}}$.

802. (a) By induction on how the function was built up. For initial functions and composition it is straightforward. For primitive recursion show by recursion on x that such a u always exists (using Theorem 10.11), and by induction on the length of u that for all such u their elements are the same.

(b) We need to show that $PA \vdash \varphi_f(\pi_{\bar{a}}, \pi_n, y) \leftrightarrow y = \pi_{f(\bar{a},n)}$. Again, it holds for the initial functions, and easy to check for composition. Suppose $f = PrRec(g, h)$. Form the defining formula we have $PA \vdash \varphi_f(\vec{x}, 0, z) \leftrightarrow \varphi_g(\vec{x}, 0, z)$. By induction, $PA \vdash \varphi_g(\pi_{\bar{a}}, 0, z) \leftrightarrow z = \pi_{g(\bar{a},0)}$, thus $PA \vdash \varphi(\pi_{\bar{a}}, 0, z) \leftrightarrow z = \pi_{f(\bar{a},0)}$. This establishes the claim for $n = 0$. For larger values of n go by induction using that we know the claim both for the function h (with arbitrary arguments) and for $n - 1$.

803. Using the idea of Solution 146, the graph $H(i, x, y)$ of the Ackermann function can be formalized using Len, Elem and the initial functions. Using Theorem 10.11 we have $PA \vdash \forall x \exists y H(0, x, y)$ and $PA \vdash \forall x \exists y H(i, x, y) \rightarrow \forall x \exists y H(i+1, x, y)$. Applying the induction scheme of PA completes the proof.

> **Remark.** The Ackermann function is not primitive recursive by Problem 111, yet it is PA-definable. In Problem 842 we construct a recursive function that is not PA-definable.

804. The coding functions are primitive recursive by Problem 128. Instances of axiom schemes Ax_1–Ax_{12} and Ex_1–Ex_3 are primitive recursive as well as the derivation rules MP and G (see Section 7.1.1). By assumption Γ is primitive recursive, thus whether the sequence $u = \langle \varphi_1, \ldots, \varphi_n \rangle$ is a correct Hilbert-type derivation from Γ can be checked by primitive recursive relations using bounded quantifiers, see Problem 104(c).

805. (a) If $\Gamma \vdash \varphi$, then $\langle u, \alpha(\varphi) \rangle \in PP_\Gamma$ for some $u \in \omega$. Since $Prov_\Gamma^\circ$ represents PP_Γ, see Problem 802(b), we have $PA \vdash Prov_\Gamma^\circ(\pi_u, \ulcorner\varphi\urcorner)$.

(b) As $Prov_\Gamma^\circ(u, x)$ is a faithful representation, it checks whether every element in the proof sequence u is correctly entered. When swapping Γ to $\Gamma \cup \Delta$ the only difference is that that instead of checking whether $Elem(u, i)$ is a member of Γ, the formula checks whether it is a member of $\Gamma \cup \Delta$. Again, it is a faithful representation, which means that PA proves that members of Γ are also members of $\Gamma \cup \Delta$.

(c) If u is a derivation of φ and v is a derivation of $\varphi \rightarrow \psi$, then it is a matter to concatenate these derivations and add an application of MP. PA proves that for any two sequences their concatenation exists (by induction on the length of the second sequence), and the append function can add a new element to the sequence.

806. (a) By Problem 805(c) this is derivable from PA. But PA is a subset of Γ, thus it is also derivable from Γ.

(b) By Problem 805(a) $PA \vdash Pr(\ulcorner\varphi \rightarrow \psi\urcorner)$. Combining with (a) gives the required implication.

(c) By induction on the number of φ_i formulas. For $n = 1$ this is (b). Otherwise $(\Phi \wedge \varphi) \to \psi$ is propositionally equivalent to $\Phi \to (\varphi \to \psi)$. Using the induction and (a) in the form

$$\Gamma \vdash \mathsf{Pr}(\ulcorner \varphi \to \psi \urcorner) \to \big(\mathsf{Pr}(\ulcorner \varphi \urcorner) \to \mathsf{Pr}(\ulcorner \psi \urcorner) \big)$$

completes the induction step.

(d) By Problem 805(b) $\mathsf{PA} \vdash \mathsf{Pr}_{\mathsf{PA}}(\ulcorner \varphi \urcorner) \to \mathsf{Pr}_{\Gamma}(\ulcorner \varphi \urcorner)$ for every φ. This together with Theorem 10.13 gives the claim.

807. Clearly, $\Gamma \vdash (\varphi \wedge \neg\varphi) \to \bot$. Use Problem 806(c).

808. By Problem 806(b), $\Gamma \vdash \mathsf{Pr}(\ulcorner \varphi \urcorner) \leftrightarrow \mathsf{Pr}(\ulcorner \neg\neg\varphi \urcorner)$. Thus let v be a fixed point $\Gamma \vdash v \leftrightarrow \neg\mathsf{Pr}(\ulcorner v \urcorner)$, then $\varphi \equiv \neg v$ works.

809. Solution 1. Take φ to be $\bot$.
Solution 2. If φ is of the form $\mathsf{Pr}(\bullet)$ then $\Gamma \vdash \varphi \to \mathsf{Pr}(\ulcorner \varphi \urcorner)$ holds by Problem 806(d). In Problem 808 we saw that for $\varphi \equiv \neg v$ we have $\Gamma \vdash \varphi \to \mathsf{Pr}(\ulcorner \neg v \urcorner)$. However, we also have $\Gamma \vdash \neg v \leftrightarrow \mathsf{Pr}(\ulcorner v \urcorner)$, thus φ it is just of the right form. We also need that $\Gamma \vdash \psi_1 \leftrightarrow \psi_2$ implies $\Gamma \vdash \mathsf{Pr}(\ulcorner \psi_1 \urcorner) \leftrightarrow \mathsf{Pr}(\ulcorner \psi_2 \urcorner)$, which is true by Problem 806(d).

810. We have $\Gamma \vdash \bot \to v$, thus Problem 806(b) gives $\Gamma \vdash \mathsf{Pr}(\ulcorner \bot \urcorner) \to \mathsf{Pr}(\ulcorner v \urcorner)$. As v is a fixed point of $\neg\mathsf{Pr}(x)$, we get $\Gamma \vdash \mathsf{Pr}(\ulcorner \bot \urcorner) \to \neg v$.

For the other direction Problem 809 showed that both $\Gamma \vdash \neg v \to \mathsf{Pr}(\ulcorner \varphi \urcorner)$ and $\Gamma \vdash \neg v \to \mathsf{Pr}(\ulcorner \neg\varphi \urcorner)$ for $\varphi \equiv \mathsf{Pr}(\ulcorner v \urcorner)$. By Problem 807 in this case $\Gamma \vdash \neg v \to \mathsf{Pr}(\ulcorner \bot \urcorner)$.

811. $(\Rightarrow)$ Let $\Gamma \vdash v \leftrightarrow \neg\mathsf{Pr}(\ulcorner v \urcorner)$. As v has contradictory consequences, $\Gamma \vdash \neg v$. By Problem 810 it implies $\Gamma \vdash \mathsf{Pr}(\ulcorner \bot \urcorner)$.

$(\Leftarrow)$ Since $\Gamma \vdash \bot \to \varphi$, Problem 806(b) shows $\Gamma \vdash \mathsf{Pr}(\ulcorner \bot \urcorner) \to \mathsf{Pr}(\ulcorner \varphi \urcorner)$. Consequently $\Gamma \vdash \mathsf{Pr}(\ulcorner \varphi \urcorner)$ for all closed φ.

812. If v is the fixed point of $\neg\mathsf{Pr}(x)$, then $\Gamma \nvdash v$ by Problem 767. By Problem 810, $\Gamma \vdash \mathsf{Con}_\Gamma \leftrightarrow v$.

813. (a) This is the case as the standard model $\mathfrak{N}$ is a model of PA. If we have $\mathfrak{N} \models \exists u \mathsf{Prov}_\Gamma^\circ(u, \ulcorner \bot \urcorner)$, then there is a $u \in \omega$ such that the pair $\langle u, \alpha(\bot) \rangle \in \mathsf{PP}_\Gamma$. Consequently this $u \in \omega$ is the code of a correct proof sequence which derives $\bot$ from Γ, thus Γ is inconsistent. PA is consistent, as it has a model $\mathfrak{N}$.

(b) As $\mathsf{PA} \nvdash \mathsf{Con}_{\mathsf{PA}}$, $\Gamma = \mathsf{PA} \cup \{\neg\mathsf{Con}_{\mathsf{PA}}\}$ is consistent, and clearly primitive recursive. Also, $\Gamma \vdash \mathsf{Pr}_{\mathsf{PA}}^\circ(\bot)$ (as this is an element of Γ). By Problem 805(b) in this case $\Gamma \vdash \mathsf{Pr}_\Gamma^\circ(\bot)$ as well.

814. By Problem 810 for theories considered in this section $\Gamma \vdash v \leftrightarrow \mathsf{Con}_\Gamma$. Thus $\Gamma \vdash \neg v$ iff $\Gamma \vdash \neg\mathsf{Con}_\Gamma$. Such a theory was constructed in Problem 813.

815. By Problem 807 in this case $\Gamma \vdash \mathsf{Pr}(\ulcorner \bot \urcorner)$. Thus take Γ from Problem 813, for this theory $\Gamma \vdash \mathsf{Pr}(\ulcorner \top \urcorner)$, and $\Gamma \vdash \mathsf{Pr}(\ulcorner \bot \urcorner)$.

816. Not necessarily. For PA yes (in general, for every Γ which has $\mathfrak{N}$ as a model), as in this case the derivation guaranteed by the formula Pr_Γ corresponds to a real derivation, see Problem 813(a).

For a counterexample consider the consistent theory in Problem 815, we cannot have $\Gamma \vdash \varphi$ and $\Gamma \vdash \neg\varphi$ at the same time.

817. (a) As $\Gamma \vdash \mu \leftrightarrow (\mathsf{Pr}(\ulcorner\mu\urcorner) \to \varphi)$, we also have $\Gamma \vdash \varphi \to \mu$ (if φ is true, then the implication $\mathsf{Pr}(\ulcorner\mu\urcorner) \to \varphi$ is true, thus μ must also be true). From here Problem 806(b) gives $\Gamma \vdash \mathsf{Pr}(\ulcorner\varphi\urcorner) \to \mathsf{Pr}(\ulcorner\nu\urcorner)$.

(b) Now $\Gamma \vdash (\mu \wedge \mathsf{Pr}(\ulcorner\mu\urcorner)) \to \varphi$, thus $\Gamma \vdash (\mathsf{Pr}(\ulcorner\mu\urcorner) \wedge \mathsf{Pr}(\ulcorner\mathsf{Pr}(\ulcorner\mu\urcorner)\urcorner)) \to \mathsf{Pr}(\ulcorner\varphi\urcorner)$ by Problem 806(c). We also have $\Gamma \vdash \mathsf{Pr}(\ulcorner\mu\urcorner) \to \mathsf{Pr}(\ulcorner(\mathsf{Pr}(\ulcorner\mu\urcorner)\urcorner)$, thus we are done.

(c) By (a), $\Gamma \vdash \mathsf{Pr}(\ulcorner\varphi\urcorner) \to \mathsf{Pr}(\ulcorner\mu\urcorner)$, thus $\Gamma \vdash (\mathsf{Pr}(\ulcorner\varphi\urcorner) \to \varphi) \to (\mathsf{Pr}(\ulcorner\mu\urcorner) \to \varphi)$. Denoting $\mathsf{Pr}(\ulcorner\varphi\urcorner) \to \varphi$ by ψ, it gives $\Gamma \vdash \psi \to \mu$ as μ is the fixed point. Consequently, $\Gamma \vdash \mathsf{Pr}(\ulcorner\psi\urcorner) \to \mathsf{Pr}(\ulcorner\mu\urcorner)$ and $\Gamma \vdash \mathsf{Pr}(\ulcorner\mu\urcorner) \to \mathsf{Pr}(\ulcorner\varphi\urcorner)$ by part (b), thus we are done.

818. The condition and Problem 806(d) gives $\Gamma \vdash \mathsf{Pr}(\ulcorner\mathsf{Pr}(\ulcorner\varphi\urcorner) \to \varphi\urcorner)$. By Problem 817 this implies $\Gamma \vdash \mathsf{Pr}(\ulcorner\varphi\urcorner)$, which gives $\Gamma \vdash \varphi$ by the condition.

819. If $\Gamma \vdash \nu \leftrightarrow \mathsf{Pr}(\ulcorner\nu\urcorner)$, then Löb's theorem gives $\Gamma \vdash \nu$.

820. (a) By Löb's theorem all fixed points of $\mathsf{Pr}(x)$ are derivable from Γ, thus $\Gamma \vdash \nu_1 \leftrightarrow \nu_2$.

(b) By Problem 810 all fixed points of $\neg\mathsf{Pr}(x)$ are provably equivalent to $\neg\mathsf{Pr}(\ulcorner\bot\urcorner)$, thus they are provably equivalent to each other.

821. By Problem 817(b), $\Gamma \vdash \mathsf{Pr}(\ulcorner\mu\urcorner) \to \mathsf{Pr}(\ulcorner\varphi\urcorner)$. Thus $\Gamma \vdash \mu \to (\mathsf{Pr}(\ulcorner\varphi\urcorner) \to \varphi)$ as μ is a fixed point. The other direction is similar using Problem 817(a).

822. $a \in \omega$ is prime iff $\mathfrak{N} \vDash a > 1 \wedge (\forall x < a)(\forall y < a)(x \cdot y = a \to x = 1 \vee y = 1)$.

823. Using the collection principle $\mathfrak{N} \vDash ((\forall x < y)\exists z\, \psi) \leftrightarrow \exists u(\forall x < y)(\exists z < u)\psi$.

824. Let $\varphi(x, \bar{y}), \psi(x, \bar{y}) \in \Delta_0$. Closedness of Σ_1 under disjunctions follows from

$$\mathfrak{N} \vDash \exists x\varphi(x, \bar{y}) \wedge \exists x\psi(x, \bar{y}) \leftrightarrow \exists x((\exists u < x)\varphi(u, \bar{y}) \wedge (\exists v < x)\psi(v, \bar{y})),$$

and similarly for conjunction. For bounded universal quantifier see Problem 823, otherwise use the equivalences

$$\mathfrak{N} \vDash (\exists y\exists z\, \varphi) \leftrightarrow \exists u(\exists y < u)(\exists z < u)\varphi,$$
$$\mathfrak{N} \vDash (\exists x < y)\exists z\varphi \leftrightarrow \exists u(\exists x < y)(\exists z < u)\varphi.$$

825. Similarly to Problem 824 Σ_n and Π_n formulas are closed under $\wedge$ and $\vee$, from which the first claim follows. Observe that the negation of a Σ_n formula is Π_n, which implies the second claim.

826. By induction on n using the definition of Σ_n and Π_n formulas.

827. (a) $\mathfrak{A}$ is an end-extension of $\mathfrak{N}$, thus $\mathfrak{N}$ is a substructure of $\mathfrak{A}$ and evaluations of the terms in bounded quantifiers are the same in $\mathfrak{N}$ and in $\mathfrak{A}$.

(b), (c) Existential formulas are preserved under extensions, while universal formulas are preserved under taking substructures.

828. As $\mathfrak{N}$ is a model of Q, the $\Leftarrow$ implication holds. To show the converse, Problem 827(b) claims that $(\exists y \varphi)[\pi_{\vec{a}}]$ holds in every end-extension of $\mathfrak{N}$. As every model of Q is such an end-extension by Problem 721, this formula holds in every model of Q. To complete to proof use the completeness theorem 10.14.

829. For any closed Δ_0-formula we have either $Q \vdash \varphi$ or $Q \vdash \neg\varphi$. This is because $\mathfrak{N} \vDash \varphi$ iff $Q \vdash \varphi$ by Problem 828.

Enumerate all the proofs from Q. Eventually, either φ or $\neg\varphi$ pops up, and thus we have a derivation of either φ or $\neg\varphi$. Let $b(\ulcorner\varphi\urcorner)$ be the code of this derivation.

830. Both A and its complement are Σ_1. By Problem 826 there are Δ_0-formulas $\varphi(\vec{a}, y)$ and $\psi(\vec{a}, z)$ such that $\vec{a} \in A$ iff $\mathfrak{N} \vDash \exists y \varphi(\vec{a}, y)$ iff $\mathfrak{N} \nvDash \exists z \psi(\vec{a}, z)$. Put

$$\vartheta(\vec{a}) \equiv \exists y \big(\varphi(\vec{a}, y) \wedge (\forall z < y) \neg \psi(\vec{a}, z) \big).$$

This is clearly Σ_1, we claim that it works. First, if $\vec{a} \in A$, then $\mathfrak{A} \vDash \vartheta(\pi_{\vec{a}})$ for each end-extension $\mathfrak{A}$ of $\mathfrak{N}$ since such an x can be found even in $\mathfrak{N}$, and Δ_0 formulas evaluate in $\mathfrak{N}$ and in $\mathfrak{A}$ equivalently, see Problem 827(a). Second, if $\vec{a} \notin A$, then

$$\mathfrak{A} \vDash \forall y \big(\varphi(\pi_{\vec{a}}, y) \to (\exists z < y) \psi(\pi_{\vec{a}}, z) \big),$$

since if $\mathfrak{A} \vDash \varphi(\pi_{\vec{a}}, y)$, then this y cannot be in N ($\vec{a} \notin A$ so $\mathfrak{N} \nvDash \exists y \varphi(\pi_{\vec{a}}, y)$), so the z witnessing $\mathfrak{N} \vDash \exists z \psi(\pi_{\vec{a}}, z)$ is smaller than y.

Therefore $\mathfrak{A} \vDash \vartheta[\pi_{\vec{a}}]$ (or $\mathfrak{A} \vDash \neg\vartheta[\pi_{\vec{a}}]$ if $\vec{a} \notin A$) for each end-extension of $\mathfrak{N}$, so we get, as in Solution 828, that these formulas can be derived from Q.

831. Let $\varphi(x, y, z) \in \Delta_0$ witness that the graph of f is Σ_1, i.e., $f(a) = b$ iff $\mathfrak{N} \vDash \exists z \varphi(a, b, z)$. Since f is a function, for each $a \in \omega^n$ there exists a unique $b \in \omega$ for which $\mathfrak{N} \vDash \varphi(a, b, c)$ with some $c \in \omega$. So if $f(a) = b$, then $\mathfrak{N} \vDash \forall y \forall z (\varphi(a, y, z) \to y = b)$, and conversely, if this holds then necessarily $b \in \omega$ is that unique element. This shows that the graph of f is also Π_1.

832. (a) $y = \beta(m, b, i)$ iff $(\exists q \le m) m = y + q(b(i+1) + 1) \wedge y < b(i+1) + 1$.

(b) The graphs of functions $K(u)$ and $L(u)$ are clearly Δ_0, thus the claim for $\mathsf{Len}(u)$ and $\mathsf{Elem}(u, i)$ follows from (a). The value of the function $u \frown z$ is v if the triplet $\langle u, z, v \rangle$ satisfies the Δ_0 formula expressing that v has length one more than u, has the same elements as u up to $\mathsf{Len}(u)$ (using a universal quantifier bounded by u) and its last element is z, while no $v' < v$ has this property (a bounded universal quantifier). This is clearly Δ_0.

833. By Problem 832(a), $\beta(m, b, i) = j$ can be treated as a Δ_0 formula, and $t_1 = t_2$ can be replaced by $\exists x (t_1 = x \wedge t_2 = x)$.

(a) This formula expresses that some sequence starts with 1 and each element is x times the previous one, and z is its last element:

$$\exists m \exists b \big(\beta(m,b,0) = 1 \ \wedge \ (\forall i < y)\beta(m,b,i+1) = x \cdot \beta(m,b,i) \ \wedge \ \beta(m,b,y) = z \big).$$

(b) Similarly to (a), but use $\beta(m,b,i+1) = (i+1) \cdot \beta(m,b,i)$ in the middle.

834. Structural induction on bounded formulas using that primitive recursive relations are closed under the Boolean operations and bounded quantification, see Problem 104.

835. By induction on the construction of primitive recursive functions using that the graphs of $\mathsf{Len}(u)$ and $\mathsf{Elem}(u,i)$ are Δ_0. The fastest way is to look at the faithful representation of primitive recursive functions (Definition 10.12), and conclude that it provides the requested Σ_1 formula.

836. Kleene's normal form theorem 4.16 says that there is a primitive recursive function $H_n(e,\vec{x},u)$ such that $f(\vec{x}) = \downarrow$ iff $H_n(e,\vec{x},u) = 0$ for some $u \in \omega$. By Problem 835 $T_n = \{\langle e, \vec{a}\rangle : \mathfrak{N} \models \exists u (H_n(e,\vec{a},u) = 0)\}$ is Σ_1.

837. We show first that every Σ_1-set is recursively enumerable. Suppose $A \subseteq \omega^n$ is a Σ_1-subset represented by the formula $\exists y \varphi(\vec{x}, y)$, where $\varphi(\vec{x}, y)$ is Δ_0. Then $\varphi(\vec{x}, y)$ represents the set $S \subseteq \omega^{n+1}$ by

$$\langle \vec{a}, b\rangle \in S \quad \Leftrightarrow \quad \mathfrak{N} \models \varphi[\vec{a}, b]$$

The characteristic function χ_S of S is primitive recursive by 834. Let f be defined as $f(\vec{a}) = \mu(x : \chi_S(\vec{a}, x) = 1)$. Then f is partially recursive and $\mathrm{dom}(f) = A$, therefore A is recursively enumerable.

For the converse assume $A \subseteq \omega^n$ is recursively enumerable. By Problem 169, A is the domain of an n-variable partial recursive function f. Take the $Sigma_1$ relation T_n from Problem 836. Now, there is an index e such that $\vec{a} \in A$ iff $\langle e, \vec{a}\rangle \in T_n$, thus A is also Σ_1.

838. A set A is recursive iff both A and its complement $\bar{A}$ are recursively enumerable (Problem 157). By 837, A is recursive iff both A and $\bar{A}$ are Σ_1. The complement of a Σ_1 set is a Π_1 set. Thus a set A is recursive iff A belongs to $\Sigma_1 \cap \Pi_1$ which is Δ_1.

839. (a) Kleene's T (Problem 836 gives us such a set U, but it is not guaranteed to be primitive recursive (only Σ_1). Another problem is the uniformity: we need the index to be the code of the formula, and not just the existence of a good index.

Mimic the proof of Kleene's normal form theorem 4.16. The sequence u is a *justified evaluation* if every element of u is a triplet $\langle \alpha(\varphi), e, v\rangle$ where φ is a Δ_0 formula, e assigns natural number to the free variables of φ, and v tells whether $\varphi[e]$ is true or not, and every element of u is justified: either it is an atomic formula whose truth is checked, or for each direct subformulas of φ

there are earlier triplets in u which compute their validity, and the validity of this formula is computed from those values.

As this is a course-of-value recursion using primitive recursive functions, being a justified evaluation is a primitive recursive relation. Given $\varphi(x)$ and $n\omega$, $\alpha(\varphi[\pi_n])$ is a primitive recursive function of $\alpha|phi)$ and n, thus we are done if we can put a primitive recursive upper bound on a justified computation of $\varphi[\pi_n]$ as a function of $\alpha(\varphi[\pi_n])$. If $b(m)$ is such a bound, then one can take $b(m) = m^m b(m-1)$ corresponding to the case when all evaluations indicated by a bounded quantifier must appear earlier. But this $b(m)$ is clearly primitive recursive.

(b) $A = \{a \in \omega : \langle a, a \rangle \notin U\}$ is primitive recursive but cannot be Δ_0.

840. $\varphi(x, y)$ can be written equivalently as $\exists z \psi(x, y, z)$ where ψ is Δ_0, see Problem 826. The assumption says that for every $n \in \omega$ there are $a, b \in \omega$ such that $\mathfrak{N} \vDash \psi(\pi_n, \pi_a, \pi_b)$. The Δ_0 set $A \subset \omega^3$ determined by ψ is primitive recursive by Problem 834, thus $h(n) = \mu\{u \in \omega : \langle n, (u)_0, (u)_1 \rangle \in A\}$ is recursive (as such a u exists for every n). Then $(h(n))_0$ is also recursive and it gives $\mathfrak{N} \vDash \exists z \psi(\pi_n, \pi_{f(n)}, z)$, as required.

841. (a) By Problem 835 the graph of a primitive recursive function is Σ_1, and, of course, the graph represents the function. For recursive functions in general observe that its graph is recursively enumerable, thus Σ_1 by Problem 837.

(b) Immediate from Problem 840 as in this case there is a unique z with $\mathfrak{N} \vDash \exists z \pi(\pi_n, z)$ for each n, and the recursive function returns such a z.

842. Since $\mathfrak{N} \vDash \Gamma$ and $\Gamma \vdash \forall x \exists y \varphi(x, y)$, we have $\mathfrak{N} \vDash \exists y \exists z \psi(\pi_n, y, z)$ for each $n \in \omega$, thus there are $k, \ell \in \omega$ such that $\mathfrak{N} \vDash \psi[n, k, \ell]$. By Problem 839 the relation

$$U = \{\langle \alpha(\psi), n, k, \ell \rangle : \mathfrak{N} \vDash \psi[n, k, \ell]\}$$

is primitive recursive. As Γ is primitive recursive, the provability predicate PP_Γ (Definition 10.8) is also primitive recursive. Define the function $f(n)$ as follows. For each $u \leq n$ check whether $\langle u, \alpha(\Phi) \rangle \in PP_\Gamma$, where Φ is of the form $\forall x \exists y \exists z \psi(x, y, z)$ with $\psi \in \Delta_0$. If no, let $k_u = 0$, otherwise take the minimal pair $\langle k_u, \ell_u \rangle$ such that $\langle \alpha(\psi), n, k_u, \ell_u \rangle \in U$, such a pair exists by the discussion above. Let $f(n) = 1 + \max\{k_u : u \leq n\}$.

This f is recursive as PP_Γ is primitive recursive, and if $\langle u, \alpha(\Phi) \rangle \in PP_\Gamma$ then $\alpha(\Phi)$ must be smaller than u. The relation U is also primitive recursive, and the minimum always exists.

To show that f dominates every provably recursive function, let g be such a function witnessed by the formula $\exists z \psi(x, y, z)$. Then $\mathfrak{N} \vDash \exists z \psi[n, k, z]$ iff $k = g(n)$ for each $n \in \omega$ (as this formula represents g in $\mathfrak{N}$), thus the only $k \in \omega$ which satisfies $\langle \alpha(\psi), n, k, \ell \rangle \in U$ is $k = g(n)$. Let $\Phi \equiv \forall x \exists y \exists z \psi(x, y, z)$. As $\Gamma \vdash \Phi$ by assumption, there is an $u \in \omega$ such that $\langle u, \alpha(\Phi) \rangle \in PP_\Gamma$. For $n \geq u$ the formula Φ is handled, ensuring $f(n) > g(n)$.

843. (a) The set $T_k \subseteq \omega^{k+1}$ form Problem 836 is such a Σ_1 set. Indeed, by Problem 837 Σ_1 sets are recursively enumerable, so let $A = \mathrm{dom}(f)$ for a k-variable partial recursive function. Then there is an $e \in \omega$ such that $\langle e, \vec{a} \rangle \in T_k$ iff $f(\vec{a}) = \downarrow$ iff $\vec{a} \in A$, as required.

(b) For if Σ_1 were closed under negation, then take $A = \{i \in \omega : (i, i) \notin U_1^1\}$. By (a) there is an integer e such that $A = \{i \in \omega : (e, i) \in U_1^1\}$. But then with $i = e$ we get a contradiction.

(c) The complement of Σ_1 is Π_1, thus $\Pi_1 \neq \Sigma_1$ by (b). But then $\Delta_1 = \Sigma_1 \cap \Pi_1$ is a proper subset of both Σ_1 and Π_1.

844. The set $\{a \in \omega : \langle a, a \rangle \notin U\}$ would also be recursive.

845. (a) By induction on n. The complement of a universal Σ_n set is a universal Π_n set, so assume, as an induction hypothesis, that V_{k+1}^n is a universal Π_n set. By definition, $A \subseteq \omega^k$ is Σ_{n+1} iff there is a Π_n set $B \subseteq \omega^{k+1}$ such that $A = \{\vec{a} : \exists b \langle \vec{a}, b \rangle \in \mathfrak{B}\}$. Consequently

$$U_k^{n+1} = \{\langle e, \vec{a} \rangle : \exists b \langle e, \vec{a}, b \rangle \in V_{k+1}^n\}$$

is a universal Σ_{n+1} set.

(b), (c) Same as for Problem 843

846. The claim for $n = 1$ was covered in Problem 839(a). In general use induction on the more general statement that for every recursive function f, the set $\{\langle f(\alpha(\varphi)), \vec{a} \rangle : \mathfrak{N} \vDash \varphi[\vec{a}]\}$ is Σ_n as $\varphi(\vec{x})$ runs over the Σ_n^* formulas.

847. Assume by contradiction, that there is a formula $\Phi(x)$ such that for every closed v, $\mathfrak{N} \vDash v$ iff $\mathfrak{N} \vDash \Phi(\ulcorner v \urcorner)$. By the fixed point theorem 10.7 there is a closed formula v such that $\mathfrak{N} \vDash v \leftrightarrow \neg\Phi(\ulcorner v \urcorner)$, a clear contradiction.

848. The idea is to enumerate the closed formulas as $\varphi_0, \varphi_1, \ldots$ Start with $\Gamma = \mathrm{PA}$. At the i-th step add φ_i to Γ only if $\Gamma, \varphi_i \nvdash \bot$. In the other case $\Gamma \vdash \neg\varphi_i$, thus φ_i will not be true in any model of Γ. At the end Γ will be a maximal consistent extension of PA, and when we encounter φ_i we know whether it holds (if it was added), or does not hold (otherwise) in any model of Γ.

Let $\mathrm{Prov}_{\mathrm{PA}}(u, x)$ be a representation of the provability predicate $\mathrm{PP}_{\mathrm{PA}}$, and $\mathrm{Pr}_{\mathrm{PA}}(x)$ be $\exists u\, \mathrm{Prov}_{\mathrm{PA}}(u, x)$. For any closed formula φ, $\mathfrak{N} \vDash \mathrm{Pr}_{\mathrm{PA}}(\ulcorner \varphi \urcorner)$ iff $\mathrm{PA} \vdash \varphi$. The formula $\Phi(x)$ will be true in $\mathfrak{N}$ if x is a code of a closed formula and it was added to Γ in the above procedure. The sequence u maintains the conjunct of formulas added so far. Thus

$$\Phi(x) \equiv \exists u \big(\mathrm{Len}(u) = x + 2 \wedge (u)_0 = \ulcorner \top \urcorner \wedge$$
$$(\forall i \leq x)\big(\text{if } i \text{ is a code of a closed formula, and}$$
$$\neg\mathrm{Pr}_{\mathrm{PA}}(\ulcorner (u)_i \wedge i \rightarrow \bot \urcorner), \text{ then } (u)_{i+1} = \ulcorner (u)_i \wedge i \urcorner,$$
$$(u)_{i+1} = (u)_i \text{ otherwise} \big)$$
$$\wedge (u)_{x+1} \neq (u)_x \big).$$

By the remarks above, $\mathfrak{N} \vDash \Phi(\ulcorner \varphi \urcorner)$ exactly when φ is a closed formula and is true in any model of Γ.

12.11 SELECTED APPLICATIONS

849. Let $\{U_j : j \geq k\}$ be relation symbols not in the finite Boolean combination $\varphi(x)$. By assumption all the combinations $\varphi(x) \wedge U_k(x)$, $\varphi(x) \wedge \neg U_k(x) \wedge U_{k+1}(x)$, $\varphi(x) \wedge \neg U_k(x) \wedge \neg U_{k+1}(x) \wedge U_{k+2}(x)$, etc. are satisfied by some element, but clearly all of them must be different.

850. Let $\mathfrak{A} = \langle {}^\kappa 2, U_i^{\mathfrak{A}} \rangle_{i<\kappa}$ be such that

$$U_i^{\mathfrak{A}} = \{x \in {}^\kappa 2 : x(i) = 1\}.$$

Then $\mathfrak{A} \vDash \Gamma$ as each finite Boolean combinations of finitely many of the U_i's are realized by 2^κ elements.

851. Solution 1. Let $\tau' \subset \tau$ contain finitely many relation symbols. Then $\mathfrak{A}{\restriction}\tau'$ is an elementary substructure of $\mathfrak{B}{\restriction}\tau'$ by Problem 483: there are $2^{|\tau'|}$ many equivalence classes in $\mathfrak{B}{\restriction}\tau'$, each is infinite, thus there is an automorphism which moves the new element into an old one and keeps finitely many other elements fixed.

For a formula φ let $\tau_\varphi \subset \tau$ contain the relation symbols in φ. Then $\mathfrak{A} \vDash \varphi[\bar{a}]$ for some $\bar{a} \in A$ iff $\mathfrak{A}{\restriction}\tau_\varphi \vDash \varphi[\bar{a}]$ iff $\mathfrak{B}{\restriction}\tau_\varphi \vDash \varphi[\bar{a}]$ (as τ_φ is a finite subtype of τ) iff $\mathfrak{B} \vDash \varphi[\bar{a}]$. Thus $\mathfrak{A}$ is an elementary substructure $\mathfrak{B}$.

Solution 2. Γ admits quantifier elimination (Problem 855) thus the condition in the Tarski–Vaught test 8.3 holds trivially.

852. Let the universe A of $\mathfrak{A}$ be the set of sequences $a : \omega \to \{0,1\}$ that are eventually zero and let

$$U_i^{\mathfrak{A}} = \{a \in A : a(i) = 1\}.$$

Then $\mathfrak{A}$ is countable and $\mathfrak{A} \vDash \Gamma$. Note that there is no element in $\mathfrak{A}$ that realizes all the $U_i^{\mathfrak{A}}$'s. Let $\mathfrak{B}$ be the structure obtained from Problem 851. Then $\mathfrak{A}, \mathfrak{B} \vDash \Gamma$ but $\mathfrak{A} \not\equiv \mathfrak{B}$.

853. For notational simplicity we give a solution for the $\kappa = \omega$ case. The general solution is analogous. We need to find models $\mathfrak{A}_i \not\vDash \Gamma$ such that $\prod_{i \in I} \mathfrak{A}_i / U \vDash \Gamma$ for some ultrafilter U over I.

For $n \in \omega$ let $\mathfrak{A}_n$ be such that for any $I, J \subset n$, $I \cap J = \emptyset$ there is an $a \in A$ such that a satisfies $\bigwedge_{i \in I} U_i^{\mathfrak{A}} \wedge \bigwedge_{j \in J} \neg U_j^{\mathfrak{A}}$, but for $k \geq n$ we have $U_k^{\mathfrak{A}} = \emptyset$. Clearly there are such structures $\mathfrak{A}_n$, and $\mathfrak{A}_n \not\vDash \Gamma$.

Let $\mathfrak{A} = \prod_{n \in \omega} \mathfrak{A}_n / U$ for a non-principal ultrafilter over ω. We claim that $\mathfrak{A} \vDash \Gamma$. Indeed, for finite subsets $I, J \subset \omega$ with $I \cap J = \emptyset$, we have

$$\left\{ n \in \omega : \mathfrak{A}_n \vDash \exists v \left(\bigwedge_{i \in I} U_i(v) \wedge \bigwedge_{j \in J} \neg U_j(v) \right) \right\} \in U$$

as the complement of this set is finite. Łoś's lemma 9.3 implies that $\mathfrak{A} \vDash \Gamma$.

854. Pick disjoint subsets $I, J \subset \omega$ and for $n \in \omega$ let

$$X_n = \{a \in A : \mathfrak{A} \models (\bigwedge_{i \in I \cap n} U_i \wedge \bigwedge_{j \in J \cap n} \neg U_j)[a]\}$$

Each X_n is infinite. Let $X = \prod_{n \in \omega} X_n / U$. The cardinality of X is continuum (see Problem 609). We claim that each $a \in X$ satisfies $\bigcap_{i \in I} U_i \cap \bigcap_{j \in J} \neg U_j$ in $\mathfrak{B}$. Indeed, for $a \in X$ and $k \in \omega$ we have that

$$\{n \in \omega : a(n) \in \bigcap_{i \in I \cap k} U_i^{\mathfrak{A}} \cap \bigcap_{j \in J \cap k} \neg U_j^{\mathfrak{A}}\}$$

belongs to U as the set is co-finite. By Łoś's lemma 9.3 we get then

$$\mathfrak{B} \models \bigwedge_{i \in I} U_i[a] \wedge \bigwedge_{j \in J} \neg U_j[a]$$

for any $a \in X$.

855. As the only closed quantifier-free formulas are the $\top$ and $\bot$, quantifier elimination implies completeness. By Problem 535 it is enough to show that each formula φ of the form $\exists y (\ell_1 \wedge \cdots \wedge \ell_n)$, where ℓ_i is a literal, is equivalent to a quantifier-free formula. Literals are of the form $x_i = x_j$, $x_i \neq x_j$, $U_i(x_j)$, and $\neg U_i(x_j)$. The formula φ can be grouped as $\beta_1(x_1) \wedge \cdots \wedge \beta_n(x_n) \wedge \beta(y) \wedge \varepsilon$, where each $\beta_i(x_i)$ is a conjunction of literals in the variable x_i only, not including $x_i = x_i$ or $x_i \neq x_i$ among its conjuncts; and ε is a conjunction whose only conjuncts are of the form $(\neg)(x_i = x_j)$ and $(\neg)(y = x_j)$. Let us write

$$\vartheta(\vec{x}, y) \equiv \beta_1(x_1) \wedge \cdots \wedge \beta_n(x_n) \wedge \beta(y) \wedge \varepsilon$$

We can eliminate quantifiers from $\varphi \equiv \exists y \vartheta(\vec{x}, y)$ as follows. If ϑ is inconsistent with Γ, then replace $\exists y \vartheta(\vec{x}, y)$ with $\bot$. If ϑ is consistent with Γ and ε contains $y = x_i$ for some i, then replace all instances of y in ϑ with x_i. Observe that $\exists y \vartheta(\vec{x}, y)$ is equivalent modulo Γ to the quantifier-free $\vartheta(\vec{x}, x_i)$. Finally, if ϑ is consistent with Γ and no $y = x_i$ appears in ε, then eliminate all conjuncts from ϑ which involve y. The resulting formula is quantifier-free and it equivalent modulo Γ to $\exists y \vartheta(\vec{x}, y)$. To see this, note that in any model $\mathfrak{A} \models \Gamma$, for any tuple $\vec{a} \in A$ which satisfies the prescriptions given by the β_i's and ε, there is always an element c distinct from $\vec{a}$ such that c satisfies the same prescriptions. (We used here that any Boolean-combinations of the U_i's are realized by infinitely many elements).

856. Suppose $\kappa = \omega$ and define $\mathfrak{A}$ and $\mathfrak{B}$ as follows. Let $A = {}^\omega 2$ and write

$$U_i^{\mathfrak{A}} = \{a \in A : a(i) = 1\}.$$

Let B be the set of sequences $a \in {}^\omega 2$ that are eventually zero and write

$$U_i^{\mathfrak{B}} = \{a \in B : a(i) = 1\}.$$

Then $\mathfrak{A}, \mathfrak{B} \models \Gamma$, thus $\mathfrak{A} \equiv \mathfrak{B}$ by Problem 855.

Observe that no element of $\mathfrak{B}$ realizes all the $U_i^{\mathfrak{B}}$'s, while there is such an element a in $\mathfrak{A}$. While playing the Ehrenfeucht–Fraïssé game $EF(\mathfrak{A}, \mathfrak{B}, N)$, I can always choose a as his first pick, blocking II to reply with an element that satisfies all the U_i's (thus blocking the existence of a partial isomorphism).

857. If $2^\lambda < \kappa$, then Γ has no model of cardinality λ because the interpretations of the U_α's are distinct sets.

Suppose $2^\lambda \geq \kappa$. By Problem 6 there is an independent family $\langle E_\alpha : \alpha < 2^\lambda \rangle$ of size 2^λ on the set $A = \lambda$. For $\alpha < \kappa$ write $U_\alpha^{\mathfrak{A}} = E_\alpha$. Then $\mathfrak{A} = \langle A, U_\alpha^{\mathfrak{A}} \rangle_{\alpha < \kappa}$ is a model of Γ and the construction in 6 ensures that $\mathfrak{A}$ has no element that belongs to all $U_\alpha^{\mathfrak{A}}$. Let $\mathfrak{B}$ be the extension of $\mathfrak{A}$ with a new element that satisfies each U_α. Then $\mathfrak{B} \vDash \Gamma$ and as Γ is complete (see Problem 855) $\mathfrak{A}$ and $\mathfrak{B}$ are elementarily equivalent. But $\mathfrak{A}$ and $\mathfrak{B}$ are not isomorphic, hence Γ is not λ-categorical.

> **Remark.** Suppose κ is finite. Then Γ is ω-categorical, but not λ-categorical for any $\lambda > \omega$ as there are models $\mathfrak{A}, \mathfrak{B} \vDash \Gamma$ with $|\mathfrak{A}| = |\mathfrak{B}| = \lambda$ such that $|U_1^{\mathfrak{A}}| \neq |U_1^{\mathfrak{B}}|$.

858. Solution 1. Define a graph as follows: the set of vertices is ω. Set $\neg E(i, i)$ for all $i \in \omega$, otherwise suppose $j > i$. Then set $E(i, j)$ if and only if the i-th digit in the base 2 expansion of j is 1. This defines a graph.

Suppose now that $x_1 \ldots x_n, y_1 \ldots y_m$ are distinct elements for some $n, m \in \omega$. Then let z be the number whose base 2 expansion is 1 at the x_i-th digits and 0 at the y_j-th digits. Clearly then z is connected to all the x_i's but none of the y_j's.

Solution 2. We start with the following statement: for each finite graph $\mathfrak{G}$ there is a finite graph $\mathfrak{G}'$ which is an extension of $\mathfrak{G}$ (i.e. $\mathfrak{G} \subset \mathfrak{G}'$) and which has the property, that whenever one picks two disjoint (finite) subsets $X, Y \subset G$ of vertices, then there is a vertex $z \in G'$ that is connected to all vertices in X but none in Y. To construct such $\mathfrak{G}'$ is easy: as $\mathfrak{G}$ is finite, there are finitely many possible ways to pick two disjoint subsets. For all such choices add a new vertex with the required property.

Now, starting from the one-vertex graph $\mathfrak{G}_0$, define by recursion $\mathfrak{G}_{n+1} = \mathfrak{G}_n'$. Then $\mathfrak{G} = \bigcup_{n \in \omega} \mathfrak{G}_n$ is as desired. For, select $X, Y \subseteq G$ with $|X| = n$, $|Y| = m$ for some $n, m \in \omega$. Then there is k such that $X, Y \subseteq G_k$ and thus one finds an element $z \in G_{k+1}$ such that $E(z, x)$ holds for all $x \in X$ and $\neg E(z, y)$ for all $y \in Y$.

Solution 3. Let $\langle M, \in \rangle$ be a countable model of set theory. Define $\langle M, E^M \rangle$ to be the graph with $E^M(x, y)$ if and only if $x \in y$ or $y \in x$.

Solution 4. For infinite κ take a bijection $f : \kappa \to [\kappa]^{<\omega}$ and for $a, b \in \kappa$ draw an edge iff $a \in f(b)$ or $b \in f(a)$. Given distinct vertices $x_0, \ldots, x_{n-1}, y_0, \ldots, y_{m-1} \in \kappa$, let $Y = f(y_0) \cup \cdots \cup f(y_{m-1}) \cup \{y_0, \ldots, y_{m-1}\}$. Y is a finite set, and there are infinitely many $z \in \kappa$ such that $f(z) \supset \{x_1, \ldots, x_{n-1}\}$ and $f(z) \cap \{y_1, \ldots, y_{m-1}\} = \emptyset$, thus there is such a $z \notin Y$. For such z we have $E(x_i, z)$ for $i < n$ and $\neg E(y_j, z)$ for $j < m$.

859. If the graph $\mathfrak{A}$ is finite, then $\varphi_{|A|,0}$ cannot be satisfied in $\mathfrak{A}$.

860. Suppose u is connected to members of the finite set B. By universality there is a v connected to all of $\{u\} \cup B$. This v cannot be in B as no vertex is connected to itself.

861. We prove the countably infinite case only, the finite case is similar. Let $A = \{a_i : i \in \omega\}$ be the universe of the countable graph $\mathfrak{A}$ and denote the i-th initial segment of A by $A_i = \{a_0, \ldots, a_{i-1}\}$. Let $f_0 = \emptyset$ and define by recursion an increasing sequence of partial isomorphisms f_n such that $A_n \subseteq \text{dom}(f_n)$. Suppose f_n has already been defined. Take a_n and let us denote the neighbors of a_n in A_n

$$N = \{a_k : a_k \in A_n, \mathfrak{A} \vDash E(a_k, a_n)\}.$$

Then N is a finite set and $\mathfrak{G}$ satisfies the formula $\varphi_{|N|,|A_n \smallsetminus N|} \in \Gamma$. Therefore there is some $z \in G$ with the property

$$\mathfrak{G} \vDash \bigwedge_{a_i \in N} E(f_n(a_i), z) \wedge \bigwedge_{a_i \in A_n \smallsetminus N} \neg E(f_n(a_i), z)$$

Put $f_{n+1} = f_n \cup \{\langle a_n, z \rangle\}$.

This construction can be carried out for all $n \in \omega$. Finally, let $f = \bigcup_{n \in \omega} f_n$. Then $f : A \to G$ is an embedding of $\mathfrak{A}$ into $\mathfrak{G}$.

862. Standard back and forth argument (cf. Solution 365). We modify Solution 861. Let $A = \{a_i : i < \omega\}$ and $B = \{b_i : i < \omega\}$. A_i and B_i denote the initial segments as before.

Build partial isomorphism f_n as before, such that $A_n \subset \text{dom}(f_n)$ and with the additional requirement that $B_n \subset \text{ran}(f_n)$. In Solution 861 we found an image z for a_n. With the same method applied backwards we can find an inverse-image z' for b_n. Then let $f_{n+1} = f_n \cup \{\langle a_n, z \rangle\} \cup \{\langle z', b_n \rangle\}$.

At the end, define $f = \bigcup_{n \in \omega} f_n$. Then $f : A \to B$ is an isomorphism between $\mathfrak{A}$ and $\mathfrak{B}$.

863. $\aleph_0$-categoricity of Γ is immediate from 862. As Γ has infinite models only (859) and is $\aleph_0$-categorical, the Łoś–Vaught theorem 8.7 implies completeness. The last statement follows form Problem 543 trivially.

864. The subgraph is also universal by Problem 860. Γ admits quantifier elimination so it is model complete (Problem 544). Thus every universal subgraph of a universal graph is an elementary submodel.

865. Enumerate the vertices of $\mathfrak{G}$ as $\{v_\alpha : \alpha < \omega_1\}$, and assume that each v_α has countable degree. Let $h_0 = \omega$, and $h_{i+1} = \sup\{\beta \in \omega_1 : v_\alpha v_\beta \text{ is an edge for some } \alpha < h_i\}$. As h_{i+1} is the sup of countably many countable ordinals, it is below ω_1. Take $h = \lim_i h_i$. If $v_0 v_\alpha$ is a vertex, then $\alpha < h$, and if $v_h v_\alpha$ is a vertex, then $h < \alpha$. Thus there is no vertex which would be connected to both v_0 and v_h, a contradiction.

866. Let K be the complete graph on ω vertices. By 861, K can be embedded into $\mathfrak{G}$, therefore we might assume that K is a subgraph of $\mathfrak{G}$. We claim that $^{\omega}K/U$ is a complete subgraph of $^{\omega}\mathfrak{G}/U$. Indeed, for any $a, b \in {}^{\omega}K/U$ we have

$$\{n \in \omega : a(n) \text{ is connected to } b(n)\} \in U.$$

By Problem 609, the cardinality of $^{\omega}K/U$ is $2^{\aleph_0}$ (for which $\aleph_1 \leq 2^{\aleph_0}$ holds).

867. (a) Take a countable universal graph $\mathfrak{G}$. The ultrapower $^{\omega}\mathfrak{G}/U$ contains a complete subgraph of cardinality $\aleph_1$ (see 866), let this be K. The Löwenheim–Skolem theorem 8.5 implies that $^{\omega}\mathfrak{G}/U$ has an elementary substructure of cardinality $\aleph_1$ that contains K.

(b) Take a bijection $f : \aleph_1 \to [\aleph_1]^{<\omega}$ and for $a, b \in \aleph_1$ draw an edge iff $a \in f(b)$ or $b \in f(a)$. The resulting graph $\mathfrak{G}_f$ is a model of Γ (see Solution 858) and $|G_f| = \aleph_1$. We claim $\mathfrak{G}_f$ does not contain a complete subgraph of cardinality $\aleph_1$. Suppose on the contrary $X \subset G_f$, $|X| = \aleph_1$ induces a complete subgraph of $\mathfrak{G}_f$. Define vertices x_n by induction on n as follows. Let $x_0 \in X$ be arbitrary. If x_i for $i < n$ has already been defined, then let x_n be any point from $X \smallsetminus (\{x_i : i < n\} \cup \bigcup_{i<n} f(x_i))$. Write $Y = \{x_i : i < \omega\} \cup \bigcup_{i<\omega} f(x_i)$. Y is a countable set, thus $X \smallsetminus Y$ is uncountable. Pick any $a \in X \smallsetminus Y$. Then a is connected to all the x_i's, but a does not belong to any of the $f(x_i)$'s, hence it must be the case that $x_i \in f(a)$ for all $i < \omega$. This contradicts $f(a)$ being finite.

868. Take the set of ordinals $\{\alpha < \omega_1\}$, and let $\mathcal{F}$ be an independent family of subsets of the first ω elements. As there is such a family of size continuum (see Problem 5), we can pick a subfamily of size ω_1 as $\mathcal{F} = \{X_\beta : \beta < \omega_1\}$. We can also assume that picking finitely many of them, taking the complement of some of them, the intersection is not only non-empty, but actually infinite (by replacing each element in the base set by a countable set). Define the graph on ω_1 such that for $\alpha < \beta$ there is an edge between them if $\alpha < \omega$ and $\alpha \in X_\beta$. As the family $\mathcal{F}$ is independent, it is clearly a universal graph on ω_1, moreover except for the first ω vertices, every other vertex has countable degree.

869. (a) Let S_n be the set of all tournaments on n vertices. Clearly, $|S_n| = 2^{\binom{n}{2}}$. Consider the uniform distribution over S_n. Call a k-element set X bad if no element dominates each member of X. Write $Y(\mathfrak{T})$ for the number of bad k-element sets in $\mathfrak{T}$. Then the expected value $E(Y)$ equals $\binom{n}{k}(1 - (1/2)^k)^{n-k}$. As $E(Y) \to 0$, by Markov's inequality $P(Y \geq 1) \to 0$. That is, a randomly chosen tournament on n vertices satisfies χ_k with probability tending to 1 as n tends to infinity. Therefore, for sufficiently large n there is a tournament on n vertices that satisfies χ_k.

(b) Take $\langle \mathbb{Z}, < \rangle$.

870. No finite $\mathfrak{T}$ can satisfy $\psi_{|\mathfrak{T}|,0}$.

Solution 1. We construct a countable model of Γ by recursion. Start with an arbitrary finite tournament T_0 and suppose T_n has already been defined.

For each pair (A, B) of disjoint non-empty subsets of T_n let $v_{(A,B)}$ be a new vertex such that $v_{(A,B)}$ dominates each element of A and is dominated by every element of B. Let T_{n+1} be the tournament that extends T_n with the new vertices $v_{(A,B)}$. Then $T = \bigcup_{n<\omega} T_n$ is a universal tournament.

Solution 2. We construct for every infinite κ a universal tournament of cardinality κ. The set of vertices is κ. Take a bijection $f : \kappa \to [\kappa]^{<\omega}$ and for $\alpha < \beta < \kappa$ draw an edge $\alpha \to \beta$ if $\beta \in f(\alpha)$, otherwise draw the edge $\beta \to \alpha$. The resulting structure is a tournament on κ vertices. We claim that it is universal. For, pick finite disjoint $A, B \subseteq \kappa$. The number of finite subsets of κ which contain A but exclude B is κ, and $A \cup B$ is not cofinal in κ, therefore there must exists an $\alpha < \kappa$ that is larger than any of the elements in $A \cup B$, and $f(\alpha) \supseteq A$, $f(\alpha) \cap B = \emptyset$. But then α dominates A and is dominated by B.

871. Same proof as in Solution 861.

872. $\aleph_0$-categoricity can be shown using the same back and forth argument presented in Solution 862. (Cf. also Solution 365).

Since Γ has infinite models only (Problem 870), completeness of Γ follows from $\aleph_0$-categoricity and the Łoś–Vaught theorem 8.7.

873. Let K be a countable transitive tournament. By Problem 871, K can be considered as a substructure of $\mathfrak{T}$. We claim that $^\omega K / U$ is a transitive subtournament of $^\omega\mathfrak{T}/U$. Pick $a, b, c \in {}^\omega K / U$ and suppose a dominates b and b dominates c. Then

$$\{n \in \omega : (a(n), b(n)) \in E, (b(n), c(n)) \in E\} \in U.$$

As K is transitive, it follows that $\{n \in \omega : (a(n), c(n)) \in E\} \in U$, hence a dominates c. By Problem 609, the cardinality of $^\omega K / U$ is $2^{\aleph_0}$ (for which $\aleph_1 \leq 2^{\aleph_0}$ holds).

874. By Problem 626 every structure can be embedded into an appropriate ultraproduct of its finitely generated substructures. Finitely generated substructures of $(X, <)$ are just finite suborderings, let us denote these structures by $(X_\alpha, <_\alpha)$. For suitable κ and U we have $(X, <) \hookrightarrow \prod_{\alpha\in\kappa}(X_\alpha, <_\alpha)/U$. By Problem 871 each $(X_\alpha, <_\alpha)$ can be embedded into $\mathfrak{T}$, therefore $(X, <)$ embeds into $^\kappa\mathfrak{T}/U$.

875. (a) Let $\mathfrak{T}_0$ be a transitive tournament of cardinality $\aleph_1$. Define by transfinite recursion a sequence of tournaments $\langle \mathfrak{T}_\alpha : \alpha < \aleph_1 \rangle$ such that the following stipulations hold.

(i) $\mathfrak{T}_\beta \subseteq \mathfrak{T}_\gamma$ whenever $\beta \leq \gamma \leq \aleph_1$.

(ii) $|\mathfrak{T}_\beta| \leq \aleph_1$ for all $\beta \leq \aleph_1$.

(iii) For each $\beta < \gamma \leq \aleph_1$ and for any disjoint finite subsets $A, B \subseteq T_\beta$ there is $z \in T_\gamma$ which dominates A and is dominated by B.

Suppose $\mathfrak{T}_\beta$ has been defined for $\beta < \alpha$. If α is limit, then let $\mathfrak{T}_\alpha = \bigcup_{\alpha<\beta} \mathfrak{T}_\beta$. If α is successor, say $\alpha = \beta + 1$, then for each finite disjoint pair of subsets

$A, B \subseteq T_\beta$ add a new vertex $v_{(A,B)}$ which dominates A and is dominated by B. $\mathfrak{T}_{\beta+1}$ is the extension of $\mathfrak{T}_\beta$ with these new vertices. As $|T_\beta| \leq \aleph_1$, there are only $[\aleph_1]^{<\omega} = \aleph_1$ new vertices, and thus the inductive hypothesis remains true for $\mathfrak{T}_{\beta+1}$. Finally, $\mathfrak{T}_{\aleph_1}$ is the desired universal tournament.

876. (a) $\text{Prob}(M_n \models \vartheta) = 1$ iff $\text{Prob}(M_n \not\models \vartheta) = 0$ for all n.

(b) Use that for all n we have

$$\text{Prob}(M_n \models \vartheta \wedge \varphi) \leq \text{Prob}(M_n \models \varphi), \text{Prob}(M_n \models \vartheta),$$

and

$$\text{Prob}(M_n \models \vartheta) + \text{Prob}(M_n \models \varphi) - \text{Prob}(M_n \models \vartheta \wedge \varphi) \leq 1.$$

877. Exactly half of the n-element τ-structures satisfy $U(c)$, thus $\text{Prob}(M_n \models U(c)) = 1/2$ for all $n > 0$, thus the limit is $1/2$.

878. $\text{Prob}(M_n \models \forall x(f(x) \neq x)) = \frac{(n-1)^n}{n^n} = (1 - \frac{1}{n})^n \to \frac{1}{e}$.

879. By completeness of Γ we have either $\Gamma \models \varphi$ or $\Gamma \models \neg\varphi$. Suppose the first is the case. By compactness, there is a finite $\Sigma = \{\gamma_0, \dots, \gamma_{n-1}\} \subseteq \Gamma$ such that $\Sigma \models \varphi$. As $\text{Prob}(M_n \models \vartheta) = 1 - \text{Prob}(M_n \not\models \vartheta)$ holds for any closed ϑ, it follows that

$$\text{Prob}(M_n \not\models \Sigma) \leq \text{Prob}(M_n \not\models \gamma_0) + \cdots + \text{Prob}(M_n \not\models \gamma_{n-1}) \to 0,$$

that is, $\text{Prob}(M_n \models \Sigma) \to 1$ as $n \to \infty$. Since $\text{Prob}(M_n \models \Sigma) \leq \text{Prob}(M_n \models \varphi)$ we obtain $\text{Prob}(M_n \models \varphi) \to 1$.

The case $\Gamma \models \neg\varphi$ is similar.

880. We prove the two statements in tandem. A pair of k and ℓ element sets (A, B) is said to be bad in the graph (tournament) G on vertex set n if there is no $g \in G$ which is connected to (dominates) every element of A and is not connected to (dominated by) each element of B. Let $N(G)$ be the number of bad pairs in G. Then the expected value of N is

$$E_n(N) = \binom{n}{k+\ell}\binom{k+\ell}{k}(1 - \frac{1}{2^{k+\ell}})^{n-k-\ell}.$$

Calculus shows $E_n(N) \to 0$ so $\text{Prob}(N \geq 1) \to 0$.

881. Take Γ from Definition 11.1 (Definition 11.3 in case of tournaments). By Problem 863 (Problem 872), Γ is complete, thus combining Problem 880 and Problem 879 completes the proof.

882. That G is a subgraph can be described by a formula ϑ. By Problem 881, ϑ is either almost surely true or false depending on whether ϑ is a consequence of the theory Γ of universal graphs. But every G is a subgraph of the countable universal graph (Problem 861), therefore $\Gamma \models \vartheta$, and thus the asymptotic probability that a randomly chosen finite simple graph contains G as a subgraph is one.

883. There are $2^{\binom{n}{2}}$ relations over an n-element set and 2^{n^2-n} of them have no loops, thus $\text{Prob}(M_n \vDash \forall x \neg E(x,x)) = \frac{2^{n^2-n}}{2^{n^2}} = \frac{1}{2^n} \to 0$.

$\text{Prob}(M_n \vDash \forall x \forall y (E(x,y) \to E(y,x))) = \frac{2^{n(n+1)/2}}{2^{n^2}} \to 0$.

884. The property expresses that for any finite subset B and a prescription a new element could be related to them, there is such an element. For $R \in \tau$ and $P_R \subseteq \{x_1, \dots, x_n, y\}^k$ (k is the arity of R) consider the formula

$$\forall x_1 \dots x_n \Big(\bigwedge_{i \neq j} x_i \neq x_j \to \exists y \big(\bigwedge_i y \neq x_i \wedge \bigwedge_{R \in \tau} \big(\bigwedge_{\bar{a} \in P_R} R(\bar{a}) \wedge \bigwedge_{\bar{a} \notin P_R} \neg R(\bar{a}) \big) \big) \Big)$$

Let Γ_τ consists of all possible such formulas.

885. Let the domain be ω and for each $R \in \tau$ and tuple $\bar{a}$ decide $R(\bar{a})$ with probability $1/2$. For any finite $B \subseteq \omega$ and prescriptions P_R there is a suitable y with probability 1.

886. Γ_τ clearly has no finite models, thus $\aleph_0$-categoricity implies completeness. Showing that any two countable models of Γ_τ are isomorphic can be done by a standard back and forth method as in Problem 863.

887. Consider the prescriptions P_R for $R \in \tau$ an $n \in \omega$ and a formula φ of the form

$$\forall x_1 \dots x_n \Big(\bigwedge_{i \neq j} x_i \neq x_j \to \exists y \big(\bigwedge_i y \neq x_i \wedge \bigwedge_{R \in \tau} \big(\bigwedge_{\bar{a} \in P_R} R(\bar{a}) \wedge \bigwedge_{\bar{a} \notin P_R} \neg R(\bar{a}) \big) \big) \Big)$$

Let $\bar{a}$ be a sequence of distinct elements, and a be an additional element. Let p be the probability that a is related to $\bar{a}$ exactly the way the sets P_R prescribe. Then $p > 0$. Let M be the class of all τ-structures. Then $\text{Prob}(M_k \nvDash \varphi) \leq \frac{k!}{n!}(1-p)^{k-n}$. Taking the limit as $k \to \infty$ this probability converges to 0.

888. Γ_τ is complete by Problem 886, thus combining Problem 887 and Problem 879 completes the proof.

INDEX

Symbols

$[X]^\kappa$, $[X]^{<\kappa}$—family of subsets, 1

$\rightleftarrows$—back and forth, 84

$\ulcorner\varphi\urcorner$, 112

μ—operator, 19, 27

Ω—set of functions, 19

Ω^*—partial functions, 27

$\wp(X)$—powerset, 1

Σ^*—set of words over Σ, 13

Σ_n, Π_n, Δ_n—arithmetical hierarchy, 120

$\alpha(o)$—code of an object, 29, 50, 63

β—Gödel's β function, 23

$\equiv$—elementary equivalence, 77

$\exists_n$—formula, 78

$\forall_n$—formula, 78

κ-almost disjoint, 2

κ-categorical, 80

κ-compact, 100

κ-saturated, 100

$\models$—semantical consequence, 42, 55

ω-complete, 76

ω-consistent, 76

ω-model, 76

ω-rule, 76

ω_1^{CK}—Church–Kleene ordinal, 37

$\varphi_i(x)$—i-index function, 32

$\prec$—elementary substructure, 78

$\vdash^R$—resolution method, 45

$\vdash$—derivation, 47

A

Ackermann function, 21, 25, 36

Ackermann model, 228

admissible substitution, 55

almost disjoint, 1

almost surely true/false, 126

alphabet, 13

append function, 24

arithmetical hierarchy, 120

arithmetical set, 120

arithmetization, 29

asymptotic probability, 126

atomic diagram, 81

atomic formula, 53

automorphism, 57

axiomatizable, 102

axioms, 46

B

back-and-forth system, 84

Beth theorem, 72

Boolean function, 40

bounded minimization, 20

bounded product, 20

bounded quantifiers, 20, 108

bounded sum, 20

C

Carroll, Lewis, 46

categorical theory, 80

chain, 63

chain (in ordered set), 1

Church theorem, 111

Church–Kleene ordinal, 37

clause, 45, 69

closed formula, 54

code, 23, 29, 50

collection principle, 82

INDEX

Printed in the United States
by Baker & Taylor Publisher Services